# Instrumentation: A Reader

This reader is one part of an Open University integrated teaching system and the selection is therefore related to other material available to students. It is designed to evoke the critical understanding of students. Opinions expressed in it are not necessarily those of the course team or of the University.

The Open University would like to thank The Institute of Measurement and Control for its co-operation and support in the publication of this book.

The Open University

Faculty of Technology

T292 *Instrumentation*

*The Course Team*

Mr R. Loxton
Dr D. Daruvala
Ms J. Dutton
Professor J.K. Fidler
Ms H. Johnson
Dr M. Meade
Dr J. Nazemi
Mr P. Pope
Mr J. Stratford (BBC)
Ms J. Younger

# Instrumentation: A Reader

Edited by R. Loxton and P. Pope
for the *Instrumentation* course
at the Open University

**CHAPMAN & HALL**
London · Glasgow · New York · Tokyo · Melbourne · Madras

**Published by Chapman & Hall, 2-6 Boundary Row, London SE1 8HN**

Chapman & Hall, 2-6 Boundary Row, London SE1 8HN, UK

Blackie Academic & Professional, Wester Cleddens Road,
Bishopbriggs, Glasgow G64 2NZ, UK

Chapman & Hall, 29 West 35th Street, New York NY10001, USA

Chapman & Hall Japan, Thomson Publishing Japan, Hirakawacho
Nemoto Building, 6F, 1-7-11 Hirakawa-cho, Chiyoda-ku, Tokyo 102,
Japan

Chapman & Hall Australia, Thomas Nelson Australia, 102 Dodds
Street, South Melbourne, Victoria 3205, Australia

Chapman & Hall India, R. Seshadri, 32 Second Main Road, CIT East,
Madras 600 035, India

First edition 1986 by The Open University
Reprinted 1993

© 1990 Chapman & Hall

Typeset by Butler & Tanner Ltd, Frome
Printed in Great Britain by The Ipswich Book Company, Suffolk

ISBN 0 412 53400 2

A catalogue record for this book is available from the British Library
Library of Congress Cataloging-in-Publication Data available

# Contents

Page

Acknowledgements     vii

Contributors     ix

Introduction     xi

1. Future instrumentation: its effect on production and research
   *R. Shaw*     1
2. Sensors for mechanical properties
   *R.S. Medlock*     12
3. Digital transducers
   *G.A. Woolvet*     24
4. Smart pressure transmitters
   *A.T. Bradshaw*     39
5. On-line measurement of liquid density
   *Conrad H. Hoeppner*     45
6. Measuring mass flow using the Coriolis principle
   *K.O. Plache*     55
7. Humidity in air and gases
   *A.K. Michell*     63
8. Development and application of a 0.14 g piezoelectric accelerometer
   *Howard C. Epstein*     68
9. The use of ultrasonic techniques for non-invasive instrumentation on chemical and process plant
   *R.C. Asher*     75
10. Level and density measurement using non-contact nuclear gauges
    *David R. Carlson*     82
11. Analytical instruments for process control
    *K.G. Carr-Brion*     91
12. Silicon sensors meet integrated circuits
    *Phillip W. Barth*     102
13. Optical measurement methods
    *Robert Jones*     114
14. Sensing with optical fibres: an emerging technology
    *Albert R. Tebo*     125
15. Feedback in instruments and its applications
    *B.E. Jones*     134
16. Communications in process control
    *P.R. Matthews*     151

17. Measurement errors and instrument inaccuracies
    *M.J. Cunningham*                                                    171
18. How to calibrate flowmeters and velocity meters
    *A.T.J. Hayward*                                                     184
19. Noise suppression and prevention in piezoelectric transducer systems
    *Jon Wilson*                                                        200
20. Process measurement in the food industry
    *D.J. Steele and I. McFarlane*                                      208
21. Some instrumental techniques for hostile environments
    *E. Duncombe*                                                       223
22. The impact of the accident at Three Mile Island on plant control and
    instrumentation philosophy
    *F. Catlow*                                                         240

# Acknowledgments

Paper 1    Reprinted from *Journal of Physics E Scientific Instruments* **15**, 1982, by permission of the Institute of Physics and R Shaw.

Paper 2    Reprinted from *Journal of Physics E Scientific Instruments* **16**, 1983, by permission of the Institute of Physics and R S Medlock.

Paper 3    Reprinted from *Journal of Physics E Scientific Instruments* **15**, 1982, by permission of the Institute of Physics and G A Woolvet, and with acknowledgement to G A Woolvet (1977) *Transducers in Digital Systems*, Peter Peregrinns, London.

Paper 4    Reprinted from *Measurement and Control* **17**, October 1984, by permission of the Institute of Measurement and Control.

Paper 5    Reprinted by special permission from *Chemical Engineering*, October 1984 © 1984, by McGraw-Hill Inc., New York, NY 10020.

Paper 6    Reprinted from *Transducer Technology* 1980 by permission.

Paper 7    Original article by A K Mitchell © Mitchell Instruments Limited (1985). Published by permission.

Paper 8    Reproduced by permission © Instrument Society of America 1973. From *ISA Transactions* **12**, Number 4.

Paper 9    Reprinted from *Measurement and Control* **15**, May 1982, by permission of the Institute of Measurement and Control.

Paper 10   Reprinted from *Measurement and Control* **10**, March 1977, by permission of the Institute of Measurement and Control.

Paper 11   Reprinted from *Measurement and Control* **10**, November 1977, by permission of the Institute of Measurement and Control.

Paper 12   Copyright © 1981 IEEE. Reprinted, with permission, from *IEEE SPECTRUM* **18** No 9, pp. 33–40, September 1981, and by permission of P W Barth.

Paper 13   Reprinted from *Engineering*, Technical File No. 110, February 1983, by permission of the Design Council.

Paper 14   Originally published in Electro-Optical Systems Design, February 1982. © Laser Focus/Electro-Optics Magazine. All rights reserved. Reprinted with permission of Pennwell Publishing, Advanced Technology Group, Littleton, Mass., USA.

Paper 15   Reprinted from *Journal of Physics E Scientific Instruments* **12**, 1979, by permission of the Institute of Physics and B E Jones.

Paper 16   Reprinted from *Measurement and Control*, November 1982, December 1982, January 1983, by permission of the Institute of Measurement and Control.

Paper 17   Reprinted from *Journal of Physics E Scientific Instruments* **14**, 1981 by permission of the Institute of Physics and M J Cunningham.

Paper 18   Reprinted from *Flowmeters — A Basic Guide and Sourcebook for Users*, Macmillan Press Limited, 1979, Chapter 11, pp. 142–163, by permission of Macmillan, London and Basingstoke, and A T J Hayward.

Paper 19   Reprinted from *Sound and Vibration* TP 270, April 1979, by permission of Acoustical Publications Inc.

Paper 20    Reprinted from *Measurement and Control* **14**, January 1981, February 1981, by permission of the Institute of Measurement and Control.

Paper 21    Reprinted from the *Journal of Physics E Scientific Instruments* **17**, 1984 by permission of the Institute of Physics and the United Kingdom Atomic Energy Authority.

Paper 22    Reprinted from *Transactions of the South African Institute of Electrical Engineers*, August 1983, by permission of the South African Institute of Electrical Engineers.

# Contributors

| | | |
|---|---|---|
| 1 | *R Shaw* | Systems Designers Ltd<br>Systems House<br>105 Fleet Road<br>Fleet<br>Hampshire |
| 2 | *R S Medlock* | Brown Boveri Kent plc<br>Biscot Road<br>Luton<br>Beds |
| 3 | *G A Woolvet* | Kingston Polytechnic<br>Penrhyn Road<br>Kingston upon Thames<br>KT1 2EE |
| 4 | *A T Bradshaw* | European Product Manager<br>Honeywell Control Systems Limited<br>Newhouse<br>Lanarkshire |
| 5 | *C H Hoeppner* | Consultant<br>320 12th Terrace<br>Indialantic<br>Florida 32903 |
| 6 | *K O Plache* | Ex: Micro-motion Incorporated<br>Boulder<br>Colorado |
| 7 | *A K Michell* | Michell Instruments Ltd<br>Nuffield Road<br>Cambridge |
| 8 | *H C Epstein* | Ex: Endevco-Dynamic Instrument Division<br>Pasadena, California |
| 9 | *R C Asher* | Section Leader, Systems Group<br>Instrumentation and Applied Physics Division<br>AERE<br>Harwell |
| 10 | *D R Carlson* | Ex: Kay Ray Incorporated<br>USA |

| | | |
|---|---|---|
| 11 | *K G Carr-Brion* | Warren Springs Laboratory<br>Stevenage<br>Herts |
| 12 | *P W Barth* | Senior Research Associate<br>Stanford University<br>California<br>USA |
| 13 | *R Jones* | Optic Group<br>Cambridge Consultants Ltd<br>Cambridge |
| 14 | *A R Tebo* | Associate Editor<br>Electro-Optics<br>Tulsa<br>USA |
| 15 | *B E Jones* | Dept. of Instrumentation and<br>   Analytical Science<br>UMIST<br>Manchester |
| 16 | *P R Matthews* | Eagleton (Saudi Arabia) Limited<br>(Ex: W S Atkins and Sons Consulting Engineers) |
| 17 | *M J Cunningham* | Electrical Engineering Laboratories<br>The University of Manchester |
| 18 | *A T J Hayward* | Moore, Barrett and Redwood Ltd<br>Windsor<br>Berkshire |
| 19 | *J Wilson* | Ex: Endevco Incorporated<br>   San Juan Capistrano<br>   California |
| 20 | *D J Steele* | Mark 1 Publishing International,<br>Weybridge<br>Surrey |
| | *I McFarlane* | Hunter Laboratory<br>Beaconsfield<br>Buckinghamshire |
| 21 | *E Duncombe* | UKAEA, Risley (Retired) |
| 22 | *F Catlow* | New Works Department<br>US Electricity Supply Commission |

# Instrumentation: A Reader

## Introduction

This book contains a selection of papers and articles in instrumentation previously published in technical periodicals and journals of learned societies. Our selection has been made to illustrate aspects of current practice and applications of instrumentation.

The book does not attempt to be encyclopaedic in its coverage of the subject, but to provide some examples of general transduction techniques, of the sensing of particular measurands, of components of instrumentation systems and of instrumentation practice in two very different environments, the food industry and the nuclear power industry.

We have made the selection particularly to provide papers appropriate to the study of the Open University course T292 *Instrumentation*.

The papers have been chosen so that the book covers a wide spectrum of instrumentation techniques. Because of this, the book should be of value not only to students of instrumentation, but also to practising engineers and scientists wishing to glean ideas from areas of instrumentation outside their own fields of expertise.

In recent years instrumentation has emerged as a discipline in its own right rather than as an adjunct to traditional science and engineering disciplines. This development has been driven partly by the needs of industries for new and improved sensing techniques, and partly by new technological developments such as microprocessors, optical fibres and integrated silicon sensors which are revolutionising sensing and signal processing practice.

The application of microprocessors to signal processing in instrumentation systems has led to an increase in digital transmission of information, and in many cases to the development of transducers which provide a digital output signal directly, rather than an analogue signal which has to be converted to digital form. Microprocessors are also appearing inside so-called 'smart' transducers where the processor is used to linearise a transducer output, to convert it from one form to another, to provide inbuilt calibration facilities, or to allow the span of the transducer to be changed *in situ*.

Optical fibres are appearing in instrumentation systems both as communication channels which are inherently safe and possess high noise immunity; and as sensors in their own right, where the ability of the fibre to conduct light is made dependent on some measurand, usually with a very high sensitivity.

The ability to create complex electronic circuits inside a chip of silicon is now being married to the technique of creating a sensing element in silicon, so opening the door to integrated sensing and signal conditioning in a very small space, and with the inherent cost advantages of mass production techniques.

The application of these new techniques is leading on the one hand to considerable improvements in sensitivity and accuracy of 'traditional' instrumentation systems, and on the other hand to the development of compact, rugged and very cheap systems. Papers have been chosen to describe some of these new technologies and their application in instrumentation.

In many cases instrumentation systems are intimately linked with closed-loop control processes which are outside the scope of this book. It is acknowledged that the need to control can dictate the form which an instrumentation system takes; this collection of papers, however, does not go beyond the point at which measurand data is ready to be displayed or recorded.

While the preponderance of instrumentation technology is currently electronic in nature, the papers do not attempt to deal with descriptions of electronic circuits. That too is considered beyond the scope of this book.

We have tried to select papers which will not become obsolete too rapidly, but it is inevitable that some specific items of hardware will cease to be manufactured during the lifetime of the book. We hope that this is more than counteracted by the quantity of unchanging basic principle which we have sought to include.

This is not intended to be a book which is read from beginning to end in the order of presentation of the material. Papers are not interdependent, and may be read in any order. Only Papers 13 and 14 have been ordered so that the material in the second is supported by, and follows logically from, the material in the first.

If you wish to have the fundamental principles used in the papers explained in detail, the Open University course T292 *Instrumentation* contains the detailed teaching of most of those principles.

We should mention that some small changes have been made to papers for the sake of uniformity of presentation. Where a substantial part of a paper has been omitted for academic reasons, the missing part is marked by [...]. In these cases editorial changes have been made to figure numbers, section numbers and references to ensure consistency in what has been reproduced.

Finally, we would like to acknowledge the help we have received from Dr J.P. Bentley and Mr G. Webb in collecting papers, and to record our gratitude to members of the *Instrumentation* Course Team of the Open University for assisting us in the selection of papers.

Paper 1

# Future instrumentation: its effect on production and research

## R. Shaw

This paper provides an overview of the industrial instrumentation scene. In any technology, commercial factors have a powerful influence on the direction of new developments. This paper gives useful perspectives on some economic aspects of instrumentation, and of the distribution of instruments within industries, techniques and measurands.

A good deal of jargon is employed, much of which is more fully explained in later papers. There is just one term, *analysis,* which could give rise to confusion and deserves explicit mention. It refers to a multitude of techniques for *chemical* analysis to determine the chemical composition of substances. Paper 11 discusses the function and typical applications of on-line analytical instruments. (Eds.)

An introductory section gives a general profile of the instrumentation and control industry in the UK, making particular mention of the increasing use of microelectronics-based systems. Subsequent sections cover in greater detail sensors, analysers, signal transmission equipment, and instrument hardware: the growing tendency for equipment to be comprised of basic hardware and adaptable software is noted.

### 1.1. Introduction

The last 100 years has seen instrumentation move from the simple principles of the dial thermometer and electrical measuring instruments of the 1880s, to the electronic digital control of the 1980s. Throughout this period the most constant aspect of the technology has been its state of rapid change (see Appendix).

To the instrumentation manufacturer the evolving technological environment has represented an opportunity for profit and growth. But for the user, in production or research, the major benefit of instrumentation is as a financial amplifier. Within the national capital investment in plant and machinery on average some 5% is spent on instrumentation and control. In exchange for this the user expects product quality, higher plant utilization, material and energy savings, better manpower productivity, and speedier accurate data reduction. There is little reason to doubt the

Originally published in *J. Phys. E Scientific Instruments* **15** (1982).

claims that for systems which are well specified and designed the pay-back can be achieved in under one year's operation.

Research departments have long been the home of sophisticated instruments because of their need to measure esoteric variables accurately. Rising staff costs have brought an increasing awareness of the need for electronic data capture and automatic sequence control of experiments. The steady penetration of data links and information networks is beginning to indicate the future organic and neural nature of instrument and information systems in the laboratory.

## 1.2. Profile of the industry

### 1.2.1. The market

To provide the users with their systems the supplier of instrumentation and control equipment competes in an aggressive and fast-moving international market. As a UK industrial sector, instrumentation is relatively healthy: of the UK market for instrumentation and control, which in total exceeds £500 million, half is supplied by imported equipment but the industry nevertheless achieves a modest positive trade balance.

The evolving commercial environment can be seen at work in Figure 1.1, which is a pie chart of sales in key market sectors together with the annual rate of growth. Oil and petrochemicals show the expected declining growth rate while new energy sources and manpower-intensive industries are the currently favoured areas.

### 1.2.2. Ownership

A significant feature of the UK instrumentation industry is the dominance of large overseas-owned companies: the bulk of UK ownership is concentrated among the medium and small firms. Another strong influence is the weighty purchasing power of the major plant and process contractors and large nationalized industries. The net effect of overseas ownership and powerful but traditional customers can be to

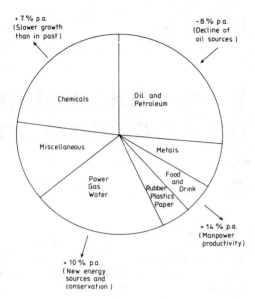

**Figure 1.1.** UK users. Total UK market ~ £500M.

encourage conservatism at the expense of innovation, and equipment developed abroad is often imposed on the UK market in preference to bright ideas conceived here.

### 1.2.3. Capabilities

Despite the harshly competitive climate (or perhaps because of it) the available variety of sensors, instruments and control systems is truly impressive. From abrasion and adsorption, through dustiness and eccentricity, past hardness and hygrometry, by way of moisture and noise, via thermography and thickness to weight and width, the product lists of the instrument industry catalogue a bewildering array of equipment to measure, control and test. Each month the list grows as new techniques probe the hitherto unmeasurable and regulate the previously uncontrollable. While the Appendix is a fascinating backward glance, Figure 1.2 shows what interests the users now.

### 1.2.4. Microelectronics

The strong growth trends in analysis

techniques and digital systems are evident in Figure 1.2, and both of these have been strongly stimulated by recent solid-state technology and microelectronics developments. The pattern of steady penetration by microelectronics is equally evident in scientific instruments where it provides significant breadth of capability, much enhanced diagnostic facilities, and flexible communication interfaces unimaginable at the price 10 years ago. Two features in particular relating to the use of microprocessors in instrument equipment are worth comment. Firstly, it is unlikely that the total power or versatility of the microchip will have been fully utilized in any unit design. Despite this, the economics of microelectronic technology will have been attractive to the supplier, especially bearing in mind the easy way in which later enhancements and design updates can be incorporated with only a small development overhead.

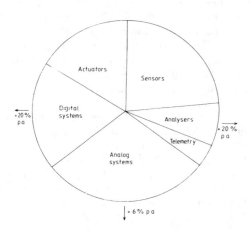

**Figure 1.2.** Equipment purchases (by value).

Secondly, so great has been the speed of advance in hardware microcircuit technology that related techniques like sensors and software now represent constraints on progress. What then is the sensor scene?

## 1.3. Sensors

### 1.3.1. Present status

Understanding rests on sound information whose foundation, in turn, is accurate measurement. In a study of its instrument and control technology a major chemical company found that over 50% of its total instrument maintenance effort was devoted to process measurements. From a range of typical plants a composite sensor population profile was developed (Figure 1.3). This profile only serves to emphasise how dominant still is the traditional type of sensor (and incidentally the traditional instrumentation knowledge and skill required to service it) and the relatively insignificant impact as yet of microelectronics on sensors. Sadly, it also indicates how very few are the examples of non-invasive measurements (categories marked with asterisks).

Discounting for the moment analysis equipment and considering total maintenance effort for each type of sensor, the lion's share of support goes into differential pressure, level and flow equipment followed by thermocouples and resistance thermometers, variable-area flowmeters and pressure gauges, in that order.

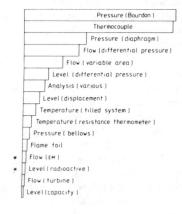

**Figure 1.3.** Sensors (relative populations). Asterisks denote non-invasive measurements.

### 1.3.2. User needs

But this is not quite the problem priority order as seen by users. They want a faster-responding and more accurate temperature device than the thermocouple, ideally using a non-contacting technique to avoid the numerous physical and heat transfer problems associated with thermosheaths. Next comes flow measurement with a call for non-invasive techniques (incidentally offering energy savings) to cope with a wide range of flows, temperatures, material density variations, viscosities and compositions. Level measurement sits in third place with special difficulties caused by dust, obscuration, coating and sticking, intrinsic safety, and again the problems of invasive measurement.

The placing of these three measurements at the top of the list does not imply that no problems exist with any others; all variables can present difficulties.

### 1.3.3. Credibility

As if these physical troubles were not enough, it appears that control-room operators have a low level of belief in their instrument readings. They make continual calls for instrument checks on the plant, less than 20% of which show any fault, but cause significant interruption to production. This finding alone points to the need for more comprehensive built-in proof-testing facilities in measurement instrumentation to facilitate instant diagnosis by the operator in the control room or even automatically.

### 1.3.4. Reliability

Analytical techniques do exist to determine at the design stage the ways in which an instrument might fail either partially or catastrophically. In the latter case the failure rate is found by summing the observed failure rates of the modules from which the instrument is constructed. Partial or potentially unrevealed maloperation analysis calls for a detailed study of the various modes of partial failure like amplifier drift or component ageing, to generate an error probability function.

Comparison of the two analyses will suggest the frequency of routine checking, calibration or replacement.

In all cases, however, the cost of designing for higher reliability must be balanced against the operational penalty of failure.

### 1.3.5. Ergonomics

However reliable the instrument, its value in use will depend on how effectively it communicates and interacts with the operator: its ergonomics.

Research studies and rules of thumb on the topic are legion. Analogue indication is best for qualitative information, and digital for precision. The height of digital characters should be 25% larger than that of associated labels and have a ratio of stroke width to height between 1:6 and 1:10 depending on contrast with background. 7 × 5 dot matrix displays are superior to seven-segment characters, and the error rate for a green display may be up to three times that for red. Analogue indicator scales should not be multiple or non-linear, and numbering should increase clockwise. Fixed-scale moving-pointer indicators with coloured markings or bands to indicate danger conditions can speed recognition of abnormal conditions up to four times.

The vast possibilities soon to be offered by time-shared flat-screen electronic displays with auto-zeroing, auto-ranging and self-diagnosis in addition to multifunction presentation will re-emphasize the golden rule 'question the need for the information and keep it simple'.

### 1.3.6. Invasiveness

In relation to the generally invasive nature of measurements (Figure 1.3), the instrumentation survey revealed that when trouble does occur, causes predominate which are associated with the sensor interacting with the process such as chokes, coatings, corrosion and damage. This suggests that Doppler flow measurement, hoop pipe-stress pressure sensing, ultrasonic process signature identification, and ultimately the new purely optical measurement techniques will be important as the basis for further development in sensors.

Meanwhile, putting several sensors together in a cluster with only one tapping point into the process could offer advantages. Cheaper installation, greater reliability through redundancy, less process energy loss and simpler maintenance would be some of the benefits. Some redundancy of this kind must also be the starting point for measurement validation checks and automatic fault-testing diagnosis procedures.

### 1.3.7. Pressures on development

The direction of future sensor development may be influenced by two strong and relatively recent pressures. The first arises from the common ground between the objectives of industrial and research users. The former emphasize high reliability and repeatability with accuracy about 1% or better, while in research applications, which can be more esoteric, the emphasis is on higher accuracy and flexibility of function. The instrument suppliers' response to both requirements is to explore and refine newer techniques, like vortex and magnetic flow metering, while extracting more data from traditional sensors using electronic techniques.

Fortunately techniques like signal correlation, comparison of multiple input signals, digital filtering and smoothing, variable-rate sampling and Kalman estimators offer much higher information content from existing input signals. These new sophistications will put pressure on the instrument makers to link more closely with manufacturers of microcircuits for the specialized low-volume customized chips and masks which they need.

It is from the solid-state technology of the chip manufacturers that the other pressure on sensor development arises.

The massive investment by vehicle manufacturers into sensor-based automobile electronics completely dwarfs the most optimistic estimates of instrument industry R & D investment. The 1981 demand by General Motors for integrated circuits and solid-state sensors will be £188 million, with an estimated world-wide 1984 market demand of £600 million. The present sensor specification in automobile and consumer applications may well be modest accuracy, strictly limited flexibility and cheap interchangeability but the huge investment and resulting steep learning curve should rapidly put this technology in a position to exert a powerful influence on industrial instrumentation.

### 1.3.8. Optical sensors

Many instrument applications are in hazardous and potentially explosive environments. Intrinsic safety and barrier techniques are well established but not cheap. Possibilities exist for pulsing the power to sensors below the explosive limit or deriving power locally from available plant energy. But in the longer term, a totally different technology based on optics may obviate the safety problem altogether and have other more general attractions.

Numerous ideas for the manipulation of optical information have lain dormant since the mid-nineteenth century but the convergence of fibre-optic transmission, new solid-state materials, and the need for yet faster data-processing circuits has led to a spontaneous interest recently in optical sensing among workers in universities, manufacturers' laboratories and one or two major instrument users.

### 1.3.9. Optical techniques

Optical sensing is based on the principle that the physical variable to be measured causes a change in the optical properties of the sensor and this change in turn affects some property of the light passing through it. Typically this could be phase, amplitude, polarization state, birefractive indices, or interference fringe effects. An example is temperature-sensitive semiconductor phosphors with spectral cut-off proportional to temperature.

Pressure sensing has been achieved by phase change detection relative to a reference wave down a long length of single mode fibre or by using optics to detect sensitively the movements of conventional diaphragms. Optical detection of vortex shedding from an immersed bluff body can be used to measure flow, while multi-reflection splitting and interference pattern detection may ultimately be a more general principle which can be applied to the measurement of many variables.

Attention is also being devoted to the development of multiport data rings and networks. Thin transparent films on a transparent substrate of lower refractive index can be prepared as low-loss film and stripe light waveguides. These act as filters, modulators, switches, power splitters and light-beam deflectors. Combinations can form complex circuits for signal processing and switching, from which optical sensors with integral computing capability may finally develop.

The Department of Industry is now taking an initiative in the formation of a UK Association to further information exchange on the whole of this developing area.

Sensors, therefore, still seem to be the Achilles heel of the technology. The users want more reliable sensors even at the expense of accuracy, and tend to focus on total life cost, while contractors are more concerned with initial capital cost. Developments described above, together with emerging computer-aided design methods, may offer the hope that sensors will soon take on an overdue new look.

## 1.4. Analysers

### 1.4.1. Evolution

Analysis in the last two decades has moved steadily from the laboratory to become a production measuring technique. Early applications tended to be little more than research laboratory techniques applied to samples of production material from some distant plant site. Usually the displacement of the sample and the measurement in both place and time made closed-loop control of quality impossible at that stage.

Gradually the introduction of electronics began to automate the laboratory techniques making them faster, more accurate and reproducible. As a result quality, which for so long has been an inferential measurement, can increasingly be directly measured. By installing specially protected cubicles on the plant, a microcosm of the laboratory environment is now provided close to the sample point, bringing a step nearer the ideal of real-time closed-loop quality control. The latest solid-state sensor materials and microelectronics make this close to fulfilment.

### 1.4.2. The market

Production-based analysis equipment is almost doubling in quantity every five years and even at the present time absorbs some 20% of instrument support effort. The UK market for analysers is a healthy £25–30 million per annum with a buoyancy derived from the increasing cost of manual analysis and the benefits which cost-conscious users get from better product consistency. Safety and reduction of environmental pollution are further strong pressures.

### 1.4.3. Support problems

Analysis instrumentation, however, does pose special difficulties. For the supplier it is, in general, a low-volume high-technology activity and to the user it represents complexity with high installation and support costs. With older equipment roughly half the support work is relatively inconsequential, involving routine checking, cleaning and chemical refilling, all of which leave the equipment in virtually the same state as before.

### 1.4.4. New materials

Newer designs offer many more built-in self-check and self-diagnostic features, and in addition have simpler and more reliable measuring elements based on solid-state materials which are highly sensitive to specific chemicals, for example, the use of lithium–tantalum solid-state detectors to measure vinyl chloride to 0.2 p.p.m. Also, many of these small-area semiconductor composites readily interface to microelectronics. Progressive surface poisoning can reduce useful life; however this is being improved using theoretical studies of surface chemistry.

### 1.4.5. Laser techniques

Both in the laboratory and on plant the increasing cheapness of lasers is steadily widening their application.

Raman spectroscopy methods like CARS (coherent antistokes Raman spectroscopy) and SIRS (stimulated inverse Raman spectroscopy) are recent techniques. One or other of these radiations is emitted when the frequency difference of two coherent beams equals the Raman frequency of the trace molecules of interest.

Another example, the laser environmental atmosphere monitor, joins together spectroscopy (measurement of environmental back-scatter at different frequencies) and electronics (microprocessor control of laser scanning and signal conditioning) and radar-imaging (computer-driven image enhancement and pattern analysis).

### 1.4.6. Microcomputer impact

Because of the ubiquitous microcomputer, standard laboratory and scientific analysis equipment now has enhanced speed, accuracy, reproducibility and flexibility. Features like easily selected automated sequence options, signal conditioning, automatic calibration and diagnostic routines, and a range of data-communication facilities to other laboratory equipment are standard. Complex microchips can perform Fourier transforms virtually in real time, and together with very-low-noise amplifier techniques imported from radio-astronomy they are now making available infrared multigas analysers with Fourier transform facilities and fast wavenumber multiplexing for real-time absorption analysis of several constituents.

Infrared analysis, which was purely qualitative until the 1960s, now offers real-time quantitative measurement covering more than 300 gases. Furthermore, attenuated total reflectance techniques have recently extended the measurement capability to rubbers, plastics and semi-solids using robust calcium fluoride or zinc selenide cells with path lengths less than 0.1 mm.

Speed improvements are particularly impressive: ultraviolet and visible spectrum analysers can now span wide-range wavebands in under one second, and medical multitest analyses which would have taken two days are printed out in as many minutes. At the other end of the size scale, huge computer-based engine test facilities are similarly reducing testing times in the car industry.

### 1.4.7. Sampling

A significant proportion of manpower used in analysis is either in the extraction and preparation of samples or the maintenance of sampling systems, and as analysis techniques become slicker, the pressure to improve productivity in the sampling area will grow. There are welcome trends towards in-line non-invasive analysis techniques and these will increasingly involve fibre optics and optical sensors on the lines described earlier.

The newer solid-state materials with high sensitivity may simplify the task of sample preparation, but clustering sample points in the plant to provide a group of analysis work stations in a protected environment can give significant productivity improvement. There are now prototype examples of sampling by the use of simple microcomputer-controlled robots in just such a plant-based group of analysers.

### 1.4.8. Product quality pressures

Finally, a further pressure towards better product quality control is evident. The increasing sophistication of micro-computer-controlled processing equipment at the factories of the customer is revealing batch-to-batch quality variations hitherto averaged out by the relative crudity of the processing machinery.

So analysis is in some respects still the Cinderella of instrumentation, although relevant and interesting techniques abound. Perhaps the awakening influence of investment money will now allow the necessary technology transfer to take place.

## 1.5. Signal transmission

### 1.5.1. The standards jungle
A healthy current buzz-word is 'digital networks'. Following a surprisingly stable period of analogue instrument transmission standards based on 3–15 p.s.i., 0–10 mA and 4–20 mA, the present digital scene is a jungle. The growing demand for more sensing, monitoring and distributed control systems has spawned equipment-based multicomputer buses, and lately PROWAY, the IEC-sponsored international standard for process control.

Small wonder that although the latter is being studied by instrument manufacturers it is without much enthusiasm. Perhaps they are watching the emerging consensus on protocol standards by the International Telegraph and Telephone Consultative Committee (CCITT) or the efforts by major systems suppliers like DEC, Intel, Xerox, and Hewlett Packard to introduce a *de facto* standard for smaller-scale local area networks (LAN).

### 1.5.2. Open systems interconnection (OSI)
The CCITT recommendations are based on the concept of open systems interconnection, in which networks of dissimilar computers, terminals or peripherals of different manufacture will be interconnected and work together without technical incompatibilities. A phased introduction of this environment is envisaged, initially at the physical layer then through the data, network, transport, session and higher application layers of protocol harmonization.

The wide availability of cheap protocol conversion microchips, like X25 for the network layer, are expected to speed the process of standardization and force digital control equipment suppliers towards compatibility.

### 1.5.3. Local area networks (LAN)
LAN arose as the huge office automation market was recognized and it was discovered that the bulk of information is used within 1 km of its source. There are nearly as many networks systems as suppliers, but Ethernet may now be gaining a lead.

Adopted by Xerox, DEC, Intel, Hewlett Packard and Siemens, Ethernet allows the interconnection of over 1000 pieces of equipment on a local site area by a coaxial cable carrying information at 10 Mbits s$^{-1}$. Fibre-optic cabling is also envisaged.

Alert research groups will not miss the significance of these local networks within a research complex for the manipulation of corporate research data together with data gathering and control at remote experimental facilities.

### 1.5.4. Industrial data links
Meanwhile on a multimillion pound plant which may have 40 000 instrument signal connections, multicore cabling can cost hundreds of thousands of pounds and make multiplexed systems economically attractive. Hazardous explosive environments add further complications and require equipment to be flame proof, or intrinsically safe or protected by individual barrier units on each measurement line. All these are expensive. Recent intrinsically safe plant-based systems therefore look interesting especially when they use cheap twisted twin cable and serial data transmission.

### 1.5.5. Fibre-optic links
Coaxial cable offers higher data rates like 24 Mbits/2 km but even this must concede advantage to fibre optic data links with 100 Mbits/5 km in the visible light range

and losses of less than 3 dB km$^{-1}$. Optical fibres also look good both for process and laboratory or office environments. They are corrosion free, tough, heat resistant, inherently safe and immune from electrical noise, electrically insulated and have enormous information capacity.

Two types of multimode fibre are used industrially. Step index has a discrete step in refractive index at the core–cladding interface and is usable up to a few kilometres before frequency dispersion causes loss of bandwidth. Graded index, with graded refractive index from core to cladding, extends the bandwidth to above 500 MHz km$^{-1}$ for a 70 $\mu$m doped silica core material.

As fibre-optic cable use grows, the possibilities offered by multicolour (frequency domain) transmission of multiplexed signals as well as high-impedance couplers to ring and network configurations, suggest that in the future data transmission will be optical, perhaps from passive optical sensors.

## 1.6. Instrument hardware

### 1.6.1. Trends

In process instrumentation the well-established trend from pneumatic equipment to electronic equipment (see Figure 1.4) has now been overtaken by the migration from analogue to digital systems, so that to talk about instrument equipment is now to talk about hardware and software. Recent price comparisons for a typical 40-loop control installation showed pneumatic equipment costing £38 000, electronic analogue at £47 000 and electronic digital at £35 000.

Not surprisingly, the use of micro-computers in the instrument industry is growing at over 35% per year. To assure their supplies and influence the design of microchips, instrument manufacturers are working with or purchasing the chip makers: Honeywell with General Instrument, Foxboro with Integrated Circuit Transducers, and Taylor with Digimetric.

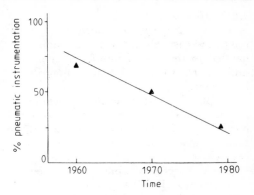

**Figure 1.4.** Trend towards electronics.

### 1.6.2. Micros everywhere

Although initial development costs can be unexpectedly high, microprocessors offer significant benefits to the manufacturer. What was complex hardware design is now a software programming task performed on a sophisticated development system with high-level languages and error correction so that design changes can be made by reprogramming. At the hardware level, because there are many fewer components and interconnections, power supplies and packaging costs are reduced and reliability improved. Finally, quite apart from add-on product features which can be accommodated at a relatively low extra cost, later product evolution can be accomplished by program alteration only.

Microprocessors are embedded in transmitters, controllers and recorders and allow many previously 'up-market' features to be offered as standard. Compensated scale adjustment, auto-calibration and -zeroing, power level compensation, range of outputs, multitest diagnostics, high-resolution displays, multimode algorithms, plug-in capability enhancement modules and digital communication interfaces are just some of the features which characterise the new generation of instrumentation. An interesting trend is the quotation of mean time between failure (MTBF) and guaranteed life performance by some manufacturers. This could become a competitive selling point.

### 1.6.3. High integrity

Steadily cheapening hardware and relatively easy electronic interconnection has led to the use of multicomponent or multi-unit systems when high integrity is important.

Apart from the normal precautions like minimum component count, proven hardware reliability history, conservative rating operation, initial burn-in and exhaustive system tests, it is possible to improve expected reliability by employing redundancy. Parallel redundancy will improve individual reliabilities from 0.9 to 0.99 for the parallel system. Standby parallel redundancy with switched-in operation of the standby improves reliability from 0.9 to 0.9948. Majority voting (for example 2 out of 3) will give enhancement from 0.9 to 0.972.

Ideally, redundant systems should be designed to avoid faults which might be common to all channels and for example should use different means of signal sensing and control output processing to this end.

### 1.7. Conclusions

For long the Cinderella of instrumentation, sensors are finally succumbing to new technological advances in sensor materials, optics and microelectronics. The effect of microcomputers in sensors and instrumentation has been to expand the range of functions available from a single device and to enable modules readily to be linked together in communicating networks.

Always responsive to commercial pressures, the instrument industry now seems to have the technological tools to approach the ideal specification of accuracy, reliability, flexibility and non-invasiveness — until the goal posts change again!

### Appendix. 50 years of non-stop development

| | |
|---|---|
| 1928 | Reset in electrical controllers |
| 1929 | Negative feedback in pneumatic amplifier |
| 1930 | Flow ratio controllers<br>Photoelectric cell and control use |
| 1931 | Electron microscope<br>Ultraviolet radiation detectors<br>Flame analysers |
| 1932 | Thermistor<br>Precision RID |
| 1933 | Variac power transformer |
| 1934 | Derivative action used |
| 1935 | Cascade control |
| 1936 | Piezoelectric elements for vibration |
| 1937 | Program controller for batch drying |
| 1938 | Strain gauge |
| 1939 | Moisture content of gases measured |
| 1940 | Three-term (proportional, integral, derivative) controllers |
| 1941 | Load cell weighing |
| 1942 | Stepless proportioning electric control |
| 1943 | Electronic strip chart recorders |
| 1944 | Valve sizing coefficient established |
| 1945 | Infrared gas analyser |
| 1946 | First digital calculating computer |
| 1948 | First transistor |
| 1949 | Random access memory used |
| 1952 | Gas chromatograph |
| 1955 | Ultrasonic flowmeter |
| 1956 | Capacitance pressure sensor |
| 1957 | Ultrasonic level sensor<br>Dead-time control used |
| 1958 | Analogue–digital converters<br>First process computer control |

| 1959 | Magnetic flowmeters<br>Intrinsic safety concept |
| 1960 | Stepping motor drives<br>Fluidics |
| 1961 | Direct digital control<br>Fibre optics<br>Integrated circuits |
| 1962 | Rotary pneumatic motor<br>Fluidic amplifiers<br>Voice recognition |
| 1963 | Executive control programs |
| 1964 | Digital flowmeter |
| 1965 | Bang–bang motor control |
| 1966 | Thermal mass-flowmeter |
| 1967 | Minicomputer control |

| 1968 | Lasers |
| 1969 | Fluidic controller |
| 1970 | Programmable logic controller (PLC) |
| 1971 | LED displays |
| 1972 | Visual display units (VDU) |
| 1973 | Solid state pressure transducers<br>Opto-isolators |
| 1974 | Microprocessors |
| 1975 | Distributed control concept |
| 1976 | Colour VDUs |
| 1977 | Microprocessor pressure sensor |
| 1978 | Fibre-optic data transmission<br>Very-large-scale integrated circuit (VLSI) |

# Paper 2

# Sensors for mechanical properties

*R. S. Medlock*

This paper presents devices to convert a wide range of physical and mechanical properties into a displacement which can subsequently generate an instrumentation signal. The application of physical principles as the basis of a practical transducer is but one fascinating aspect of instrumentation. The reader should not expect to find the operating principles of all the devices explained in detail, but rather should use the pictorial presentation as a pot-pourri of physical ideas which can be dipped into at a first reading and used later as an *aide-mémoire* to the range of available techniques. In many cases, for example the differential manometer, the principle of operation is quite elementary while in others, for example the Coriolis flowmeter, the principle of operation is hidden in the underlying physics. (Paper 6 describes the principle of operation of a Coriolis flowmeter in some detail.)

Later papers provide greater detail of ultrasonic sensors (Paper 9), nucleonic sensors (Paper 10) and optoelectronic sensors (Papers 3, 13 and 14). (Eds.)

The range of mechanical measurands is defined for seven areas of application. Attention is drawn to the problem of terminology and classification. Transduction techniques used in a large number of sensors are described by numerous line diagrams each with a brief explanation. Finally, some indications are given of trends in future development from which it is concluded that electronic and opto-electronic microchips could play an important role in sensor technology of the future.

## 2.1. Introduction

Measurement technology embraces a wide vocabulary with poorly defined meanings. A device which can perform a measurement conversion is variously called one of the following: sensor, sensing element, transducer, transmitter, converter, detector (e.g. infrared detector), cell (e.g. load cell), gauge (e.g. strain gauge), pick-up (e.g. vibration pick-up), probe (e.g. temperature probe), transponder, Xmeter where X is any measurand. The word 'sensor' in the title of this paper is intended to be interpreted widely and can be defined simply as 'a device which provides a usable output in response to a specific measurand'. However in this paper, the

Originally published in *J. Phys. E Scientific Instruments* **16** (1983).

word 'sensor' will not be used exclusively and deference will be made to convention by using popular terminology.

The definition of 'mechanical properties' is also subject to interpretation. Table 2.1 provides a list of measurands which will be regarded as having a 'mechanical' classification although it must be accepted that a more precise description would be 'physicomechanical'. Sensors are also used for the primary measurement of five other classes of variables, namely electrical, magnetic, thermal, radiation and chemical but these will be excluded from the discussion.

### Table 2.1. List of mechanical measurands

1. *Displacement and dimensional*
   Linear, angular, position, size, level, area, thickness, shape, strain, volume, surface finish
2. *Velocity*
   Linear, angular, speed, flowrate
3. *Acceleration*
   Vibration
4. *Mass*
   Weight, load, density
5. *Force*
   Load, weight, pressure, differential pressure, absolute pressure, dynamic pressure, torque, stress, power
6. *Other physicomechanical qualities*
   Viscosity, hardness, phase concentration

## 2.2. Applications

Seven areas of application can be recognized (Table 2.2). The transduction principles embodied in sensors for these areas have much in common but the completed sensors have quite different physical forms; e.g., a pressure sensor for physiological research bears little resemblance to its counterpart used in the process industries even though the principle of operation is the same.

### Table 2.2. Areas of application

1. Process control
2. Laboratory, test and scientific
3. General industrial
4. Medical and biological
5. Environmental and survey
6. Automobile and consumer markets
7. Military and aerospace

## 2.3. Sensor anatomy

Sensors are often regarded as black boxes with 'input' and 'output' terminations, and some users may not be too concerned with what lies between the terminals. This may be satisfactory for well-established applications but the discriminating user will need to understand the transduction principles and characteristics of the black-box components before applying a sensor to a new problem. Several common themes can be recognised amongst the large number of sensors available today. Some of these classification descriptions are given below.

*Passive sensors* are those requiring a source of energy which can be modulated by the measurand to provide the output signal. The alternative title is 'modulating'. Example: a capacitor sensor for the measurement of displacement by altering the distance between the capacitor plates.

*Active sensors* are those which draw their energy from the measurand. The alternative title is 'self generating'. Example: a piezoelectric vibration sensor.

*Simple sensors* are those which have a single stage of transduction. Example: a liquid manometer measuring static pressure.

*Compound sensors* can have two or more stages of transduction. Example: a pressure convertor can incorporate a pressure-to-displacement primary sensor such as a diaphragm connected to a secondary system to convert the displacement to an electrical signal.

*Feedback sensors:* these incorporate

primary detectors whose output signals are amplified and fed back to the input in polarity opposition. The amplified signals then provide the output signals of the sensor. The complement is the 'feed-forward' sensor which has no feedforward arrangement.

*Displacement, velocity* and *acceleration sensors*: the first-named are sometimes referred to as 'steady-state' sensors and the other two types as 'dynamic' sensors.

*Analogue* and *digital*: these terms generally refer to the nature of the output signal.

*Invasive* and *non-invasive*: these terms are used mainly in the process industries. An invasive sensor is one which has to contact and interact with the measured fluid in pipes or tanks and is therefore subject to difficult accessibility and to a harsh environment.

## 2.4. Sensor output signals

Whilst most available sensors provide an electrical output signal, the process industries still employ some sensors with pneumatic outputs.

The output signals can take many forms but most will fall into one of the first three of the following categories.

*Analogue:* voltage, current, displacement (linear, angular), pressure, ratio amplitude modulation.

*Frequency:* various waveforms, pulse rate, frequency modulation.

*Digital encoded:* serial (pulse-coded modulation), parallel (optical, electrical or magnetic encoders, using binary or Gray code).

*Other pulse coding:* amplitude, width, position

*Hybrid:* combined analogue and digital.

## 2.5. Sensor classification

Various systems of classification have been proposed. The significant features in any system appear to be as shown in Table 2.3.

**Table 2.3.**

| Feature | Example |
|---|---|
| The primary measurand | pressure |
| The sensing element | diaphragm to convert pressure to displacement |
| The sensing means | conversion of diaphragm movement to an electrical signal by a capacitance sensor |
| Output signal | standardized 4–20 mA DC |
| Range | 0–100 bar (0–10 MPa) |
| Special features | suitability for specific applications |
| Applicational area | industrial process control |

## 2.6. Transduction techniques

Figures 2.1–2.13 illustrate the main techniques used for measuring the mechanical properties referred to in Table 2.1. A significant proportion of these properties can be measured by sensors responding to a force or a displacement but as force can be transformed into a displacement through an elastic element, the dominant measured variable is linear or angular displacement.

Although an attempt has been made to include the better-known sensors, no claim is made that the listing is comprehensive. Each sensor described or illustrated covers only the basic form and it should be appreciated that each form has many variants to meet price, application and manufacturing requirements. For example a strain gauge may be of the unbonded, bonded, thin-film, semiconductor or piezoresistive type and can be designed for attachment to diaphragms, cantilevers or other mechanical structures.

Furthermore, the majority of sensors incorporate two or more sensor elements depending on the type of output signal required. Thus a Bourdon pressure gauge may have a mechanical output in the form of an indicating pointer or it may incorporate any one of a number of displacement sensors to give a pneumatic, electrical analogue or digital output signal.

## 2.7. Technological trends

Sensor technology has of late benefited considerably from advances made in other technologies particularly associated with semiconductors, microprocessors, fibre optics, opto-electronics, ferroelectrics, acoustics and material science. Some of the technological areas which promise to contribute to the birth of new sensors are briefly referred to below.

### 2.7.1. Silicon technology and integrated sensors

Silicon and other semiconductor materials offer great potential for research and development in new types of sensors. The existing excellent facilities for batch fabrication of silicon structures is inviting. These offer micromachining capability plus on-board sensing and signal conditioning with the result that a new generation of sensors could be small, cheap and thoroughly reliable. Already, silicon sensors are available for mechanical measurements such as pressure, force and acceleration. Sensors for other measurands, particularly in the electro-chemical field, are either available or in the course of development. Another important advantage of silicon sensors is the potential for creating intelligence or smartness on the chip. Very few sensors available today have microprocessor compatibility but there is a good possibility that silicon sensors of the future could incorporate interface circuitry and thus truly become integrated.

Because silicon has such excellent mechanical properties (high tensile strength, Young's modulus, hardness and strength to weight ratio) it is an excellent material for mechanical measurements. It also has a high fatigue strength, negligible hysteresis, consistent quality, high corrosive resistance and the ability to accept a high surface finish. In addition to all these virtues there is a vast knowledge of techniques to combine silicon with other materials to add to its mechanical and sensing virtues.

Sensors based on silicon need not be restricted entirely to the application of its semiconductor properties. An example of a sophisticated integrated sensor is described by Ko *et al.* (1982) as an absolute pressure sensor combining a silicon diaphragm as one plate of a parallel-plate capacitor with an evacuated and sealed cavity and on-board circuitry to give a pulse-modulated output.

However, it would be wrong to give the impression that the development of accurate silicon sensors for a wide range of mechanical parameters is near at hand. There are many problems and techniques to be resolved such as bonding, eliminating temperature effects, fabrication and packaging. The cost of incorporating on-board circuitry will be high so that integrated sensors will initially only be commercially viable for the consumer and automobile markets. Nevertheless, the medical, scientific and process control markets could well benefit by adapting the sensor chips to meet their specific needs. In the meantime integrated sensors are being developed for mechanical measurands including displacement, pressure, flow, acceleration and humidity. An additional advantage conferred on an integrated sensor is the capability of including compensating circuitry for the removal of non-linearities and secondary parameter sensitivities — particularly temperature. On account of developments in sensor integration, it is probable that the existing piezoresistive silicon pressure sensors will be superseded by greatly superior silicon devices which operate in the capacitance mode. Apart from a substantial improvement in sensitivity over piezoresistive devices, capacitance sensors are potentially capable of negligible temperature sensitivity and a high degree of linearity.

In summary the major advantages that might accrue from integrated sensors could be:

reduction in external leads
signal conditioning
lower signal-to-noise ratio
standardization of output signals compatible with future data highways and computers
compensation for secondary parameters
small physical size
high reliability
low cost if quantities are sufficient
incorporation of diagnostic routines and alarm generation
greater sophistication
higher performance

### 2.7.2. Piezoelectric sensors

Piezoelectric devices have been in common use for many years in the measurement field, e.g. in accelerometers, distance measurements by sonar, doppler devices, velocity measurements in fluids and viscosity measurements. The piezoelectric materials selected for these applications have been quartz, and ceramics of the lead titanate zirconate type. More recently polymer materials such as polyvinylidene difluoride (PVDE) have become available in thin films having a high compliance and low permittivity relative to other piezoelectric materials.

An interesting phenomenon based on the properties of piezoelectric material is the surface acoustic wave generator (SAW). If interdigital electrodes are deposited on a piezoelectric plate, surface waves can be propagated in the surface plane normal to the overlap of the electrodes at a frequency dependent on the propagation velocity and spacing of the electrodes. Its use as a selective filter is obvious but possibilities exist of using the device as a sensor element. For example its centre frequency can be changed by stress and some development has taken place to use a SAW device as a pressure sensor. These devices can also be applied to provide acoustic beams which can diffract light in the Bragg mode. This technique hints that the combination of optics, acoustics and electronics could play an important role in the future of sensor technology.

### 2.7.3. Optical sensors for mechanical measurements

These sensors can be classified into three groups:

opto-electronic with an integral light source and electrical output circuit;
as above but with optical fibres conveying the light input energy and transmitting a light-modulated output signal;
true fibre-optic sensors in which the measurand modulates light in the fibre by intensity, phase, wavelength, polarization or time.

Some simple optical sensors have been available for some years. One example is the Fotonic displacement sensor (Figure 2.12(d)) in which light is transmitted down one fibre and reflected back into an adjacent second fibre from a displaceable target. A second example is the level monitor (Figure 2.12(g)). Meanwhile considerable effort is being made to produce a whole new range of optic, fibre-optic or electro-optic sensors. This is a subject which has been discussed in more detail in other conference proceedings. Future developments will depend on several factors, particularly materials science:

*Development of fibres with special properties*
Polarization-preserving fibres.
Heavy metal halide fibres having special optical properties in the long-wavelength region.
Fibres coated with piezoelectric or magnetostrictive material which can provide a conversion from mechanical stress to optical modulation.
Specific doping of fibres to provide specific measuring capability.

*New materials and components*
Electro-optic filters which make use of the dispersive properties of birefringent materials (e.g. ternary metal sulphides) which become isotropic at a specific wavelength. Isotropism can be controlled by electrically rotating the principal optic axes. Thus, such material, placed between crossed polarizers can provide an electrically controlled optical filter.

Composite piezoelectric materials with high piezoelectric activity. The potential for producing new sensors combining piezoelectrics and optics is promising.

*Integrated optics*
When the main functions of optical components can be performed on commercially available semiconductor chips a whole new field of optical measurement will be opened up. The main functions will need to be switching, coupling, polarizing, modulating, reflecting, diffracting, amplifying and filtering.

## 2.8. Conclusion

In this broad survey of sensors for mechanical measurements an attempt has been made to illustrate the great variety of disciplines, technologies and inventions associated with measurement. Although a comprehensive review would be impossible, it is hoped that sufficient representative examples have been chosen in this paper to give a clear picture of the current situation and probable future trends. Acknowledgments are made to the authors of some hundreds of books, papers and articles, too numerous to mention singly, from which the examples in this paper were sifted. However, in the reference list six books are mentioned which were particularly helpful and can be recommended for further reading.

## References

Herbert, J. M. (1982) *Ferroelectric Transducers and Sensors* (London: Gordon and Breach).

Jones, B. E. (1977) *Instrumentation Measurement and Feedback* (London: McGraw-Hill).

Ko, W. H., Bao, M. H and Hong, Y. D. (1982) A high sensitivity integrated-circuit capacitive pressure transducer. *IEEE Trans. Electron Devices* **29**, 48–56.

Morris, N. M. (1973) *An Introduction to Fluid Logic* (London: McGraw-Hill).

Norton, H. N. *Sensor and Analyser Handbook* (New Jersey: Prentice Hall).

Sydenham, P. H. (1980) *Transducers in Measurement and Control* (Bristol: Adam Hilger).

Woolvet, G. A. (1979) *Transducers in Digital Systems* (Stevenage: Peter Peregrinus).

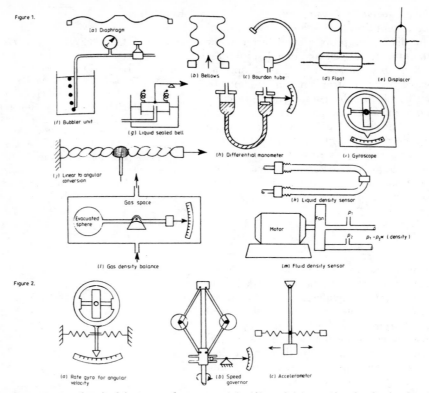

**Figure 2.1. Sensors: mechanical input and output.** (*a*), (*b*) and (*c*) are simple elastic elements for measurement of pressure: (*d*) measures liquid level and the float has a density less than the liquid: (*e*), measures liquid level but the 'float' has a density greater than the liquid: (*d*) and (*e*) can also measure density: (*f*), measures level by bubbling a gas through the liquid: level is indicated by a pressure gauge: (*g*), an inverted liquid sealed bell whose buoyancy is controlled by pressure: the range is adjusted by spring rate: (*h*), differential ($p_1 - p_2$) is measured by a float on the liquid: (*i*), measures angular rotation of gyroframe pivoted in a vertical plane: (*j*), double twisted elastic member converts linear motion to angular: (*k*), weight of liquid in elastically pivoted U-tube produces deflection in vertical plane proportional to density: (*l*), gas density balance measures buoyancy of evacuated sphere: (*m*), constant speed centrifugal fan creates pressure difference proportional to density.

**Figure 2.2. Dynamic sensors: mechanical input and output.** (*a*), Gyroscope frame is spring controlled to a centre position instead of having free rotation: (*b*), diagrammatic arrangement of centrifugal device based on the speed governor design: (*c*), acceleration is measured by the displacement of a mass positioned by springs.

**Figure 2.3. Flow sensors.** (*a*), (*b*), (*c*) Measure flowrate by differential pressure measurements which are proportional to (Flow)$^2$; (*d*), Measures flowrate by the height of the float which has a linear relationship and a constant pressure drop. (*e*), The hinged gate swings through an angle which is a function of flowrate; (*f*), The flow drives the conical body against the reaction of the spring and causes the annulus between the cone and orifice to increase. The position of the cone, or the pressure differential across the orifice, is a measure of flowrate quantity; (*g*), This is a positive displacement meter for measuring volumetric quantity. The sliding blades provide sealed volumetric compartments conveying the measured fluid; (*h*), Gas flow enters at the centre hub and drives the spiral vanes in an anticlockwise direction. Precise volumes of gas are held in chambers sealed with water; (*i*),

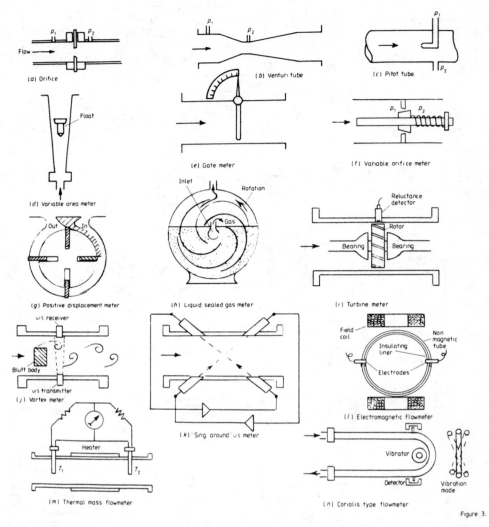

(a) Orifice

(b) Venturi tube

(c) Pitot tube

(d) Variable area meter

(e) Gate meter

(f) Variable orifice meter

(g) Positive displacement meter

(h) Liquid sealed gas meter

(i) Turbine meter

(j) Vortex meter

(k) 'Sing around' u/s meter

(l) Electromagnetic flowmeter

(m) Thermal mass flowmeter

(n) Coriolis type flowmeter

Figure 3.

The turbine meter is used as a rate of flow meter and as a quantity meter. The turbine rotor speed is proportional to flowrate and is measured by a detector which senses the passage of the rotor blades and causes an electric pulse to be generated. Summation of the pulses gives a measure of flow quantity; (j), The vortices are generated by the interaction of the flow with the bluff body and develop in an alternating manner from the two sides of the bluff body. Various detectors are used to measure the frequency of vortex generation. The method shown relies on modulation of an ultrasonic beam; (k), Ultrasonic beams are transmitted across the flowmeter body and are regenerated by feedback amplifiers. The frequency difference of the two feedback loops is proportional to the flowrate and is independent of the speed of sound in the fluid; (l), This meter generates a voltage across the electrodes which is a function of flowrate and magnetic field strength and is a form of a liquid dynamo. Both AC and DC fields are employed in commercial designs; (m), Provided the thermal characteristics of the fluid remain constant, this meter measures mass flow. $T_1$ measures the initial fluid temperature and $T_2$ the temperature after a constant heat input. ($T_1 - T_2$) is a function of flowrate; (n), This is one version of a Coriolis type of flowmeter. The U-tube is vibrated at its resonant frequency. The Coriolis component of the fluid motion causes the two legs of the U-tube to depart from parallelism by an amount proportional to mass flow. The deflection is measured photoelectrically as a pair of variable-duration pulse signals.

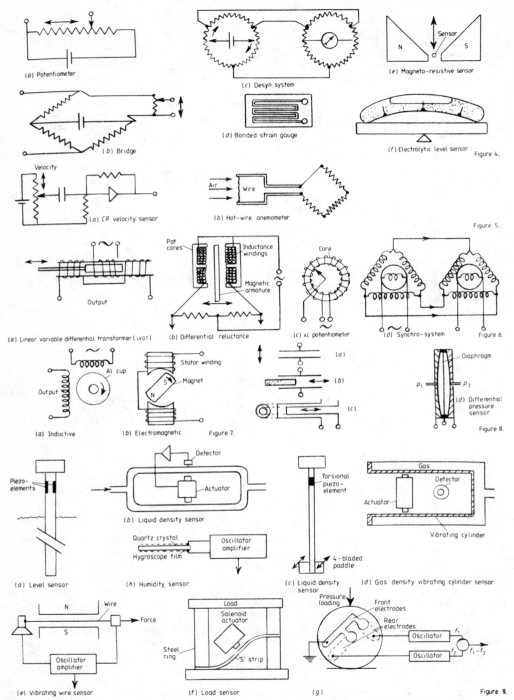

**Figure 2.4. Sensors based on resistance change.** (*a*), A simple potentiometric displacement sensor with an EMF output; (*b*), A simple bridge circuit with EMF or current output; (*c*), The desyn system for converting angular rotation into an electrical output signal; (*d*), A bonded type strain gauge whose resistance varies linearly with strain. Other types of strain gauge include unbonded,

film, semiconductor and piezoresistive; (e), A magnetoresistive sensor which is a resistor whose coefficient varies with magnetic field strength; (f), This is a special type of 'spirit level' in which the spirit is replaced with a conductive fluid which is in contact with three electrodes connected into an AC bridge circuit. The bubble position differentially affects the resistance of each arm.

**Figure 2.5. Velocity sensors utilizing resistance change.** (a), The velocity of the contact arm of the potentiometer is measured through a differentiating $CR$ network; (b), The velocity of a fluid is measured by its cooling effect (and hence resistance change) on a fine heated wire.

**Figure 2.6. Sensors utilizing inductance change.** The techniques illustrated in this section can also be applied to the measurement of thickness of coatings, level and hardness of ferrous materials. (a), The core of magnetic material differentially links the flux generated by the centre winding and the two outer windings which produces a resultant voltage whose phase and magnitude is a measure of displacement; (b), Two opposing coils are embedded in magnetic cores and form two arms of a reluctance bridge. This bridge balance and hence its output is a measure of the differential air gap between the armature and each coil; (c), This is the inductive equivalent of the circular resistive potentiometer: (d), The AC-energized transmitting rotor creates three voltages in the stator windings. The output of the receiving rotor is zero when in the same angular relationship as the transmitting rotor but angular displacement creates proportional AC output signals.

**Figure 2.7. Tachometers.** (a), (b), Two forms of tachogenerator. (a), Has an AC-energized stator winding and an aluminium cup rotor which induces a velocity output signal in the second stator winding; (b), Creates an AC signal in the stator windings whose magnitude and frequency is dependent on the angular velocity of the permanent magnet rotor.

**Figure 2.8. Capacitance sensors.** (a), (b), (c), Three forms of capacitance sensor in which displacement effects a change in impedance by (a), Separation of the plates; (b), Displacement of the dielectric; (c), Axial displacement of one element of a coaxial capacitor. (d), A differential capacitance sensor which provides a convenient way of measuring differential pressure across an elastic diaphragm.

**Figure 2.9. Vibration-type sensors.** (a), Two piezoelectric transducers are fixed to the inside of a hollow tube. These are employed to generate flexural vibrations in the tube at its resonant frequency. A phase-locked loop circuit tracks an oscillator to follow changes of frequency as the liquid level moves up and down the tube. The change in resonant frequency is dependent on the degree of submersion. If the liquid level is constant then the resonant frequency is a function of liquid density — see also (c); (b), Liquid or slurry flows through the two parallel arms of the sensor. An electromagnetic actuator, a detector and a feedback amplifier maintain the two parallel tubes in a resonant vibratory mode, the frequency of which is a function of fluid density. (c), A piezoelectric transducer is fixed internally to a hollow tube and operates in the torsional mode. At the bottom of the tube a four-bladed paddle is fixed. This device, when in resonance, generates a frequency proportional to density. It is possible to combine (a) and (c) to produce a sensor which can simultaneously measure level and density; (d), By combining an electromagnetic actuator, detector and feedback amplifier the steel tube can resonate in an elliptical mode. Changes in gas pressure or density outside the tube alter the resonant frequency. (e), A vibrating wire has a resonant frequency = $0.5\,L^{-1}(Tm^{-1})^{1/2}$ where $L$ is the length of wire, $T$ is the tension, and $m$ is the mass per unit length. The wire is maintained at its resonant frequency by an oscillator amplifier. This frequency is thus a root function of the force stretching the wire. (f), This is another resonating device, the 'S-sensor'. It is maintained in resonance by an oscillator amplifier and a solenoid actuator. The S-shaped metal strip is clamped diagonally across a high-tensile steel ring. Compressive loading of the ring alters the natural frequency of the strip; (g), This is a pressure or load sensor in which the stressing of a quartz AT cut crystal is achieved by a diametral force in a compressive mode. Two pairs of electrodes are fixed to the crystal, one at the centre and one near the edge as shown. Each active area between the electrodes is maintained in oscillation. The centre pair is relatively insensitive to the applied force and its function is to compensate for temperature. The output is the frequency difference of the two crystal oscillators and is relatively unaffected by ambient temperature.

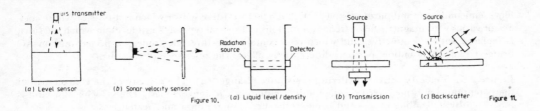

(a) Level sensor   (b) Sonar velocity sensor

Figure 10.

(a) Liquid level / density   (b) Transmission   (c) Backscatter

Figure 11.

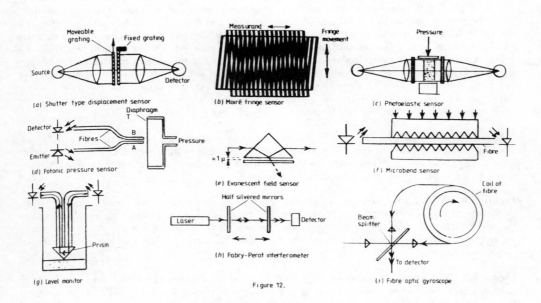

(a) Shutter type displacement sensor

(b) Moiré fringe sensor

(c) Photoelastic sensor

(d) Fotonic pressure sensor

(e) Evanescent field sensor

(f) Microbend sensor

(g) Level monitor

(h) Fabry-Perot interferometer

Figure 12.

(i) Fibre optic gyroscope

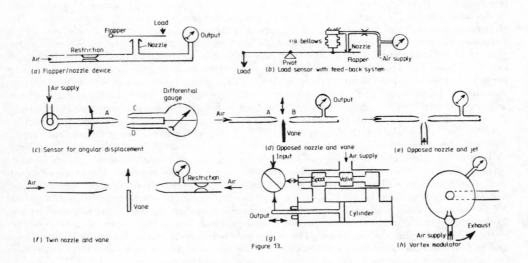

(a) Flapper/nozzle device

(b) Load sensor with feed-back system

(c) Sensor for angular displacement

(d) Opposed nozzle and vane

(e) Opposed nozzle and jet

(f) Twin nozzle and vane

(g)

Figure 13.

(h) Vortex modulator

**Figure 2.10. Ultrasonic sensors.** (a), Level, displacement and distance can be measured by measuring the time taken by an ultrasonic pulse of energy to be reflected from the surface of the liquid or some other target. In order to compensate for any change in velocity of sound, a ratio method can be devised in which a target at a fixed position from the ultrasonic transducer acts as a reference; (b), If the target is moving, its velocity can be determined by the Doppler shift in frequency.

**Figure 2.11. Nucleonic sensors.** The measurement of thickness and density can be made either by absorption or scattering of alpha, beta, or gamma radiation. (a), Represents non-invasive detection of liquid or solid level in a tank; (b), Represents a thickness level gauge; (c), A thickness gauge using the back-scatter effect.

**Figure 2.12. Optoelectronic sensors.** The line diagrams represent only a selection of recent developments in optical and electro-optic sensors. The main theme is the modulation of light by amplitude, phase, wavelength, polarization and time. (a), Uses occultation to modulate amplitude. Light is modulated by relative displacement of two line gratings; (b), Relative movement of two similarly inclined gratings generates moiré fringe movements which can be measured photoelectrically. (c), Two crossed polarizers are oriented to provide light extinction. An elastic material such as polyurethane rubber when stressed becomes birefringent and allows light to reach the detector; (d), Fibre (A) transmits light on to a target (T) and some of the light is reflected into fibre (B) and thence to a photodetector. The reflected light has a characteristic relationship to the distance between the fibre ends and the target; (e), Light can undergo total internal reflection under certain conditions (as in glass fibres) when travelling in a medium having a higher refractive index than its boundary medium. There is an evanescent field which extends exponentially from the surface of the denser medium and a suitable absorbing surface can frustrate total internal reflection when positioned in the evanescent field, i.e. within approximately 1 $\mu$m; (f), Loss of light can occur, particularly with graded index fibre when the fibre is subject to bending. Maximum sensitivity occurs when the spacing of the mechanical grating is equal to the focal length of light propagation; (g), This level monitor operates on the principle that double total internal reflection in the prism is destroyed when the two sides are immersed in liquid in which event, light disappears from the return light guide; (h), This Fabry–Perot interferometer can measure small displacements of the order of half a wavelength by interference of two light beams; (i), A fibre–optic gyroscope operates on the Sagnac effect. Light beams circulating in clockwise and anticlockwise directions in a long length of coiled fibre become phase-shifted proportional to rotational velocity of the coil and this can be measured by an interferometer.

**Figure 2.13. Pneumatic sensors.** (a), This device can measure load. The upward force due to air pressure in the vertical nozzle is balanced by downthrust due to the load; (b), The device shown in (a) is non-linear but by applying a pneumatic feed back as in (b) the system can be made linear. The nozzle in this case acts as a position detector of balance; (c), Angular movement of nozzle A creates a differential pressure across nozzles C and D; (d), A displaceable member modulates the stream of air or fluid issuing from nozzle A to B; (e), This is similar to (d) but the horizontal stream is modulated by a cross jet of air; (f), This is a combination of (d) and (e); (g), An input displacement $X$ operates a spool valve which positions an air cylinder proportional to the displacement; (h), A jet of air discharging into a cylindrical chamber can be swivelled to vary the fluid discharge conditions from normal turbulence to vortex motion and thereby modulate the pressure in the chamber.

# Paper 3

# Digital transducers

## G.A. Woolvet

This paper discusses a number of transducers having an output signal in digital form. A digital signal can be transmitted with better inherent noise immunity than an analogue electrical signal, and when received is in a convenient form for input to a microcomputer for signal processing, recording or for use in a computer-controlled system. The author mainly discusses digital displacement transducers, commonly used in numerically controlled machine tools and increasingly in robot control.

Paper 4 discusses the use of a microprocessor inside a transducer, but producing an analogue electrical output signal. The two papers clearly contrast different philosophies in the application of the ubiquitous microprocessor to instrumentation systems. (Eds.)

This article introduces some of the techniques used in transducers which are particularly adaptable for use in digital systems. The uses of encoder discs for absolute and incremental position measurement and to provide measurement of angular speed are outlined. The application of linear gratings for measurement of translational displacement is compared with the use of Moiré fringe techniques used for similar purposes.

Synchro devices are briefly explained and the various techniques used to produce a digital output from synchro resolvers are described.

The article continues with brief descriptions of devices which develop a digital output from the natural frequency of vibration of some part of the transducer. The final section deals with descriptions of a range of other digital techniques including vortex flowmeters and instruments using laser beams.

## 3.1. Introduction

The increasing use of digital systems for measurement, control and data handling leads naturally to a need for transducers which provide a digital output. A digital output from a transducer enables direct acquisition of the output by a digital system and simplifies processing for a digital readout or for control purposes.

Unfortunately, nature has not provided any phenomena that give a reasonably detectable output in directly digital form. The only possible exceptions to this are those devices in which the frequency of free vibration varies in response to some change in a physical characteristic experienced by the device.

Most transducers used in digital systems are primarily analogue in nature

Originally published in *J. Phys. E Scientific Instruments* **15** (1982).

and incorporate some form of conversion to provide the digital output. Many special techniques have been developed to avoid the necessity to use a conventional analogue-to-digital conversion technique to produce the digital signal. This article describes some of the direct methods which are in current use of producing digital outputs from transducers.

### 3.1.1. Transducers in digital systems

Systems based on a central processor and using a number of transducers can use the processor as an intelligent interrogating system and can, therefore, make use of conventional transducers each providing a DC output. For example, the system can be programmed to identify the transducers in sequence and use a single analogue-to-digital converter to provide a digital signal representing the transducer output. The data can be stored for further use or programmed to provide a digital readout.

Such systems are becoming progressively more economical with the relatively decreasing cost of microprocessors and integrated interface devices.

An alternative approach is to use compact electronic packages that can be housed within the transducer. The package can provide all the sampling and control circuits for the primary analogue signal developed by the transducer element and convert this directly to a digital output. Such instruments will have the appearance of digital transducers. The overall accuracy and resolution of the instruments will now be associated with the characteristics of the transducer element itself and with the characteristics of the digitizing electronics.

Use of the digital output signal from any instrument containing its own analogue-to-digital conversion normally presents no difficulties if a digital readout only is required. However, to access the digital output by another digital system, such as a microprocessor, will generally require some 'handshake' control. This will be necessary to prevent updating of the digital information from the analogue-to-digital converter during the time that transfer of information between the systems is actually taking place.

Much greater flexibility can be introduced by using a dedicated microprocessor-based system in each instrument. This method has its particular uses where the instrument may be working remotely from the central control or data acquisition system but where a local independent digital readout is also required.

For systems involving a large number of reasonably local transducers it is more economical, at the present time, to have all the conversion and control organized centrally by a master processor. The transducers can then be conventional analogue types and the central processor used for some intelligent assessment of the incoming signals. The computer could be programmed to make allowance for non-linearities and for signal recovery where the connections from the transducers have given rise to noisy signals.

Although microprocessor systems can accomplish a great deal, the obvious and most direct method in any systems involving digital control, or digital readout, is to have transducers that develop an output which is directly available in some binary coded form or requires the minimum of additional electronics to provide such an output. Devices which meet this criteria are generally referred to as digital transducers and some of these are described below.

## 3.2. Angular digital encoders

Shaft encoder discs of the type shown in Figure 3.1 were originally developed for direct electrical contact. The dark areas represent conducting material through which a brush could be made to complete an electrical circuit. A separate brush is required for each of the concentric rings or tracks, each track representing a separate 'bit' of the digital output with an additional brush as the 'common' connection on the energizing track.

Most of the shaft encoders in use to-day, however, make use of optical and photoelectric devices. The discs are basically transparent with opaque areas arranged in concentric rings. A common light source on one side of the disc illuminates a stack of photocells on the other side, usually with one photocell for each ring. Some optical elements are always included so that the tracks are viewed by the system through a narrow radial slit (Figure 3.2). The output of the cells will be a parallel binary coded signal which represents the absolute shaft position according to the opaque or transparent areas of the rings along the radial slot.

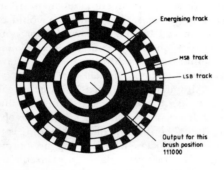

**Figure 3.1.** An absolute encoder disc.

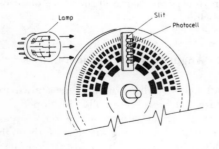

**Figure 3.2.** Optical shaft encoder.

In the normal absolute encoder the resolution is a function of the number of tracks since each produces one 'bit'. To obtain a resolution of small angles of arc, therefore, requires a large number of tracks which results in a relatively large-diameter transducer. The electrical ouput of each cell must be conditioned and converted to a square pulse for positive interrogation purposes.

In a natural binary output each track represents one bit of the output, with the inner track representing the most significant bit and the outer track the least significant bit. This has the disadvantage in that at some positions many bits are required to change simultaneously for a small angular displacement of the disc. For example, changing from 00111111 to 01000000, representing a minimal detectable angular change, requires seven of the eight bits to change simultaneously. Since it is physically impossible to manufacture an encoder disc or assemble its optical system such that all bits change simultaneously, then any reading taken during a change-over could be false. This problem can be overcome by using a disc having a cyclic code, e.g: Gray (Woolvet 1977) code or for a continuously rotating shaft, a memory which stores the reading and changes only at the midpoint of each least significant bit of the outer track.

Absolute encoders of the type described above are available with resolutions up to 14 bits representing approximately 0.02 of a degree of circular arc or 1.3 minutes of arc. Changes in position can be detected at rates up to 0.5 Mwords per second. For rotating discs, the maximum permissible speed for accurate position measurement depends upon the resolution and the switching speed of the associated electronics. Generally the greater the resolution the lower is the maximum rotational speed that can ensure an accurate read-out. For example, a 14-bit resolution with electronics capable of detecting 0.5 Mwords per second has a limiting speed of about 1500 r.p.m. before changes in position occur faster than can be detected. Only when the shaft speed is less than this can an absolute encoder of this type output the true absolute value of the shaft position.

### 3.2.1. Optical resolver

By suitable modification to the optics and/or the electronics, the output of the cells of the outer track of an angular digital encoder can be made to produce a sine-wave output. By suitable resolving, this output can be made to yield additional resolution. In practice an additional track is usually added radially outside the outer track of the absolute encoder and used exclusively as part of the resolver system. This additional track would normally have twice the number of bits as the outer track of the absolute encoder.

One sensor of this additional track is positioned such that the electrical output will vary sinusoidally in magnitude as the disc is displaced by one complete 'opaque–transparent–opaque' cycle. A sinusoidal output can be achieved by careful design of the optics of the detector, particularly, for example the shape of the aperture through which the sensor is illuminated. A second sensor is positioned such that its electrical output is displaced electrically by 90° from the first sensor. It represents a cosine output relative to the sine output of the first sensor. In practice pairs of sensors are often used to produce each of the two signals.

The sine and cosine outputs can be processed by an interpolator to provide up to 16 or more intermediate sine waves, that is 16 sine waves differing in phase equally over the 360 electrical degrees represented by one digit of the least significant bit of the absolute encoder. Each separate sine wave can be processed to provide two output pulses.

Sixteen interpolated sine waves can, therefore, provide 32 additional divisions of the least significant bit. This is equivalent to an additional resolution of five bits. Figure 3.3 illustrates the arrangement of a transducer of this type. This has an encoder disc with 14 inner tracks providing a 14-bit absolute output. The total resolution is increased by the interpolator to 19 bits representing a shaft displacement of less than 2 seconds of arc. The angular displacement represented by the last five bits cannot provide a truly absolute output.

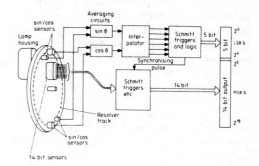

**Figure 3.3.** 19-bit shaft encoder.

A counting system is necessary to count the number of the 32 generated divisions displaced by the rotating disc from the last change in the least significant bit of the 14 inner tracks of the absolute encoder. However, once the system is operating, i.e. the shaft has rotated, at least by the amount represented by the least significant bit of the 14-track encoder, then the total system should always output a total absolute value of the shaft position whether it is stationary or rotating. Some further detail on interpolating systems is given in section 3.3.3.

### 3.2.2. Incremental shaft encoders

An incremental encoder disc uses a single track, usually with optical detectors to provide sine and cosine outputs, similar to the resolver output described above. In addition, a single mark on the disc with an associated optical pick-up is used as an angular datum from which the actual angular position of the shaft can be determined. One of the outputs of the sine/cosine pair is used to count the number of bits to or from the datum. Together the sine/cosine pair are used to determine the direction of rotation (Woolvet 1977) and hence an up or down count of the counter.

The current counter output represents the absolute value of the angular position of the shaft from the datum. When starting, the encoder disc must pass the datum position before the digital output has any validity. The two separate outputs, sine/cosine, also provide opportunity for increasing the resolution, either by interpolation as described above or by similar techniques. The simplest method is to use the leading and trailing edges of both outputs to develop a count pulse, thus increasing the resolution by a factor of 4. A disc with 5000 segments, which is about the maximum currently available, on a disc 150 mm in diameter, can, therefore, provide 20 000 pulses in one revolution without any resolver techniques, which represents a resolution of approximately 1 minute of arc. In incremental encoders as in any other position encoder using counting and/or triggering circuits, there are limitations on the speed of rotation due to the maximum switching speeds of the electronics. However, for rotational speeds found in industrial situations, suitable electronics is available.

### 3.2.3. Digital tachometers

Any device which can generate an electrical pulse or a series of pulses for each revolution of a shaft can be used to initiate a digital ouput of the shaft's speed of rotation. The simplest device is one using a photocell to detect the passing of a white or bright mark on the shaft from reflected light. A toothed wheel using an electromagnetic pick-up, or a capacity sensitive network and a simple probe, can also provide the necessary pulses. Greater accuracy can often be obtained by using an incremental optical encoder disc with a single photocell detector. Using two detectors and the techniques employed in shaft position encoders the direction of rotation can also be determined. In all cases a counter/timer circuit is required and two basic techniques are currently used. The selection of the method to be used is governed by the speed range and accuracy required.

The first method uses a clock as a timer to count the pulses developed by the tachometer over a given period of time. This can only provide the average speed over the measurement time period. Further, the digital output available represents the average speed over the previous time period, since the counter must be used to count the pulses during the current time period. The accuracy of this method depends upon the accuracy of the clock providing the time period and the length of the time period chosen relative to the actual shaft speed. The resolution of the output increases linearly with the shaft speed and therefore very low-speed measurements are not really possible.

The second method uses the time between successive tachometer output pulses as the time period over which a high clock rate is counted. In this method the resolution is inversely proportional to the shaft speed and high-speed measurements are often not possible. It has the advantage, however, that it gives the average speed between each tachometer output pulse and if there are a number of pulses per revolution the angular speed at various shaft positions can be determined.

A third method, not often used in practice, is to make use of the output of an absolute encoder. For this method it, is necessary to sense the time interval between two or more shaft positions. The direction of rotation is also required. A separate counter is necessary to accumulate clock pulses at a known rate between the two shaft positions. This enables the average shaft speed, between these positions, to be determined.

## 3.3. Linear displacement transducers

The digital measurement of linear (or translational) displacement can be achieved by mechanical conversion of the linear motion to rotary motion and then using a rotary digital transducer such as a rotary encoder. Various methods are currently used but precautions must be taken to avoid errors arising from backlash and non-linearities which will reduce the overall accuracy compared to the digital transducer itself. Many computer-controlled machine tools use ball-screw drives to move the horizontal slides with rotary encoders on the lead screws. These devices are extremely accurate and backlash problems have been reduced to levels representing less than 0.002 mm of linear travel.

### 3.3.1. *Optical gratings*

A more positive approach for measurement of linear motion is to use a direct linear encoder track or scale providing either absolute or incremental output. The most popular types in use are those using optical techniques, many with tracks of opaque and transparent areas. Basically the scale is a straightened version of a rotary shaft encoder. Whilst it is possible to have multitrack absolute encoders the scales become very wide and difficult to install and maintain. Most scales are of the incremental type and consist of finely divided optical gratings. Manufacturing techniques have been developed which enables scales up to 10 m in length to be made with a grating pitch of 0.01 mm with a very high degree of accuracy.

Two measurement systems are in current use; in both cases, the scales move relative to the optical system. One uses transparent scales illuminated one side with light-sensitive cells on the other as shown in Figure 3.4. The other method uses a reflective technique with the illumination and the cells on the same side and the scales usually polished steel with the grating engraved or etched on the surface. The optical system is arranged such that it is the reflected light from the scale which is detected by the cells. Simply counting pulses generated by the grating scale leads only to a coarse resolution. The systems actually in use, therefore, usually involve some form of resolver/interpolation to increase the resolution similar to that described above for the subdivision of the outer track of a shaft encoder. This requires multiple outputs electrically phased to provide at least sine and cosine outputs. The phasing is relative to the sinusoidal output derived from one photocell output. This type of output (sine/cosine) is also necessary in order to derive the direction of motion. In addition, there must be some form of extra marking on the scale to act as a position datum. This datum usually is at one end of the scale and is the point where

the counter would normally be set to read zero. In some systems the counter can be set to zero at any given position along the scale. This position then becomes the datum from which all measurement is determined.

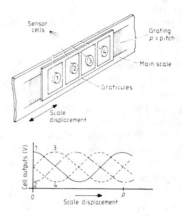

**Figure 3.4.** Linear grating assembly.

In most linear transducers the optical system is arranged to have four simultaneous separate outputs. In one the output will vary sinusoidally by one cycle as the optical system moves one 'grating' distance. A second output is arranged to produce the inverse of the sine, i.e. ' – sine'. The third output is so positioned to produce the 'cosine' output, that is displaced electrically by 90° from the sine output. The fourth provides a negative cosine, i.e. a ' – cosine' output.

Each of the four outputs is obtained by detecting the light passing through the scale and a graticule. The graticule consists of a short length of transparent grating having the same pitch as the main scale (Figure 3.4). The photocell will have maximum output when the ruled lines of the scale and graticule coincide and minimum output when the lines of the scale coincide with the spaces of the graticule. The photocell behind the graticule averages the light received over a short length of the scale, i.e. over a number of grating lines of the main scale. This further minimizes any small errors that might exist in the spacing of the grating. Four separate graticules are used on a single indexing frame, positioned relative to each other, to provide the four phased outputs as the graticule and optical assembly moves relatively to the scale as shown in Figure 3.4. These outputs are used in interpolator networks to provide four or more subdivisions of the main scale division.

### 3.3.2. Moiré fringe techniques

Moiré fringes are produced by a transparent grating and graticule, both having the same pitch, but with lines of the graticule inclined at a very small angle to the lines on the scale. Figure 3.5 shows the effect produced when the graticule is inclined by one grating pitch ($p$) over the width of the scale. The angle $\alpha$ is given by:

$$\alpha = \tan^{-1}(p/y).$$

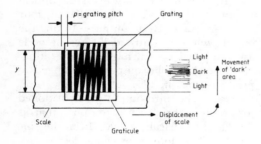

**Figure 3.5.** Moiré fringe.

In this case there is one horizontal dark area which moves vertically the distance $y$, for a horizontal displacement of the scale by a distance $p$.

Two horizontal dark areas are produced if the angular incline of the graticule is $2p$. Similarly, three horizontal dark areas are produced by an inclination of the

graticule of $3p$ and four horizontal dark areas by an inclination of $4p$.

A single light-sensitive cell at a fixed vertical position across the width of the main grating will sense one complete dark–light–dark cycle as the graticule is displaced by a distance $p$.

By careful positioning of two separate light cells the output can be made to represent sine and cosine of one cycle of displacement, that is, the two outputs are separated by 90 electrical degrees. These two outputs can then be used to increase the resolution by a factor of 4 in a similar manner to the methods used in incremental shaft encoders. Operating on a grating with 100 lines for each linear millimetre, this will provide a resolution of 0.0025 mm. The direction of movement can also be determined from these two outputs, in order to direct the counter to count up or down.

Four light cells can be positioned to provide the sine/cosine outputs and their negative values as described in section 3.2 in relation to shaft encoders. Interpolation of these four outputs can use the techniques used for many other incremental techniques where similar output can be produced.

### 3.3.3. Interpolation systems

The simplest interpolation network consists of a resistor chain as shown in Figure 3.6 to which the sine/cosine outputs are connected. The outputs A, B, C, D and E produce a set of quasi-sinewaves, differing in phase according to the tapping position on the resistance chain. Each separate waveform from the pattern is then used to generate a square wave, by the use of level detectors and these differ in phase relationship as shown in Figure 3.7. By using a relatively simple logic system, 10 separate pulses can be generated. This method provides a tenfold increase in resolution over the scale grating.

## 3.4. Synchro-resolver conversion

Synchros are used extensively in military control systems and are finding increasing use in numerically controlled machine tools and other industrial applications. Although most of the various synchro devices are more suited to AC control systems, a number of different conversion techniques have been developed to provide a digital output of a shaft position.

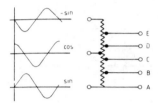

**Figure 3.6.** Simple interpolation network.

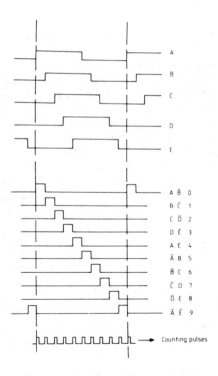

**Figure 3.7.** Interpolator logic.

Resolvers are effectively rotary transformers in which a rotor represents the primary and is fed through slip rings by a single-phase reference supply. The stator, which is in effect the secondary, contains two windings which are arranged to produce two separate voltages, at the same frequency as the reference supply. The magnitude of the voltages on the two stator windings varies as the angle of displacement of the rotor relative to a datum on the stator. The magnitude of one voltage varies as the sine of the angle of the shaft position and the other as the cosine. The two outputs are, therefore:

$$v_1 = V_\mathrm{m} \sin \omega t \sin \theta \qquad (3.1)$$

$$v_2 = V_\mathrm{m} \sin \omega t \cos \theta \qquad (3.2)$$

where $\omega$ is the frequency of reference supply, $V_\mathrm{m}$ is the maximum output voltage and $\theta$ is the angular position of the rotor. The supply or reference frequency may be 50 Hz but is more usually 400 Hz in instrumentation systems and higher for special purposes.

The two resolver outputs are in effect amplitude modulated signals at the reference frequency. These signals are demodulated or 'converted' by one of the following methods:

> phase-shift converters
> function-generator converters
> tracking converters
> successive approximation converters.

The aims of each are similar, that is to provide a digital output proportional to the rotor position. The information contained in the two resolver signals is sufficient to define uniquely the position of the rotor relative to the stator over the full 360° of rotation. The various conversion techniques all use the two analogue signals to produce a digital output. It is the converter that makes the resolver into an incremental transducer producing pulses which have to be counted to provide digital position data. The differences between the various converter methods are in the resolution available, the speed at which the shaft can be rotated and still maintain the designed resolution and the sensitivity of the system to the unwanted distortion of the resolver signals.

### 3.4.1. Phase shift converters

If the two resolver output equations (3.1) and (3.2) are fed into the circuit shown in Figure 3.8 the voltage $v_3$ will be given by

$$v_3 = \sin(\omega t + \theta).$$

By changing this sinusoidal voltage and the reference voltage to square waves, the two waveforms can provide a start and stop signal for a counter system as shown in Figure 3.9. The number accumulated in the counter then represents the phase shift of $v_3$ relative to the reference supply and therefore the angle of rotation of the rotor, $\theta$. It is important that the reference frequency and the count rate are synchronized since the real time between the start and stop pulses is a function of the reference frequency. The resolution depends upon the ratio of count rate to reference frequency. However, to maintain this resolution the speed of rotation of the shaft is limited since the counter update takes place only at the reference frequency.

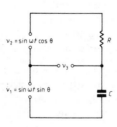

**Figure 3.8.** Phase-shift converter. $\omega RC = 1$, where $\omega$ is the reference frequency.

Generally, therefore, this method is used only for slow rotational speeds; 20 r.p.m is a typical maximum speed for a resolution of about 1 degree of arc. Special techniques have been developed to over-

come the synchronization problems and to increase the resolution and the operating speeds, but for significant improvement other conversion techniques are usually adopted.

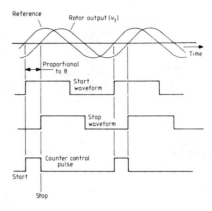

**Figure 3.9.** Phase-shift converter waveforms.

### 3.4.2. Function-generator converters

These converters use a feedback technique in which the digital output is fed back to a function generator and a comparator which develops a signal proportional to the error between the shaft position and the digital output. The error signal is used to drive the digital output towards a value which reduces the error to zero (Figure 3.10).

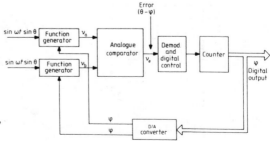

**Figure 3.10.** Function generator converter.

Even in its simplest form this system provides a higher degree of accuracy and resolution ($\pm5$ seconds of arc) although more complex and more expensive than phase-shift converters. It is also less sensitive to unwanted components in the resolver signals and to variations in the reference supply.

There are a number of methods of achieving the 'function generation', but most are effectively hybrid multipliers which generate an analogue output signal which is the product of analogue inputs, (representing $\sin \theta$ and $\cos \theta$) and a function of a digital input representing the encoder output $\phi$).

The outputs of the function generators are therefore:

$$v_a = \sin \omega t \, \sin \theta \, \cos \phi$$

$$v_b = \sin \omega t \, \cos \theta \, \sin \phi$$

These two voltages are fed to a comparator circuit producing an error voltage $v_e$, a function of $(\theta - \phi)$.

$$v_e = v_a - v_b$$
$$= \sin \omega t (\sin \theta \, \cos \phi - \cos \theta \, \sin \phi),$$

therefore

$$v_e = \sin \omega t \, \sin(\theta - \phi).$$

The voltage $v_e$ is demodulated to produce an analogue signal which is a function of $(\theta - \phi)$ only and is used to control a counter to produce the digital output. When $\theta = \phi$, the error $(\theta - \phi)$, and therefore $v_e$, becomes zero and the counter value remains constant and the output is then the digital equivalent of the encoder shaft position $\theta$.

The simplest form of function generator uses a linear resistor network similar to that shown in Figure 3.11. The output voltage achieves the required function by the digital input selectively switching the voltage divider network.

### 3.4.3. Tracking converters

These are function generator converters which adopt a particular technique for generating the digital output. The demodulated error signal in analogue form is integrated with respect to time and the result used to drive a voltage-to-frequency

converter. The resulting serial output is fed to a counter whose output represents the value of $\phi$. This digital signal is fed back to the function generators as previously described. When $\theta = \phi$ the error voltage is zero, the frequency output also zero and, therefore, the counter and hence the digital output remains constant. A tracking converter produces a digital output which remains equal to the rotor position whilst the shaft is stationary or rotating at constant velocity.

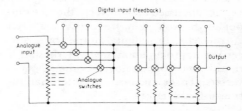

**Figure 3.11.** Function generator.

The block diagram, Figure 3.12, shows the counter as an integrator and indicates the feedback aspects of the system. The double integral in the forward path formed by the analogue integration of the error signal and the counter, defines the steady state and transient characteristics of the complete converter loop. These are:

no steady-state error, i.e. $\theta = \phi$ when $\theta$ is unchanging
if the input $\theta$ is changing, (i.e. the rotor shaft turning) $\phi$ will also be identical with $\theta$ and there is no error due to velocity of $\theta$
if the input $\theta$ is accelerating the digital output $\phi$ will lag behind the input $\theta$

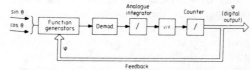

**Figure 3.12.** Tracking converter block diagram.

Typical tracking encoders can maintain a 14-bit resolution (1.3 minutes of arc) up to speeds of 240 r.p.m. Higher speeds will limit the resolution and vice versa. If the encoders are given a step input, e.g. the input–output error is $180^{\circ}$ of shaft rotation, the time taken for the output to represent the input depends upon the maximum frequency of the voltage to frequency converter and the time taken could be up to 0.5 s. Special techniques are used for increasing the resolution and adding other refinements to the overall performance.

### 3.4.4. Successive approximation converters

These are very similar to the tracking converters discussed above except that the output of the voltage-to-frequency converter is converted to a digital output by a successive approximation method similar to those used in analogue-to-digital converters. This system can be faster than the tracking converters.

### 3.4.5. Harmonic oscillator converters

The central feature of this type of converter is an harmonic oscillator which is an analogue circuit (Figure 3.13) with two integrators, and if left free-running would oscillate at a frequency in the range of 100 to 200 Hz. The outputs of the integrators represent the sine and cosine of the oscillator waveform. The two resolver outputs, proportional to $\sin \theta$ and $\cos \theta$, are demodulated and the resulting two DC outputs, whose magnitudes represent the $\sin \theta$ and $\cos \theta$ respectively, are used to set the initial conditions of the two integrators of the harmonic oscillator.

The converter includes FET switches to switch these initial conditions to the integrators, a clock, a counter and control logic. The complete measurement cycle starts by stopping the harmonic oscillator and switching the analogue values of $\sin \theta$ and $\cos \theta$ to set the initial conditions of the two integrators. The programming logic then sets the harmonic oscillator running at its natural frequency with the analogue outputs at the initial conditions previously set, simultaneously isolating the $\sin \theta$ and $\cos \theta$ inputs. At the same time the counter is initiated and clock pulses are counted until either of the harmonic oscillator

outputs cross zero. This has the effect of providing a count equal to the initial value set on the integrators and therefore equal to the magnitude of $\theta$, the resolver shaft angle.

The counter information is updated periodically at the harmonic oscillator frequency, e.g. approximately 200 conversions per second. With a clock rate of 1 MHz the output could have a total resolution of 12 bits including those bits which are derived from addition logic which identifies the quadrant (Woolvet 1977) in which $\theta$ exists. Overall accuracy depends upon accurate clock rate, stability of the harmonic oscillator and a number of other characteristics which can usually be improved by further sophistication of the system.

Synchro-encoders of any type are effectively absolute devices always giving the angular position of the rotor over $360^{\circ}$, relative to an electrical datum of the stator. If the encoder is used for continuous rotation then an additional counter system must be used to take account of complete revolutions, and the $\sin \theta$ and $\cos \theta$ outputs used to determine the direction of rotation.

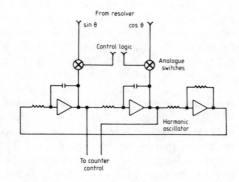

**Figure 3.13.** Harmonic oscillator converter.

## 3.5. Variable frequency devices

There are a number of techniques in which the variation of a parameter to be measured can be used to cause the variation in the natural frequency of vibration of some part of the transducer. The actual measuring method will be similar in many ways to simple incremental encoders in that a counter/timing circuit is required. Either the periodic time of the vibration frequency may be measured by counting pulses from a high-frequency clock or alternatively the frequency output can be converted to pulses which can be counted over a given time period.

The common feature of the type of transducer described in this section is that the change in frequency does not depend upon relative motion between parts of the transducer as in encoder systems. Almost any parameter which can be measured by a change in DC potential or change in resistance can be used by a voltage-to-frequency converter, or be formed as part of an oscillator circuit, to provide a variable frequency output which can then be used to develop a digital output. Examples of these systems are, a strain-gauge bridge on a diaphragm for measuring pressure, potentiometers for measuring displacement and thermistors for measuring temperature.

### 3.5.1. *Vibrating strings and beams*
Vibrating string transducers have taken a variety of forms, the most popular being used as strain gauges, and have been used to measure both force and strain. The string or wire, usually steel, is fixed at one end and the force to be measured applied at the free end. As a strain gauge both ends would be fixed to the structural member whose strain is to be measured. Any change in tension in the wire caused by a change in the force applied will change the natural frequency of free vibration. The frequency change is measured by a variable-reluctance pick-up which is amplified to provide the frequency output and also to provide a feedback to an electromagnetic exciter to sustain the wire in free oscillation as shown in Figure 3.14.

The frequency of free vibration of a taut

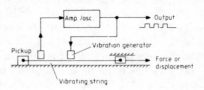

**Figure 3.14.** Vibrating string transducer.

wire or string is given by

$$f = \frac{1}{2L} \left(\frac{T}{m}\right)^{-1/2} \text{Hz}$$

where $L$ is the length of wire between supports, $T$ is the tensile load and $m$ is the mass per unit length.

A thin beam, that is, a small steel tape, can be used in a similar manner. An alternative application has used a twisted beam or tape stretched across the faces of the electromagnetic pick-up and exciter. The frequency changes as the beam is twisted and the transducer therefore acts as the measurement of angular displacement.

### 3.5.2. *Vibrating cylinders*
However, the most popular of transducers under this heading are those incorporating vibrating cylinders which have been commercially developed for the measurement of fluid pressure, density and mass flow. In the most common design the measuring element consists of a steel or alloy cylinder closed at one end approximately 25 mm diameter, 50 mm long with the side wall only 0.075 mm thick. The natural frequency of vibration selected is usually in one of its most stable modes, which is sustained and detected by iron-cored solenoids across the diameters of the cylinder. An amplifier and filter system and the position of the exciter ensures that the cylinder vibrates only in the mode selected.

The frequency of vibration depends upon the dimensions and the material of the cylinder and upon any mass caused to vibrate with the cylinder walls. Since the gas in immediate contact with the walls of the cylinder also vibrates, the frequency depends upon the density of the gas. In this case the transducer will output a frequency and hence a digital output which is a function of the gas density. In this application the gas pressure must be the same on both the outside and the inside of the cylinder. Any differential pressure across the cylinder wall will create a tension in the cylinder and change the frequency of free vibration. This factor allows the cylinder to be used as a pressure transducer providing one side of the cylinder is maintained at constant pressure, (e.g. a vacuum).

A typical pressure transducer measuring up to 300 atmospheres will have a natural frequency varying from 1.5 kHz to 5 kHz over its full working range. The cylinders will be of different physical size for different ranges. Special precautions must be taken to minimize errors due to temperature change and also non-linearities in the pressure and the frequency output.

Another type of vibrating tube transducer, larger in physical size than the pressure transducer, is used for measuring liquid density. The liquid is caused to flow through two parallel pipes whose ends are secured together and to a rigid base plate. The tubes are coupled by flexible couplings to the main flow. Between the tubes are the drive and pick-up electromagnetic coils which cause the tubes to vibrate in a simple lateral mode. The frequency of free vibration is a function of the density of the liquid filling the tubes.

Other applications of vibrating cylinders have been suggested, including the measurement of force and the measurement of torque.

## 3.6. Other techniques

### 3.6.1. Depth transducers

A direct digital output of fluid depth can be obtained either by using transducers to detect the change of inductance or change of capacitance. For the inductance types, the transducer elements consist of pairs of coils located either side of a vertical tube. The coils are located outside the tube. The inductance across the tube between a pair of coils is influenced by the presence or absence of liquid in the tube separating the coils. Nearly all fluids will cause sufficient change of inductance to produce a signal change. A number of such pairs of coils may be located along the length of the tube, each pair designed to provide a trigger when the fluid reaches a level adjacent to the coils. The resolution is only equal to the number of such pairs of transducers. Also, the system can only sense large increments in depth due to the minimum vertical separation of the coils necessary by their physical size. However, since the system is in effect a digital manometer the sensitivity can be greatly increased by inclining the tube towards the horizontal. This leads to a reduced detectable range. The method could also be used as a pressure gauge by using a capsule whose displacement caused by a change of pressure is used to displace the fluid along the tube. Similar limitations apply to the capacitance sensors since they are used in a similar way. In place of pairs of coils, there would be pairs of plates representing a capacitor, the actual capacity depending upon the dielectric. In some cases the capacitors may be completely immersed in the fluid whose depth is to be measured. Both inductive and capacitive systems require an AC modulation and demodulation system which can add considerably to the overall cost.

### 3.6.2. Magnetic effects

The magnetic recording of computer data, on tapes and disks, has been developed to densities up to 200 bits $mm^{-1}$ and greater densities are theoretically possible. Since it is possible to record eight or more tracks simultaneously it would seem that there are considerable possibilities for direct digital readout of any parameter which can be measured as a displacement.

Two magnetic systems are currently available, both used for the measurement of translational displacements and both working in a similar way. One method employs a flexible metal tape and the other an alloy rod. The rod or tape is prerecorded, with a continuous track of bits, which is moved relative to a fixed replay head or heads. The replay heads detect the passing of the recorded bits which provide an output which can be summed incrementally. Two or more heads may be used to provide the sine/cosine outputs for interpolation to improve the resolution (see section 3.3.3). The maximum displacement velocity is limited by the quality of the recording on the tape or rod. For example a tape or rod velocity of $50 \text{ mm s}^{-1}$ (corresponding approximately to the standard domestic cassette tape speed of 1.875 i.p.s) and prerecorded to a density of $200$ bits $\text{mm}^{-1}$ is equivalent to a recording frequency of 10 kHz. Higher velocities would therefore require good quality recording on the tape or rod and very high-precision reading heads.

### 3.6.3. Radiation transducers

The random radiation from a radioactive source can be detected by a photomultiplier or other type of detector. The random series of electrical pulses generated can be counted over a given period and provide a digital output proportional to the radiation received by the detector. A counting period in the order of 1 s has been found in some practical cases to give a repeatable resolution of 0.1%. The radiation received by the detector depends upon the strength of the source, the distance of the source and the area of exposure. The long life of a properly selected source and a fixed distance between source and detector reduces the detector output as a function of the area of exposure only and this can very conveniently be made to vary as a function of linear displacement or angular rotation. Practical transducers working on this principle are very rugged and more independent of many environmental effects than many other types of transducer.

In some installations it is possible to use radiation techniques for measurement of fluid flow. A neutron source located upstream of a fluid flow will cause some of the water particles to become radioactive and some of these can be sensed by a detector further down stream. For a given fluid, pipe size and given distance between the source and detector, the detector count gives a direct measurement of the flow rate. As with any counting technique, resolution and accuracy is dependent on the time period of the count.

### 3.6.4. Vortex transducers

There have been a number of techniques used for determining flow by the natural vortices caused by interference in the smooth flow of a fluid. The method now successfully developed commercially measures the flow rate in pipes by measurements of the vortices developed by a thick strut placed across the diameter of the pipe. The flow is disturbed and above a minimal velocity a regular series of vortices will be shed, the frequency depending upon the velocity.

The sensing element is required to detect the frequency of the vortices as they pass a detector located immediately downstream of the strut. The most successful method uses an ultrasonic transmitter radiating across the flow and a detector which senses when a vortex passes, by a change in the level of the signal received. This change can be used as a trigger to start and stop a counting circuit to provide a digital output measuring the time between successive vortices and therefore the fluid flow rate.

Other means of detecting the vortices such as sensitive piezoelectric pressure transducers to detect the pressure change in a vortex have also been used.

### 3.6.5. Laser techniques

The collimated beam produced by a continuous-wave laser can be used in many instrument systems, particularly for measurement of the relative displacement of large structures and for measurement over long distances. A digital output of measurement of the shorter distances commonly associated with metrology and instrumentation generally requires the use of a laser interferometry system. The fringes produced by a laser interferometer are similar to the moiré fringes referred to earlier but in the laser system they are due to interference of the laser beam and a reflected beam made to follow the same path as the original beam. The effect is to produce a dark fringe when the two coincident waves are out of phase and a light fringe when in phase. Any movement of the reflecting surface which changes the distance travelled by the reflected wave will produce a fringe. The fringe pattern is focused on to photocells in a similar manner to that used with moiré fringes. One complete cycle of interference corresponds to a movement of half a wavelength and this would be in the order of 0.001 mm.

For measurement purposes a counter would be required to count the number of fringe changes. A second sensing element would also be required to produce a quadrature signal in order to determine the direction of movement in order to generate a signal to initiate a count-up or count-down. Some compensation may also be required to overcome the variations caused by changes in temperature and air pressure. Instruments of this type are expensive but have been used to measure minute changes in displacements over long distances and also for very small displacements caused by adhesive films or thermal expansions.

A second method uses two laser beams and is, therefore, even more expensive.

The two lasers must emit beams of slightly different wavelength and in opposite polarizations. The two beams are reflected from the surface whose displacement is to be measured and compared in an interferometry circuit with the beams directly from the lasers. The beat signal of the reflected laser beams will vary with the motion of the reflecting surface and the difference between the beat frequency of the reflected beams and the beat frequency of the direct beams is a measure of the velocity of the reflecting surface. The difference is a Doppler shift and if a counter is used to count the difference in frequencies a measure of the reflector displacement is obtained.

## 3.7. Conclusion

The principles described above cover the majority of the techniques which have been adapted for use in digital techniques. Whilst it is difficult to suggest future developments it is safe to forecast that other devices will be developed. Most of these will undoubtedly be adaption or combinations of methods already existing. The most valuable advance will no doubt be made by the invention of relatively simple devices to provide the digital outputs required.

## Reference

Woolvet, G. A. (1977) *Transducers in Digital Systems* (London: Peter Peregrinus).

Paper 4

# Smart pressure transmitters

*A. T. Bradshaw*

This paper describes an approach to the measurement of pressure using microprocessor technology. The approach taken is to utilize the flexibility inherent in a microprocessor to increase the accuracy and versatility of the transducer, while retaining the traditional output signal format of a 4–20 mA analogue current signal representing the total input pressure span. This retention of a standard signal output allows direct replacement of a traditional pressure transmitter with a 'new technology' device without any redesign of control room or cabling. This is in direct contrast to the approach implied in the previous paper, where the transducer is designed to produce an output signal in digital form for transmission to a computer situated external to the transducer.

Readers unfamiliar with microcomputers will not find Figure 4.3 comprehensible, but this will not impair the accessibility of the major messages present in the paper. (Eds.)

## 4.1. Introduction

Microprocessors are now being applied to pressure and differential pressure transmitters to enable the resulting 'smart' transmitters to perform their tasks more intelligently. A number of instrument companies in the USA, Europe and Japan have now released these third-generation pressure transmitters for sale.

The characteristics of these three generations are:

*first generation:* force-balance type electromechanical requiring spanners and screwdrivers to alter pivot points for range changes: typical accuracy 0.5% of span.

*second generation:* electronic type using screwdriver adjustments of potentiometers for fine zero/span changes and switches/links for coarse adjustments: typical accuracy 0.25% of span.

*third generation:* digital type with push-button settings, selectable linear/square-root characterization, temperature compensation, built-in diagnostics and simplified calibration procedures: typical accuracy 0.1% of span.

While it is known that other innovative instrument companies are developing their own smart transmitters, two of the best-known models in the marketplace today are the H-Series (Toshiba) and the ST-3000 (Honeywell). Both of these offer improved accuracy, ambient temperature compensation, selectable linear or square-root output, direct or reverse electrical output and diagnostics facilities.

The subject of this paper is to

Originally published in *Measurement and Control* **17** (October 1984).

highlight how the use of modern technology can transform traditional products like pressure transmitters and provide a wide variety of features and benefits.

First- and second-generation pressure transmitters give a 4–20 mA signal output on a two-wire connection, corresponding to 0–100% of the measured variable range, and all electrical power in the unit is derived from this 4 mA live zero. They have limited adjustability for different pressure ranges. Performance varies with temperature changes, time and static pressure variations. Frequent recalibration is needed to maintain acceptable performance. This recalibration is time-consuming and difficult to carry out in an open environment with, usually, hazardous conditions. Most transmitters are used for fluid flow measurement and the remainder for pressure or level applications.

Flow measurements are primarily made with an orifice plate and differential pressure unit whose output can be processed by a square-root auxiliary to provide a signal directly related to flow. Orifice plates are highly repeatable, well-documented, readily available items and both now and in the forseeable future represent the preferred way to measure flow. Where higher accuracy is required, magnetic flow meters or turbine meters are used in spite of their higher costs, but are limited to conducting fluids or clean fluids. Vortex and ultrasonic approaches for measuring flow are becoming more common and will establish themselves in certain niche applications. They lack the simplicity of an orifice plate and differential pressure meter which can be easily zeroed and spanned, and disconnected from the process, are price independent of pipe size, and operate with a wide variety of fluid and gases, at even high temperatures.

The concept of the third-generation transmitter is shown in Figure 4.1. It can be a replacement for conventional units and is connected via ordinary two-wire 4–20 mA wiring but with the added capability of digital communication from a hand-held interface connected anywhere the 4–20 mA is accessible. This enables remote adjustment of the transmitter database and acquisition of diagnostic information to minimize loop downtime. Even more significantly the transmitter concept provides vastly improved performance and rangeability over present-day offerings. All of these are explained and expanded on below.

Inside a conventional-looking meter body is a three-variable sensor measuring differential pressure, static pressure and temperature. This meter body is pre-programmed in manufacturing to characterize the unit for linearity, static pressure and temperature effects, and it computes a highly repeatable and accurate pressure measurement. These characteristics are held in PROM memory and, being specific to one meter body, are kept with the meter body.

The combination of characterized meter body and digital electronics has enabled a quantum leap forward in performance and benefits.

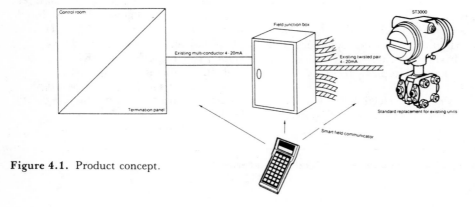

**Figure 4.1.** Product concept.

## 4.2. Performance

The major improvements made available by third-generation transmitters in accuracy, temperature performance, static pressure effects and time stability give an overall improvement in performance of three to five times in typical conditions over one year.

This three to five times improvement is a conservative figure for typical spans of 250 mbar and 500 mbar when the combined effects of accuracy, temperature and static pressure errors and time drift are considered. Figure 4.2 shows the combined effects of accuracy, temperature changes of 55 °C, static pressure changes of 70 bar one one-year time drift.

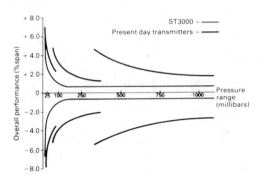

**Figure 4.2.** Overall performance comparison, showing the combined effects of accuracy, temperature changes of 55 °C, static pressure changes of 70 bar and one-year time drift.

Since a control system is only as good as the inputs it receives, the unit provides significant advantages in control accuracy, product yield and reduced wastage. It obviates the need to install high-cost magnetic flow meters or the equivalent.

The actual value of savings depends on many factors. e.g. throughput volume, value of material, and whether the measurement is used for cost purposes. The overall performance improvement of 1–2% can be very significant in the full spectrum of applications, for example:

*alarms:* enables closer working to plant limits
*control:* more repeatable product: closer control in varying conditions
*balancing/costing:* improved mass/energy balancing
*applications:* more accurate recovery of costs.

An immediate saving is available by not having to build a temperature-stabilized enclosure around the transmitter to minimize one of the contributors to overall unrepeatability. These enclosures are used for all transmitters on many plants to bring transmitter performance to acceptable levels. The cost of purchasing and installing these enclosures together with the temperature-controlled heating element is normally more than the cost of the transmitter itself, and they eliminate only one of the factors giving unacceptable overall performance — temperature. They have no effect on time and static pressure variations. Impulse lines and manifolds may still need heat tracing for high viscosity fluids.

## 4.3. Rangeability

Rangeability is increased to 400 : 1 for the typical DP/I unit (from 6 : 1). i.e. 2.5 mbar minimum span to 1 bar maximum. A typical unit can now cover the requirements of virtually all flow and level applications in a single unit. Only three sensors are required to cover the complete range, 2.5 mbar to 700 bar.

This greatly extended rangeability reduces costs of spares stocking, increases interchangeability of units with plant alterations, and provides other less tangible benefits such as freedom from having to define exact ranges when ordering units.

An obvious example of real saving through increased rangeability is in applications where two or more conventional units are applied across a single orifice plate to improve accuracy over a wide range of flows. The increased rangeability/performance of the third-generation transmitter means that only one unit is required, thereby saving not only the other units and the cost of installing and maintaining them and also avoiding the inherent low reliability of two or three loops against one. In the case of the single smart transmitter the range can be changed either manually through the communicator for different plant operating conditions, as discussed in the next section, or automatically via a dedicated-flow computer package.

## 4.4. Remote adjustability

The unit range, damping etc., can now be adjusted from up to 1500 m away by communication over the 4–20 mA signal line. The adjustment is made by connecting a smart field communicator (SFC) across the two signal lines at any point along their path. For control security, the first message in any communication is verification of identity. Then the unit can be interrogated and adjusted within very wide limits, for example:

  output may be characterised linear or square root
  damping time constant may be adjusted in ten steps
  lower range value/upper range value may be adjusted to any value in the pressure range
  adjustments are made in engineering units selectable from mbar, bar, kPa. p.s.i. and all other pressure units
  direct or reverse electrical output
  electrical output may be held constant at any value in the 4–20 mA range for loop integrity checks

The SFC is a portable adjustment and maintenance device with rechargeable batteries, 16-character alphanumeric LCD display and 32-key membrane keyboard which is ergonomically colour-coded. The SFC is also intrinsically safe for use with the transmitter.

The ability to re-range the transmitter without applying pressure is possible because of the stable, accurately known pressure characteristics which are in digital format. Thus the transmitter can be instructed to change its range by specifying the pressure equivalent to 4 mA and 20 mA, and it will accurately obey this instruction. Calibration corrections can be made for small errors due to mounting angle effects or variations between standards. The greatly improved stability of the sensing and measurement approach means that zero checks, and corrections if necessary, can be performed from the control room at plant shut-downs which typically occur for yearly maintenance. Use of digital techniques also gives improvements in stability. For example, digital memories, unlike resistors, do not suffer from time drift or changes in value with temperature.

This feature is absolutely unique and provides major savings in installation, operations and maintenance. No longer need impulse pipes be brought down to accessible levels or expensive catwalks be constructed to provide access to measuring point. Avoiding impulse lines minimizes condensation/degassing problems. Re-ranging and remote diagnostics can now be performed back at the control room, thus taking a fraction of the previous maintenance time which usually requires two men for several hours to do any field check or change.

Out of the wide range of real savings and benefits, the one highlighted here is the sequence of events at plant start-up. Key activities at this time are belling out wires, coordinating control room and plant signals, calibrating the control loop, calibrating the transmitter, fine-tuning the control loop, adjusting damping. Anyone who has witnessed this event at a plant's birth knows the problems of coordinating activities at both ends of a loop. With the remote adjustability of the third-generation transmitter, the identity of the transmitters connected to the wires can be confirmed, the unit can be switched to constant output to calibrate the loop, the transmitter can be ranged and the input loop/controller can even be exercised, alarm setting can be verified and damping can be adjusted without any process running. All of this can be carried out by one man from the convenience of the central location.

## 4.5. Reliability

The typical third-generation transmitters have been designed with reliable and stable operation as the dominant requirement. The minimum number of components is used, and protection is provided against all foreseeable damaging influences, e.g. radio frequency interference, reverse polarity, over-pressure, surge voltage and lightning. The unit has also been designed to meet Cenelec Intrinsic Safety and Explosion-proof Standards as well as North American and other International Safety Approval requirements. The enclosure complies with classification IP67 which prevents any corrosion, contamination or moisture problems with the working of the transmitter. Since there are no switches, links or adjustments in the electronics compartment the integrity of the seal need not be disturbed and the likelihood of a cap being left

loose is greatly reduced.

In spite of all precautions, failures could occur either through misuse of equipment or component breakdown. Diagnostic aids are included in both pressure transmitters and SFC to avoid misuse, e.g. by identifying excessive process temperature or, in the event of component breakdown, to identify the fault, advise the SFC operator and alert the control room by giving an upscale alarm signal.

In the event of a transmitter or loop problem the diagnosis via the SFC from the control room area will identify the problem and corrective action so that maintenance staff know what tools/replacement parts are needed. Loop downtime is minimized.

The overall logic diagram of a typical modern transmitter is shown in Figure 4.3. Key blocks of interest are given below.

### 4.5.1. Sensor
This is based on silicon ion implant technology and combines advances in silicon technology with the use of a perfectly elastic drift-free material to provide an accurate, stable sensing element. For

**Figure 4.3.** Logic diagram.

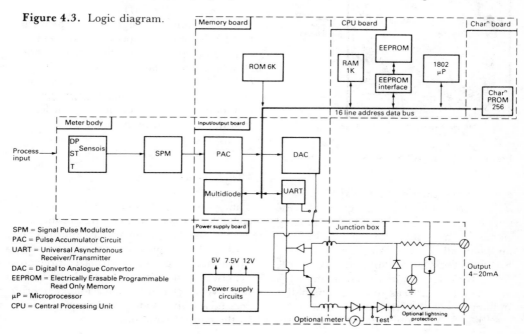

SPM = Signal Pulse Modulator
PAC = Pulse Accumulator Circuit
UART = Universal Asynchronous
        Receiver/Transmitter
DAC = Digital to Analogue Convertor
EEPROM = Electrically Erasable Programmable
         Read Only Memory
µP = Microprocessor
CPU = Central Processing Unit

example, total transmitter drift is specified as ±0.06% upper range limit/year.

The sensor mentioned earlier actually senses three variables. Figure 4.4 shows the front face of the sensor with the following sensing elements pointed out. The differential pressure across the silicon chip is sensed by a Wheatstone bridge arrangement of four piezoresistive elements, two of which increase in value with applied pressure and two of which decrease. With a constant voltage across the bridge this gives a quadrupling effect on the electrical change and hence a high-millivolt output.

The static pressure is also sensed by a Wheatstone bridge network of resistances, situated this time on the section of silicon in close proximity to the bond to the supporting tube of glass. The different coefficients of compression of silicon and glass result in a piezoresistive measurement of applied static pressure. Temperature is sensed by a single resistive element.

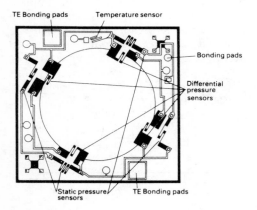

**Figure 4.4.** Ion implant sensor

### 4.5.2. Signal pulse modulator (SPM)
This converts the analogue signals from the sensor into a pulse-width modulated digital signal. The gain of the SPM is automatically changed to provide the very high input resolution required, that is, 400 times turn-down with better than 0.1% accuracy (i.e. at least one part in half a million repeatable resolution).

The SPM is time-shared between the three variables sensed and two zero check resistors. Since the temperature and static pressure values are slower moving than differential pressure and are only being used for compensation, they are sampled less frequently. Typically for any 20 s period the DP will have been sampled 120 times, static pressure 12 times and temperature once.

### 4.5.3. Microprocessor
The microprocessor carries out its task on less than 4 mA total power. Two hardware multiply/divide chips are included to speed up some computations. The input is sampled six times per second and a new output value is computed at the same rate.

### 4.5.4. Universal asynchronous receiver/transmitter (UART)
This is the communications module which looks for messages on the input wires and, if they are present and valid, responds with the correct action. The response is given by overriding the normal 4–20 mA signal with 16 mA digitally coded pulses.

### 4.5.5. Electrically erasable programmable read only memory (EEPROM)
The transmitter database is held in EEPROM, i.e. the specific loop identity, range and damping are held in non-volatile memory. This enables units to be factory or workshop customized to a specific loop without memory loss. Subsequent changes of database are automatically transferred into this non-volatile memory.

## 4.6. Summary

The application of well-established technology has enabled significant advances in the performance and overall application of pressure transmitters. As this range of products displaces previous generations it is predictable that similar improvements will be made for other process measuring devices and their interface to the control equipment.

# Paper 5

# On-line measurement of liquid density

## *Conrad H. Hoeppner*

This review paper on the measurement of liquid density is included to illustrate the wide range of techniques available for sensing a single measurand. The techniques described range from some of very long standing (the pneumatic bubbler) to some which are very modern (the fibre-optic interferometry density gauge). (Eds.)

Liquid density may be measured in many ways. Here is a rundown of the various techniques, and their advantages and disadvantages in various situations.

Chemical engineers often need to know the density of liquids in a process stream. For example, the density of a liquid is frequently an indication of dissolved-solids concentration or the liquid's composition; in the case of fuels, density may indicate the heat energy available on burning.

By definition, density is mass per unit volume. The mass of a liquid can be found in two ways: by measuring its gravitational attraction, or by determining its inertia. Since volume is a function of length ($l^3$), the dimensions of density are $M\,l^{-3}$, where $M$ is mass and $l$ is length.

The density of a liquid is often compared with that of water; and the term 'specific gravity' is defined as the ratio of the density of a particular liquid to that of water.

## 5.1. Measuring liquid density

Instruments for measuring liquid density are used in two ways: with flowing liquids, or with liquids in storage tanks. Not all instruments are suitable for both applications.

Many times, in process control, it is desired to know the total mass of liquid rather than just its density. As we will discuss later, mass can be measured directly without the need for separately determining volume and density.

The 'mass per unit volume' class of density meters comprises instruments in which mass and density may be measured interchangeably. An instrument such as a tank with a pressure gauge and a height gauge yields density by simple computation. Pressure (lb in$^{-2}$)/height (in) = density (lb in$^{-3}$). To obtain total mass, we find

$$\text{Mass} = d \int_{0}^{h_1} A \; \mathrm{d}h$$

where $d$ is density and $A$ is the cross-sectional area of the tank at height $h$.

Originally published in *Chemical Engineering* (1 October 1984).

In a tank, the density of a liquid often changes with height. This may occur if the tank is exposed to the sun, which may heat the top layers, thereby reducing their density. Variations may also occur if petroleum products such as gasolines of different densities are pumped into a tank (the lighter liquids remain on top).

The method of measurement described above provides *average* density readings, but if the density at each level is multiplied by the volume element at that level, total mass is obtained. It should be noted that it is difficult to obtain the density at each level, and thus it is difficult to measure total mass. If an alternative measurement is made — simply weighing the tank — then total mass is easily and accurately obtained and neither height nor density measurements are required.

### 5.1.1. *Instruments that use gravity*

Figure 5.1*a* shows a simple way of measuring average density in a tank, a principle that can also be employed to measure the density of a flowing liquid. The method also yields total mass, because the device's sidewalls are vertical. (If the sidewalls are not vertical, it is necessary to know the density at all levels — rather than just average density — to find total mass.)

Figure 5.1*b* represents another method of measuring density or total mass in a standing tank. In this case, two pressure sensors are employed. Density is found by

$$D = (P_2 - P_1)/H$$
$$M = (P_2 - P_1)A$$

where $A$ is the cross-sectional area of the tank.

### 5.1.2. *Flowing liquids*

An obvious extension of the method of using weight of a known volume to control the density of flowing liquids (by adding a liquid of different density) is shown in Figure 5.2. Here the known volume is weighed, and the balance is used to activate a control valve that proportions the mixing fluid, using more or less of it to

change the density of the flowing fluid. This completes the feedback control loop, and keeps the density of the flowing liquid constant.

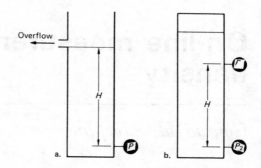

**Figure 5.1.** Ways of measuring density and total mass.

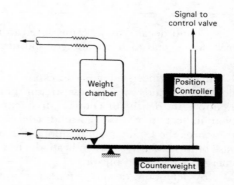

**Figure 5.2.** Controlling density of a flowing fluid.

The setpoint may be changed by moving the fulcrum. Moving it to the right decreases density and moving it to the left increases density. Figure 5.3 shows another arrangement of a similar system.

In both systems, it should be noted that restoring forces in the flexible couplings will affect the setpoint. (This error may be calibrated out. However, if the flexure force changes with temperature or pressure, this may result in erroneous control.)

With many instruments, the velocity of flow may also be a disturbing factor in the measurement of density. Consider the common laboratory hydrometer, shown in Figure 5.4*a*. It consists of an element that floats in the fluid to measure its density — the higher it floats, the more dense the fluid.

In this form, the instrument is useful for sight readings. In order to take a reading, a sample of the liquid is usually drawn off. Modifications are required if the hydrometer is to be used in flowing liquids, or if it is to provide a signal for process control.

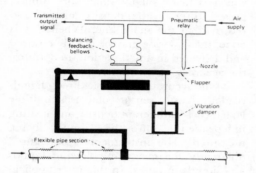

**Figure 5.3.** Pneumatic control of density of a flowing fluid.

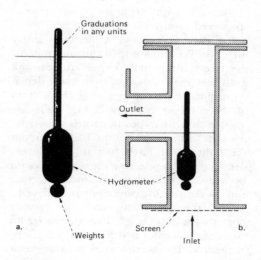

**Figure 5.4.** The laboratory hydrometer, and its use in a flowing fluid.

Figure 5.4*b* shows the hydrometer in a flowing liquid. The measurement may be satisfactory if both the rate of flow and the viscosity of the liquid are low. With viscous liquids or high flowrates, friction lifts the hydrometer, yielding an erroneous reading. If the float is completely submerged and held down by a torque arm, it displaces an amount of liquid equal to its volume, and when it is less dense than the liquid, it exerts an upward force, which is

$$F = (D_L - D_F)V_F$$

where $D_L$ is the density of the liquid, $D_F$ is the density of the float and $V_F$ is the volume of the float.

The float remains essentially in a fixed position relative to its housing. This is an advantage in compensating for the effect of flowing liquid. Figure 5.5*a* shows an arrangement using a piezometer ring to introduce the liquid radially inward, equally from all directions, causing the fluid to flow in equal amounts at equal velocities in opposite directions that are parallel to the upward force vector exerted by the float.

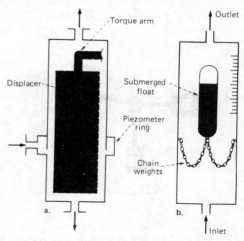

**Figure 5.5.** Two types of density sensors that use floats.

In this manner, the effects of flow-velocity and viscosity are cancelled and a correct density signal is obtained. (Note that this method can be used in a pressurized system.)

Another method of using the submerged float is shown in Figure 5.5*b*. Here the float moves, so it is not as well compensated for flow and viscosity changes by the piezometer ring (if one is used), but will give a visual reading of a pressurized flowing fluid.

Still another method of reducing the effects of flow and viscosity changes is shown in Figure 5.6. In this instrument, the flow is made parallel to the axis of rotation and does not affect the density measurement. The instrument produces an angular rotation proportional to liquid density, over a limited range. Since the rotor is rigid, it may be considered that all of the unbalanced mass is in Unit $D_3$. The buoyant force is proportional to the volume of the floats, Unit $D_1$ having the greatest volume. If just Units $D_1$ and $D_3$ were present, the instrument would be operative but quite non-linear. Unit $D_2$ may be considered as producing a torque opposing that of Unit $D_1$.

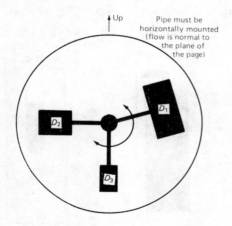

Up
Pipe must be horizontally mounted (flow is normal to the plane of the page)

**Figure 5.6.** Angular-position density sensor.

The force vector being caused by gravity is vertical on each. (The shaft about which rotation occurs must be horizontal.) As the unit rotates counterclockwise, the moment arm on each, which is the length of the float arm projected on the horizontal (i.e., the horizontal distance from the

shaft), decreases. So while the torque produced by $D_1$ decreases with counterclockwise rotation, the opposing torque produced by $D_2$ also decreases, producing greater linearity of rotation vs. density. The restoring torque is that produced by $D_3$, which must be present in the densest liquid to be measured. In order for this to be the case, the ratio of the unbalanced mass to the volume of $D_3$ must always be greater than the densest liquid.

It is often possible to locate the density meter in a sample of the flowing liquid, which is taken off by small-bore tubing, passed through the density meter, and returned to the main line. When this is done, the flow-rate may be greatly reduced and delivered in any direction, to optimize the accuracy of the meter.

One accurate and stable direct-digital-readout density sensor comprises multiple floats, each of different density and floating independently in the liquid being measured. The instrument is configured differently for measuring flowing liquids and liquids in tanks. In the latter application, it adapts well to tanks in which the pressure undergoes change as the tanks tip and tilt and are subject to varying acceleration and varying *gravity* — which diminishes with altitude — as is encountered in aircraft installations.

Figure 5.7 shows a digital density meter for flowing fluid. In this instrument, the free-floating balls either rise or sink until they reach liquid of their same density, or float to the top or sink to the bottom of the fluid or container. The balls are opaque and their positions are sensed by the optical fibres where the balls interrupt the light path. Each ball is of different density, and by selecting the dividing point between the balls that float and those that sink, a density measurement is achieved.

If the balls are made of opaque borosilicate glass (e.g., Pyrex), the coefficient of expansion is 0.5 p.p.m. °F. Consequently, the accuracy of measurement over a temperature range of 600 °F is better than 0.1%, without temperature compensation.

It is not necessary that the instruments

be vertically aligned if the shape of the float is spherical. Shapes other than spherical should not be used because they may stick in the tubes — even if mounted vertically.

In the case of flowing fluids, or even in filling or emptying tanks, the flow of the liquid should be maintained at approximately right angles to the gravity vector, to prevent the liquid from lifting or depressing the balls.

**Figure 5.7.** Floating-ball fibre-optic density sensor for flowing fluid.

When a long vertical tube is used, filling and emptying should be done from sideholes to prevent the tube from filling from the bottom — with, of course, only the most dense fluid rising in the tube.

For use in a controller, the light signals may be converted to electrical or pneumatic ones, either digitally or from a digital-to-analogue converter.

To permit its use in lighted areas, the optical source is usually modulated on and off at a high frequency, and a tuned filter is used to enable discrimination of the signal from the ambient light.

It should be noted that these sensors are immune to electromagnetic pulse effects. The extremely powerful electromagnetic pulse that is generated from a nuclear explosion in space is destructive to electrically wired equipment for hundreds of miles from the explosion.

While the sensors above must be used with a light-transmitting liquid, the distance is so short that most liquids are acceptable. A similar sensor has been developed for completely opaque liquids. It has larger float balls, each of different density, and each operating a separate electrical switch when it floats.

### 5.1.3. Instruments that use inertia

The foregoing examples were all of density meters using earth's gravity in their operation. Consequently, with the exception of the free-floating balls, they must be aligned properly in the gravitational field.

The inertial meter does not have this requirement; it may be aligned in any direction. It operates by vibrating a known volume of fluid, with the equivalent of spring coupling between the force and mass. The resonant frequency is obtained by employing the spring-coupled mass as part of an oscillator. The frequency is a function of the mass in the resonator cavities. The cavities being of known volume, density can be determined by frequency measurement.

Figure 5.8 shows examples of a torsional density meter using piezoelectric activators and piezoelectric sensors as the electromechanical transducers.

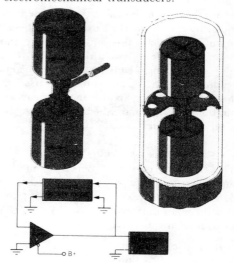

**Figure 5.8.** Torsional inertial-mass density sensor.

Another example of an inertial density meter is shown in Figure 5.9. Here, an electromagnetic transducer provides the excitation for, and a piezoelectric crystal detects the vibration of, a plate immersed in the liquid. The volume of the liquid being vibrated is not as well defined as in the previous example, but it is constant and, thus, the meter lends itself to calibration.

A very popular density meter is shown in Figure 5.10. Here, an open, straight-through tube is used, which minimally restricts flow. It is usually used in mainline flow. An electromagnetic transducer provides excitation, and another electromagnetic transducer detects vibration. Its resonant frequency changes with density and it is calibrated for direct density reading.

The electromagnetic transducer usually contains a permanent magnet as a force multiplier. This can be a serious source of instrument drift and error if there are magnetic particles in the liquid. Iron and the oxides of iron ($Fe_2O_3$ and $Fe_3O_4$) are magnetic and are attracted to, and attach themselves to, the magnet or the container's walls near the magnet. Practically all fuels and hydrocarbons are pumped through iron pipes. Water condenses in the fuel, the pipes rust and the iron flakes off, to be carried to the density meter by the liquid. The iron oxides, being more dense than the liquid, change the resonant frequency when they build up on the instrument, causing an erroneous reading. When the instrument is used in flowing liquid, the oxides may be carried off, but if they remain, the build-up may be a serious problem.

### 5.1.4. Inferential-type density meters

The inferential-type density meters measure a property of the liquid other than density, and from that measurement the density is inferred. The properties measured are usually:

dielectric constant
index of refraction
radioactive particle absorption
speed of sound.

These meters are generally useful for a specific class, or composition, of liquids and will vary greatly as the composition changes, or if small amounts of contaminants are present in the liquid. Their principal use is in the metering of hydrocarbons, as used in fuels. The density of a hydrocarbon can be employed as an inferential measurement of the heat content of the fuel, when burned; this measurement can be used quite accurately (with an error of less than 1%) to determine the fuel content of the liquid.

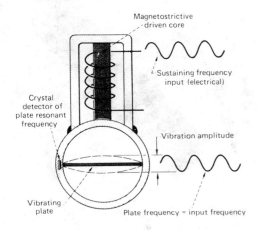

**Figure 5.9.** Vibrating-plate inertial-mass density sensor.

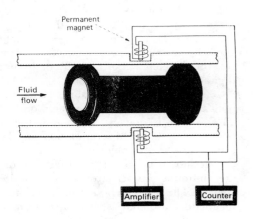

**Figure 5.10.** Vibrating-spool density meter.

A density meter that makes use of the dielectric constant of the liquid as an indirect measurement of density is shown in Figure 5.11. The capacitance between the inner and outer conductors of the sensor is measured electrically, usually by means of a bridge circuit.

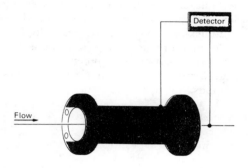

**Figure 5.11.** Dielectric-constant (or capacitance-gauge) density meter.

Another method of measuring density by inference is to measure the index of refraction of the liquid. This technique also tracks density closely in families of similar liquids. A typical instrument is shown in Figure 5.12.

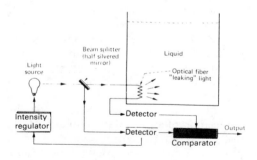

**Figure 5.12.** Measuring density by inference from index of refraction.

Here, optical fibres are used, which are bent or shaved to allow light to leak out. The more closely that the index of refraction of the liquid approaches the index of refraction of the fibre, the more light leaks out of the fibre and the less light reaches the detector. The detector output

will then be inversely proportional to the density of the liquid.

Since most sources of light will vary with time, current or voltage, a beam splitter is often used in the light path to conduct some of the light through another fibreoptic path, not immersed in the liquid, to a matching detector. In this manner, compensation is provided for the side effects that are not due to index-of-refraction changes of the measured liquid. Figure 5.12 illustrates a means by which the compensation may be accomplished.

Another inferential method of measuring density in specific families of fluids is that of radiation absorption. Figure 5.13 gives an example of this method. Here again, the radiation of the source changes with time — a half-life is known for each source. Thus, compensation must be used, either as a function of time or by parallel path outside the liquid (as described above).

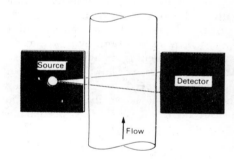

**Figure 5.13.** Radiation-absorption density meter.

It is also possible to infer density by measuring the speed of sound through a fluid. The projecting transducer and/or the receiving transducer may be immersed in the liquid or located outside the tank or pipe. The method is easily adaptable for measurements of static liquids in tanks, but must be used with care in moving liquids. One method of measuring density by measuring speed of sound is shown in Figure 5.14.

The speed of sound in a liquid is a function of other parameters besides density. Furthermore, there is a mismatch in the acoustic coupling between the transducers and the liquid, which varies with liquid density. For a restricted family of liquids, only temperature need be measured in addition to density; but if a wide range of liquids are to be measured, compressibility and viscosity must be considered. The simplest method is to calibrate the density instrument in each of the liquids, finding an output at the same time that the density is measured by immersing the solid standard in the liquid (see section 5.3). The effect of temperature change is different for each family of liquids; hence, calibration should be performed over the range of temperatures expected.

When the speed-of-sound method is used to measure density in moving liquids, an additional complication results from the change in frequency due to the Doppler effect. At low speeds of liquid motion, this may be compensated for by using a sound path in opposite directions through the same part of the fluid. Even such a path is disturbed by vortices and by flow that has a velocity gradient that changes with time. At high speeds of flow, the apparent bidirectional speed of sound actually changes with flow velocity, so a separate measurement of velocity is needed for adequate compensation.

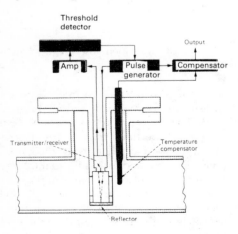

**Figure 5.14.** Speed-of-sound density meter.

## 5.2. Transmitting the measurement

In order to use the measurement made by an instrument, it usually must be transmitted to a controller, a display, or both. Measurements other than density may also need means of transmission and are multiplexed on the transmission carrier. Some carriers are not easily adapted to multiplexing and are unusable if multiplexing is a requirement. Other considerations are also important and sometimes restricting. First, let us examine the available methods. Typically these include:

> electrical
> pneumatic
> mechanical
> magnetic
> electromagnetic
> radio telemetry
> ultrasonic telemetry
> light and optics
> light and fibre optics
> nuclear radiation

A usual requirement is first to transmit the measurement through, or away from, the tank or pipe containing the fluid. Following this, either a short or long transmission distance may be desired. There may be unfavourable environmental factors, such as electrical interference, smoke, fog, snow, iron housings, high sound-levels, etc. Each of these may have an effect on the choice of transmission method.

Figure 5.15 is an example of the application of a pneumatic bubbler. Here, dried and filtered air is delivered by a pump, through a constriction and a long tube, to the liquid to be measured. The far end of the tube is immersed in the liquid unit its end is near the bottom. The air is allowed to bubble freely, automatically regulating the pressure in the tube to that of the liquid. The height of the liquid is known, so the pressure is proportional to density. Hence, back at the intake end, or wherever a side-tube leads to a pressure gauge or controller, the air pressure is proportional to the liquid density.

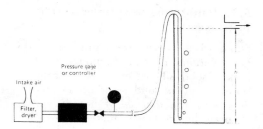

**Figure 5.15.** Pneumatic bubbler density meter.

It is sometimes not desirable to bubble the air through the process liquid. In such a case, instead of releasing the air to the liquid, an instrument with a diaphragm-controlled needle-valve sensor regulates the pressure, delivering the bleed air to a tube that carries it off without contacting the process liquid. Very often, the sensor also contains an adjustable constrictor, and then another tube following the constrictor delivers the gas pressure to the indicator and controller.

### 5.2.1. Fibre optics

Combining the sensor with the transmission means is likely to create a future family of fibre-optic systems. Passing the fibre through the liquid, so as to lose light into the liquid — as a function of index of refraction of the liquid — is one method of combination sensing, which may be called remote sensing. Using two fibres, and combining the light outputs for constructive and destructive interference (interferometry) is another method of remote sensing. For example, if a float is suspended between two stretched glass fibres, as shown in Figure 5.16, and the liquid becomes more dense, the upper fibre increases in length and the lower fibre decreases. A length differential of $10^{-7}$ is full scale!

This means that light travels farther in the upper fibre than in the lower one, producing a phase difference at the end. When the light is in phase, a strong signal is received from the detector. When the light is out of phase, there is darkness, or no signal, from the detector. Between full

scale and zero is the range of density measurement. Obviously this system may be extended to measure pressure, strain, torque, turbidity, liquid level, etc.

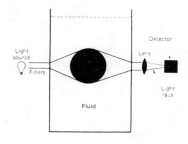

**Figure 5.16.** Fibre-optic interferometer density gauge and transmission system.

Another feature of fibre-optic light transmissions is that light may be transmitted concurrently in both directions along the fibre. A reflective instrument and transmission line as shown in Figure 5.17 may be used. When the density of the liquid increases, more light reaches the detector. If the reflector is specular it must be accurately aligned. If a diffuse reflector is used, accurate alignment is not necessary, but there is considerable loss of reflected light.

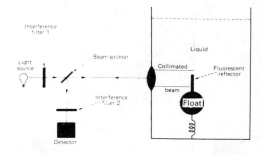

**Figure 5.17.** Fibre-optic reflective density gauge and transmission system.

While the reflective system can be used without the two filters shown in the figure, this configuration has limitations because of fibre imperfections, bends and stresses that may reflect some of the transmitted light.

The filters circumvent this limitation. Light is transmitted to the density gauge through Filter No. 1, which removes long-wavelength light. This light impinges on the fluorescent reflector and is converted to long-wavelength light, which is returned to the detector through Filter No. 2, which selects only the long-wavelength light that the fluorescent reflector generates. Only this light reaches the detector; light reflected by the fibre is blocked out. The reflector is only diffuse, hence higher-intensity light levels must be transmitted.

Nuclear radiation is presently useful only for short transmission distances. It may be used with opaque liquids and opaque containers, however. Figure 5.18 shows a typical tank application. This is an inferential system and liquids of different composition absorb nuclear energy in different amounts. Hence, it is useful only with families of fluids such as hydrocarbons, where absorption of energy tracks the density closely. In addition, material properties other than density affect the absorption of nuclear radiation; for example, small amounts of impurities may cause large errors.

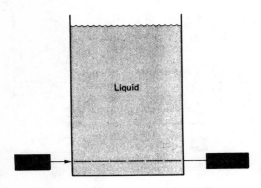

**Figure 5.18.** Nuclear density-detection-and-transmission system.

## 5.3 Calibrating density-measuring instruments

Standards of density are maintained at the U.S. National Bureau of Standards, which uses a solid silicon mass as its primary standard. This solid standard is used to determine the density of liquids and to calibrate liquid-density meters.

The Bureau also sells blocks of silicon, as secondary standards, for a price of $441 (as of June 1984). Either a 100 g or a 200 g secondary standard can be purchased at that price. These are described as follows:

Part No. 1840; 100 g;
    Dens. = 2.329074 ± 0.000019
Part No. 1841; 200 g;
    Dens. = 2.329075 ± 0.00017

These secondary standards are solid homogeneous pieces of silicon, the density of which is not affected by scratching, abrading or even breaking. If the piece is broken in two, then two secondary standards are obtained. (The size, shape and total mass of the standard are not important when it is used to determine the density of a liquid.)

Liquid density is determined by weighing the standard suspended in air and then weighing it again suspended in the liquid. The density of the liquid is calculated by:

$$D_L = [D_s(W_a - W_L) + W_L D_a]/W_a$$

where $D_L$ = density of the liquid; $D_s$ = density of the standard; $W_a$ = weight of the standard in air; $W_L$ = weight of the standard in the liquid; and $D_a$ = density of the air.

If, then, the liquid density (without change in temperature) is also measured by an instrument, a calibration of that instrument, traceable to the U.S. National Bureau of Standards is thereby obtained and the instrument is ready to go on-line.

Paper 6

# Measuring mass flow using the Coriolis principle

*K. O. Plache*

The paper describes the construction and theory of operation of a flowmeter which directly senses the mass flow rate of the fluid passing through the sensor. The paper is included as a good example of the way in which a description which could have been extremely mathematical and difficult to comprehend has been made accessible by a careful step-by-step approach to the underlying theory. (Eds.)

Described below is a unique mass flow meter that monitors the motion of each molecule of flowing material and sums all of the flow forces to generate a total force that is an accurate, linear measurement of mass flow. The meter, which can be used to measure the flow of gases, liquids, or non-homogeneous substances, uses electro-optical techniques to sense the forces.

## 6.1. Introduction

To measure the rate of mass passing through a pipe, it is first necessary to fully comprehend how mass itself can be measured. Newton's second law ($F = Ma$) provides the means of measuring mass. The law states that when an unbalanced system of forces acts on a body, it produces an acceleration in the direction of the unbalanced forces that varies inversely and proportionally to the mass of the body. The important concept is that mass cannot be measured without applying a force on the system and then measuring the resulting acceleration.

There are obvious difficulties in measuring the acceleration of a fluid due to a given force. Therefore, most flow meters on the market today measure a quantity that implies mass — instead of measuring that mass directly. One example of an inferred mass flow meter is the thermal-type meter which measures the implied mass rate of flow under certain restrictions such as constant specific heat. Another inferred method is by use of a volume-type meter with proper corrections for temperature, specific gravity, and pressure. The mass flow rate is based on a computation using these several parameters.

Editorial Note
In section 6.4 the author has used upper case M to represent both moment and mass. In reprinting the paper we have used roman M for moment and italic *M* for mass.

Originally published in *Transducer Technology* (March/April and May/June 1980).

There is a need to measure mass flow directly, and devices have been developed to accomplish this difficult task. The angular momentum mass flow meter consists of an impeller to impart a swirl to the flowing fluid and a means to measure the amount of torque required to remove the swirl that is generated. The concepts of measuring Coriolis forces and measuring gyroscopic precession have been investigated and development work has been going on for many years. The major problem associated with such meters is the difficulty in measuring the extremely small forces produced.

This article describes a mass flow meter that successfully uses the gyroscopic or Coriolis principle in directly measuring the mass flow rate. It also describes the development of a new and simple method of applying electro-optical techniques to make use of these concepts practical.

## 6.2. Mass flow meter

The new gyroscopic mass flow meter employs a C-shaped pipe and a T-shaped leaf-spring as opposite legs of a tuning fork (Figure 6.1). An electromagnetic forcer excites the tuning fork, thereby subjecting each moving particle within the pipe to a Coriolis-type acceleration. The resulting forces angularly deflect the C-shaped pipe by an amount that is inversely proportional to the stiffness of the pipe and proportional to the mass flow rate within the pipe.

The angular deflection of the pipe is optically measured twice during each cycle of the tuning-fork oscillation. The output of the optical detector is a pulse that is width modulated proportional to the mass flow rate. An oscillator/counter digitizes the pulse width and displays a numerical indication of mass flow rate.

The total mass flow over a given time interval is obtained using a digital integrator to sum the pulses of the flow rate indicator. In this way totalized flow is updated each oscillation of the tuning fork.

This mass flow meter eliminates the many problems associated with mass flow measurement. The output of the meter is directly proportional to mass flow rate; consequently, there is no need to measure the critical parameters of pressure, velocity, temperature, viscosity, or density. There are no parts in the flowing fluid and accuracy of the meter is unaffected by erosion, corrosion, or scale build-up in the flow sensor.

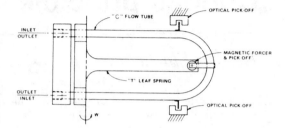

**Figure 6.1.** Mechanical configuration of the inside of the mass flowmeter including the C-shaped tube, the electromagnet which vibrates the tube, and the optical pick-offs to detect the amount of twist the mass imparts on the tube.

In laboratory tests on prototype units, accuracies in the order of $\pm 0.2\%$ full scale have been recorded using fluids ranging in specific gravity from 0.5 to 2.5 and in flow rates from 0.1 to 25 lb min$^{-1}$ (Figure 6.2). The flow meter is linear to within $\pm 0.2\%$ full scale over the total flow range. Tests have been conducted using both Newtonian and non-Newtonian fluids with a wide range of viscosities; the results indicate the meter is totally insensitive to viscosity. With the meter it is now possible to measure pulsating flow and two-phase flow.

In discussing the concepts for measuring mass flow with a meter based upon the principles of Coriolis force or gyroscopic precession, an explanation of Coriolis force, the force that causes a gyroscope to precess, is in order.

Coriolis force is generally associated with a continuously rotating system. Coriolis force due to the earth's rotation causes winds from a high-pressure area to spiral outward in a clockwise direction in the northern hemisphere and counterclockwise in the southern hemisphere. Depending on location, a projectile fired from a gun in the northern hemisphere will appear to veer slightly to its right, and to the left in the southern hemisphere. Figure 6.3 illustrates how the effect of Coriolis force can be experienced. A body moving on a rotating frame of reference such as a turntable or a

merry-go-round will experience a lateral force and one must lean sideways in order to move forward when walking outward along a radius.

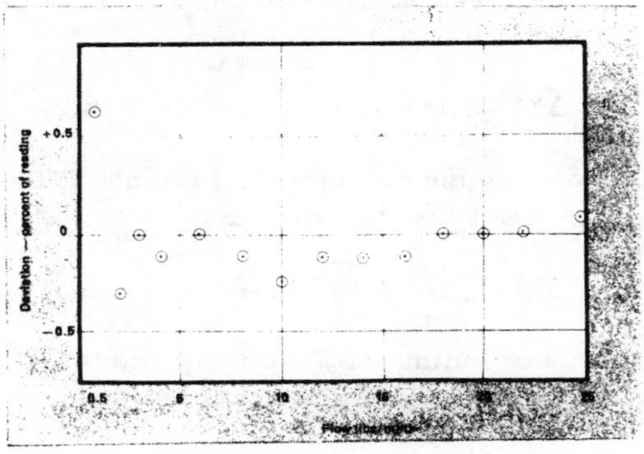

## 6.3. Gyroscopic precession

Gyroscopic precession is a property of gyroscopes that is exhibited when a torque is applied at right angles to the spin axis. The torque will produce a rotation at right angles both to the spin axis and the applied torque axis. To better understand this, imagine a gyroscope with its flywheel in a vertical plane, spinning about its horizontal axis, and supported at one of its ends. The spinning flywheel is apparently resisting the force of gravity and at the same time it moves around its point of support in the horizontal plane. This movement is called precession.

In essence the Coriolis force and gyroscopic precession are a result of the same principle. Viewed in a simplified manner, Coriolis force involves the radial

**Figure 6.2.** Laboratory tests using both Newtonian and non-Newtonian liquids with a wide range of viscosities indicate accuracies within ±0.2% of full-scale readings.

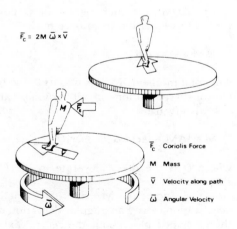

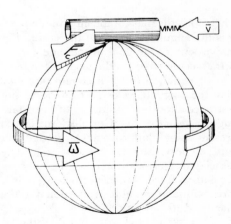

**Figure 6.3.** If a person were standing at the centre of a merry-go-round and tried to walk in a straight line toward the edge, he or she would have to lean sideways against the Coriolis force to stay on the line. The Coriolis force $F_c$ can be calculated using the mass of the person's body, $M$, the velocity of travel $\bar{V}$, toward the edge, and the angular velocity of the merry-go-round, $\bar{\omega}$

$$F_c = 2M\bar{\omega} \times V.$$

**Figure 6.4.** If a section of pipe is placed on the earth as shown with a mass, $M$, moving through it at a velocity, $\bar{V}$, the angular velocity, $\bar{\omega}$, of the earth would create a force, $\bar{F}_c$, that would twist the contents of the pipe eastward. Three things are necessary for the Coriolis force $F_c$ to appear: a mass, a rotating vehicle, and a motion relative to the vehicle.

movement of mass from one point on a rotating body to a second point. As a result of such movement, the peripheral velocity of the mass changes, which means the mass is accelerated. The acceleration of the mass, in turn, generates a force in the plane of rotation and perpendicular to the instantaneous radial movement. Such forces are responsible for precession in gyroscopes.

## 6.4. General operation

Operation of the Coriolis or gyroscopic mass flow meter is most easily explained by referring to Figure 6.4, which depicts a section of pipe located on a revolving earth with a north-south orientation. As a mass travels longitudinally through the pipe with a velocity $V$, a Coriolis force $F$ is present that tends to rotate the pipe about an axis that is parallel to the axis of the earth's rotation. The magnitude of this force $F$ is extremely small and can be calculated as follows:

$$\bar{F} = 2M\bar{\omega} \times \bar{V} \qquad (6.1)$$

where $M$ is the mass of the fluid in the pipe, $\bar{\omega}$ is the angular velocity of the earth, $\bar{V}$ is the velocity of the fluid in the pipe and $\times$ is the vector cross product operation.

In equation (6.1), it should be noted that the angular velocity term $\omega$ is not restricted to the domain of constant angular velocity. The Coriolis force is also present if the platform angularly oscillates with a peak angular velocity $\omega_p$. The associated Coriolis force $F_p$ is now an oscillatory force, but nevertheless, it is a force with a peak value proportional to the mass and its velocity.

The sketch in Figure 6.5 is of the C-shaped pipe and shows both the axis of oscillation ($\omega$) and a unit length of fluid within each leg of the pipe.

It is seen that the velocity vectors $V_1$ and $V_2$ are perpendicular to the angular-rotation vector ($\omega$), and that the Coriolis force vectors $f_{c1}$ and $f_{c2}$ are opposite in direction since the velocity vectors, $V_1$ and

$V_2$, are in opposite directions. When the flow meter is in operation, the angular velocity ($\omega$) is a sinusoidal function, as with any tuning fork, and the forces $f_{c1}$ and $f_{c2}$ are therefore sinusoidal and $180°$ out of phase with each other. Forces $f_{c1}$ and $f_{c2}$ create an oscillating moment, $\Delta M$ about axis 0, as illustrated in Fig. 6.5. The moment can be expressed as a force times a distance:

$$\Delta M = f_{c1}r_1 + f_{c2}r_2 \qquad (6.2)$$

If we assume a symmetrical geometry, the two terms are the same, and

$$\Delta M = 2f_{c1}r_1 = 4M_1V_1\omega r_1 \qquad (6.3)$$

by substituting for $f_c$ from equation (6.1). If the units of $M_1V_1$ are examined,

$$M \frac{\text{lb mass}}{\text{unit length}} V \frac{\text{unit length}}{\text{s}}$$

$$= \frac{\text{lb mass}}{\text{s}} = \Delta Q$$

where $\Delta Q$ is the incremental mass flow rate. Thus equation (6.3) becomes

$$\Delta M = 4\omega r_1 \Delta Q \qquad (6.4)$$

The total moment, M, about axis 0 due to Coriolis acceleration on all moving particles is given by

$$M = \int \Delta M = \int 4\omega r_1 \Delta Q = 4\omega r_1 Q \qquad (6.5)$$

where $Q$ is the mass flow rate in the C-shaped pipe.

The moment, M, due to Coriolis acceleration causes an angular deflection of the C-shaped pipe about the central axis. This angular deflection can be examined using an end view of the C-shaped pipe, (Figure 6.6) which shows the resultant twist-type motion.

The deflection angle $\theta$ due to the moment M is determined by the spring constant of the C-shaped pipe system. This spring constant is a function of the cantilever stiffness of each longitudinal pipe and also the torsional stiffness of each longitudinal pipe. For any given pipe system

$$\text{torque} = K_s\theta \qquad (6.6)$$

where $\theta$ is the pipe-system deflection angle and $K_s$ is the pipe system angular spring constant.

Using this equation, one can relate the mass flow rate, $Q$, to the pipe deflection angle as follows:

$$Q = \frac{K_s\theta}{4\omega r_1} \qquad (6.7)$$

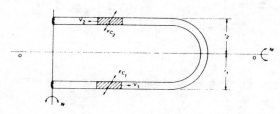

**Figure 6.5.** In operation, the mass flowmeter forces, $f_{c1}$ and $f_{c2}$ creating an oscillating moment, $M$, about axis, $O$.

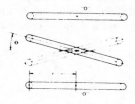

**Figure 6.6.** End view of the C-shaped tube showing the parameters for calculating the torque and its relationship to the mass flow rate. The torque depends on the deflection angle, $\theta$, of the pipe and its spring constant, $K_s$.

Thus, the mass flow rate, $Q$, is directly proportional to the deflection angle, $\theta$, and inversely proportional to the angular velocity, $\omega$, of the C-shaped pipe system. In considering the optical pick-off system (Figure 6.7) one can express $Q$ as a function of $\Delta t$, the time interval between optical pulses. The excursion of the C-pipe can be expressed as its velocity times the time interval, $\Delta t$, between photopulses $P_1$ and $P_2$:

$$V_p\Delta t = 2r_1\theta \quad \text{or} \quad \Delta t = \frac{2r_1\theta}{V_p} \qquad (6.8)$$

where $V_p$ = the velocity of pipe at the position of optical pickoffs, $\Delta t$ = the time interval that photo pickoff $P_1$ leads or lags $P_2$.

$V_p$ the vertical velocity of the end of the C-pipe, depends on the angular velocity, $\omega$:

$$V_p = L\omega \qquad (6.9)$$

where $L$ = the length of the C-shaped pipe.

By combining equations (6.8) and (6.9) we have

$$\theta = \frac{L\omega\Delta t}{2r_1} \qquad (6.10)$$

and by combining equations (6.7) and (6.10),

$$Q = \frac{K_sL\omega\,\Delta t}{8r_1^2\omega} = \frac{K_sL}{8r_1^2}\Delta t \qquad (6.11)$$

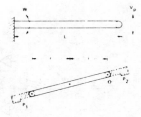

**Figure 6.7.** The excursion of the C-pipe can be expressed as its velocity times the time interval $\Delta t$ between photo-pulses, $P_1$ and $P_2$

The mass flow rate is seen to be a function of pipe geometry constants and $\Delta t$, the time interval between photo-pulses. It can also be shown that, if the pipe has a zero-flow deflection angle $\theta_0$, this error is easily removed by comparing the time interval $\Delta t_1$ of the downward pipe movement with time interval $\Delta t_2$ of the upward pipe movement. If a no-flow condition exists, these time intervals subtract out, whereas a fluid flowing in the pipe causes different time intervals (depending on the direction of angular travel) that are detected as flow-induced pipe moments.

An important feature of the optical detection system is that deflection-angle measurements are made near the centre position of the C-pipe travel. This is important because this is the time when the velocity and the deflection angle are the greatest. Also, this centre position of the C-pipe is the position where the angular acceleration of the pipe is near zero, and any imbalance between plates is least likely to cause an angular deflection that could be interpreted as a flow signal.

## 6.5. Mechanical configuration

The mechanical configuration of the gyroscopic mass flow meter is shown in Figure 6.1.

The T-shaped leaf spring is clamped or welded to the stationary inlet/outlet end of the C-shaped flow pipe. A magnetic sensor/forcer coil is mounted on the leg of the leaf spring. A permanent magnet, suspended from the center of the C-shaped pipe, passes through the middle of the sensor/force coil. In operation, the velocity of the leaf spring relative to the C-pipe generates a voltage in the sensor coil that is amplified and used to drive the concentric force coil. The force-coil amplifier is gain-controlled using a peak-detector circuit that compares the peak-velocity signal from the sensor coil with a reference voltage, Figure 6.8.

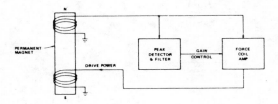

**Figure 6.8.** Block diagram of sensor circuitry.

### 6.5.1. Mass flow rate logic
Figure 6.9 illustrates a slightly warped C-pipe with two photo pickoffs and some circuitry.

### 6.5.2. Photo pickoff logic circuit
Figure 6.10 is a timing diagram illustrating the photo pick-off waveforms $P_1$ and $P_2$ and also the flipflop waveforms $f_1$ and $f_2$ for both a flow and a no-flow condition of the unit.

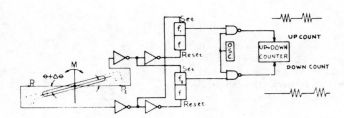

**Figure 6.9.** Mass flow rate meter logic circuit.

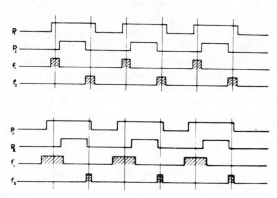

**Figure 6.10.** Timing diagram showing the photo pick-off waveforms $P_1$ and $P_2$ and the flip-flop waveforms $f_1$ and $f_2$ in both the flow (bottom diagram) and no-flow (top diagram) conditions. The photo pick-offs are staggered so that the system can easily detect subtle changes in mass flow rate. Under no-flow conditions, the areas under the waveforms $f_1$ and $f_2$ are equal and the areas under $P_1$ and $P_2$ are unequal. Under flow conditions, the flip-flops $f_1$ and $f_2$ show a discrepancy in areas and the up-and-down counter will display a count proportional to the difference in their pulse widths. The accumulated number in the up-down counter after a given number of pipe cycles will be proportional to the time interval and the mass flow rate in the pipe.

## 6.6. Mass flow meter timing diagram

On the timing diagram, Figure 6.10, the continuous vertical lines represent time intervals where photo pick-offs $P_1$ and $P_2$ would switch if the deflection angle $\theta$ of the C-pipe was equal to zero. In the first sequence, that of no flow, photosensor $P_1$ switches prior to the vertical time reference on the downward pipe stroke because of a constant deflection angle $\theta_0$ in either the photosensor alignment or C-pipe warp angle. On the upward stroke, photosensor $P_1$ switches late with respect to the time reference for the same reason. Photosensor $P_1$ switches late on the downward stroke of the C-shaped pipe and early on the upward stroke because of the no-flow angle $\theta_0$. If one sets and resets flip-flop $f_1$ with the up-going edges of photosensor $P_1$ and $P_2$ the

waveform shown is easily derived. Similarly, flip-flop $f_2$ is set and reset with the up-going edges of the inverted photosensor signals $P_1$ and $P_2$. As illustrated on the timing diagram, the positive areas of waveforms $f_1$ and $f_2$ are equal, and if we use these signals to gate an oscillator into an up-down counter, the net count will be zero following each pair of oscillator bursts.

If the same waveforms are now examined for a flow condition, the following situation is found to exist. Photosensor $P_1$ switches even earlier due to a counterclockwise Coriolis torque on the downstroke of the C-pipe and $P_1$ switches slightly earlier on the upstroke due to a clockwise Coriolis torque on the C-pipe. The same Coriolis torque causes photosensor $P_2$ to switch a little later on the downstroke and slightly later on the upstroke. The waveforms of flip-flops $f_1$ and $f_2$ are modified due to the flow forces and as illustrated in the lower portion of figure 6.10 the positive area of flip-flop $f_1$ is now much larger than that of $f_2$ and, hence, the up-down counter will display a positive count that is proportional to the difference in pulse widths of the two flip-flops. As is obvious, the number accumulated in the up-down counter after a given number of C-pipe cycles will be directly proportional to the time interval, $\Delta t$, and, therefore, directly proportional to the mass flow rate in the C-pipe. This number is not dependent upon pipe oscillation frequency or amplitude; it is not dependent upon cross-sectional area of the pipe; it is only a function of pipe geometry, spring stiffness, and mass flow rate $Q$, as seen in equation (6.11).

The total mass flow between time $t_1$ and $t_2$ is given by the equation:

$$\text{total flow} = \int_{t_2}^{t_1} Q \, dt \qquad (6.12)$$

where $Q$ = the mass flow rate during the time interval $dt$

If one examines the lower portion of the timing diagram of Figure 6.10 it is easy to see that the above integration can be performed digitally if we take the up count pulses from the first burst of flip-flop $f_2$, and subtract those pulses from the first burst of flip-flop $f_1$, and then multiply this difference by the $dt$ time interval that represents one total pipe oscillation cycle. Since the time interval of one C-pipe cycle is the inverse of the pipe frequency, it is an easy matter to simply divide the pulse difference by the pipe frequency and then continuously accumulate the quotient in a resettable counter that then provides a real time display of the totalized flow following any preset zero reference time.

## 6.7. Density measurement

The natural frequency of the C–T tuning fork (a function of the pipe geometry and materials) is also related to the density of the material within the pipe. Thus, for a given pipe geometry and material, the specific gravity (density) of the fluid within the pipe may be determined by measuring the natural frequency of the tuning fork.

The pipe and leaf spring oscillate $180°$ out of phase with each other in the same manner as the tines of a tuning fork oscillate and, as in the case of the tuning fork, little or no vibration is coupled into the base. The frequency of oscillation is determined by the natural frequency of the pipe/leaf spring; the amplitude is controlled by the peak detector circuit. The period of oscillation of a spring/mass system is given by

$$\tau = \sqrt{(2\pi M/k)} \qquad (6.13)$$

where $\tau$ = the period of oscillation, $M$ = the mass of the system being oscillated, and $k$ is the system spring constant.

In the case of the pipe/leaf spring tuning fork of the gyroscopic/Coriolis flow meter, the mass being oscillated can be divided into two parts: the mass of the pipe, $M_p$ and the mass of the fluid in the pipe, $M_f$. In this case the period of oscillation is

$$\tau_1 = k_1 \sqrt{(M_p + M_f)} \qquad (6.14)$$

where $\tau_1$ is the period of oscillation of pipe/leaf spring and

$$k_1 = \sqrt{(2\pi/k)} \qquad (6.15)$$

Since the mass of the fluid, $M_f$ is equal to the volume multiplied by the mass density, one can restate equation (6.14) as

$$\tau = k_1 \sqrt{(k_2 + k_3 D_f)} \qquad (6.16)$$

where $k_1$, $k_2$, $k_3$ are all fixed constants, defined by the geometry of the flowmeter $D_f$ = the density of the fluid in the C-pipe.

## 6.8. Conclusion

The Coriolis/gyroscopic mass flow meter can be used to measure the flow of gases, liquids, or non-homogeneous substances. It is insensitive to temperature and pressure and can be constructed using a variety of materials to provide the necessary corrosion resistance. The unit provides a solution to the problems associated with the mass flow measurement of multiphase flow and cryogenic mass flow.

# Paper 7

# Humidity in air and gases

## A. K. Michell

This paper defines the most common measures of humidity, and the modern implementation of long-standing methods of humidity measurement, as well as some recent electronic humidity sensors. It does not attempt to review all available techniques for humidity sensing. (Eds.)

## 7.1 Introduction

Humidity is a measure of the water vapour present in air or in other gases; the discipline of humidity measurement is called hygrometry.

Water vapour is present in quantities of up to about 2% by volume in ambient air and at least in trace quantities in industrial gases. The water vapour in a gas can be treated as a gas itself, behaving in accordance with the gas laws. Therefore, the total pressure $P_t$ of a gas mixture composed of two gases a and b, and water vapour w can be expressed as:

$$P_t = P_a + P_b + P_w$$

The fundamental measure of humidity is therefore water vapour pressure (or partial pressure of water vapour).

## 7.2. Hygrometric definitions

*Water vapour pressure:* that part of the total pressure of a gas which is contributed by the water vapour component.

If a sealed vessel containing moist gas is halved in volume, then both the total pressure of the gas and the water vapour pressure are doubled. However, if the total pressure is doubled by introducing totally dry gas into the vessel, the water vapour pressure is unaltered.

*Saturation water vapour pressure:* the maximum attainable water vapour pressure of a gas.

If a sealed vessel contains a quantity of water with an air space above it, then at constant temperature and pressure evaporation will take place from the water until the air is 'saturated' and cannot accept further water. The water vapour pressure of the gas is then known as the saturation water vapour pressure. This pressure will vary with temperature.

*Dewpoint:* the temperature at which condensation would occur if the gas were to be cooled (at constant pressure).

Although below $0\,^\circ C$ the appropriate term is, strictly, 'frostpoint', the practice

An unpublished article from Michell Instruments Ltd.

has developed of using the term 'dewpoint' irrespective of the temperature.

Dewpoint is measured by finding the temperature at which dew or frost forms and then obtaining the saturation water vapour pressure from Smithsonian tables.

*Absolute humidity:* the ratio of the amount of water vapour to the amount of dry carrier gas.

Absolute humidity can be expressed in parts per million by volume (p.p.m.V) or by weight (p.p.m.W), or as a percentage by volume (%V) or by weight (%W).

*Relative humidity:* the ratio of the water vapour pressure (w.v.p.) to the saturated water vapour pressure (s.w.v.p.) at the same temperature expressed as a percentage. It is also the ratio of absolute humidity to the saturation absolute humidity at a given temperature expressed as a percentage.

$$r.h. = \frac{w.v.p.}{s.w.v.p.} \times 100\%$$

$$= \frac{\%H_2O}{(\text{saturation } \%H_2O)} \times 100\%.$$

Relative humidity varies with changes in absolute moisture content of a gas *and its temperature.*

Figure 7.1 illustrates the relationship between some of these units. The horizontal axis shows temperature and the vertical axis shows absolute humidity, expressed as a percentage. The curve is the graph relating absolute humidity to dewpoint temperature.

Point A represents the humidity conditions on a typical British summer day. The air temperature is 22 °C, the absolute humidity is 1.6%V (B) and the saturation absolute humidity is 2.7% (C). The dewpoint is 14 °C (D). The relative humidity is

$$r.h. = \frac{1.6}{2.7} \times 100\% = 59\%.$$

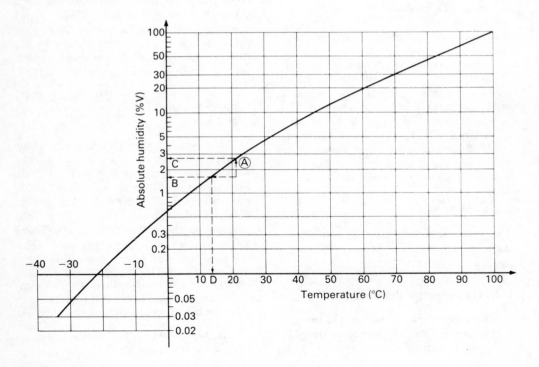

**Figure 7.1.** Relationship between absolute humidity and dewpoint.

## 7.3. Methods of humidity measurement

Absorption, evaporation and condensation are the three basic principles used for practical humidity measurement. The dimensions of natural and synthetic materials change in response to changes in humidity. The human-hair hygrometer utilizes a bundle of hairs, held under tension in a mechanical arrangement. Expansions and contractions of the bundle due to changes in humidity and temperature are transmitted through a linkage to a dial pointer or recording mechanism. A readout is provided in terms of relative humidity. This method continues to be widely used, and instruments are available which provide a potentiometric output for automatic signal transmission.

Wet-and-dry bulb psychrometry also finds much favour as a simple, low-cost method of determining atmospheric humidity. The instrument depends on the use of two identical thermometers, one of which has a moistened wick surrounding its bulb. The temperature recorded by the 'wet bulb' thermometer is depressed, owing to the energy required to evaporate water from the wick. The degree of cooling is related to the water vapour pressure in the atmosphere and the latent heat of evaporation. Relative humidity may be determined from readings from the two thermometers by referring to psychrometric charts. There are modern implementations of this principle, which use thermocouples in place of mercury-in-glass thermometers. Instruments which rely for evaporation on a fixed forced draught rather than natural draughts are more accurate.

The dewpoint or frostpoint of a gas can be measured directly by a Regnault hygrometer (Figure 7.2). The device is installed with the thimble immersed in the gas to be studied. Air is bubbled through the ether, venting to the atmosphere. The latent heat of evaporation of the ether is provided by the thimble, which gradually cools until a dew or frost forming on the outer surface indicates local saturation con-ditions for the gas. The thermometer immersed in the ether gives the dewpoint temperature.

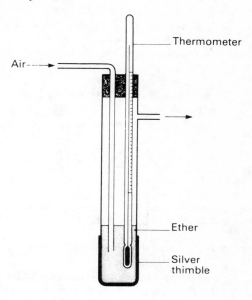

**Figure 7.2.** Regnault hygrometer.

Many variants of the Regnault hygrometer have been produced, utilizing a variety of cooling media. Recent advances in electronics, however, have enabled the production of a fully automatic instrument based on this principle, which can follow changes in humidity rapidly and continuously.

Figure 7.3 shows the measuring principle of the Michell Series 3000 Dewpoint Hygrometer sensor. It consists of a thermoelectric heat pump capable of developing large temperature gradients, on which is mounted a thermally conductive mirror. A thermocouple or platinum resistance thermometer is embedded in the mirror. Light from a source is normally reflected into a photoelectric cell. The photocell and heat pump operate in a closed-loop control arrangement in such a way that the power driving the heat pump is controlled by the amount of light arriving at the photocell, which in turn is controlled by the presence of dew or frost on the mirror surface. The mirror surface is therefore continuously maintained at the dewpoint temperature of the gas. The thermometer measures the mirror temperature, which is the dewpoint. Changes in humidity have an immediate effect on the optical system, and the instrument quickly finds the new dewpoint.

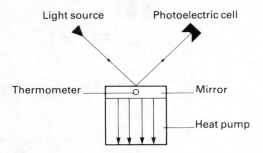

**Figure 7.3.** Dewpoint hygrometer sensor.

Very sophisticated instruments utilizing more complex optical systems have been designed. The Michell Series 4000 Dewpoint Hygrometer detects both reflected and scattered light from the mirror surface and utilizes the resultant signals in a differential mode in the control electronics. This instrument, a model of which has been installed in the National Physical Laboratory as part of the UK Humidity Standard, is capable of detecting dewpoints as low as $-75\,^{\circ}$C, the equivalent of an absolute humidity of 0.5 p.p.m.

Faraday's law of electrolysis forms the basis of the 'electrolytic hygrometer'. A bifilar platinum winding on a quartz tube coated with phosphorous pentoxide provides the measuring cell (Figure 7.4). The moisture contained by the stream of the gas being measured is absorbed by the hygroscopic phosphorous pentoxide and continuously electrolysed into its constituent gases, hydrogen and oxygen. The electrolysing current required is directly proportional to the amount of water electrolysed. Measurement of the current therefore provides a continuous measurement of the moisture content of the gas.

A wide range of instruments is in regular use which relies on changes of elec-

trical properties (capacitance, resistance etc.). These provide readouts in a measure of absolute humidity such as dewpoint, or have deliberate temperature coefficients programmed into them so that the readout is a measure of relative humidity. Unlike the methods described previously, these transducers all require calibration against a standard.

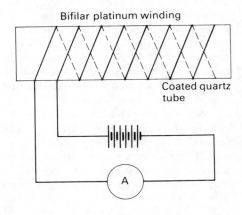

**Figure 7.4.** Electrolytic hygrometer. The ammeter displays humidity in p.p.m.

Typical of these devices is the 'Al-Ox' thin-film aluminium oxide capacitance sensor manufactured by Michell Instruments. Figure 7.5 shows its construction. A high-purity aluminium substrate is anodized to form an insulating layer of hygroscopic aluminium oxide containing a dense pore structure. A very thin layer of gold is evaporated over the oxide film to make a sandwich of the aluminium oxide layer. Water vapour is rapidly transported through the permeable gold layer and equilibrates on the pore walls in a manner functionally related to the water vapour pressure in the atmosphere in which the transducer resides. The changing capacitance is measured by suitable electronic circuitry and the signal is converted into one of the hygrometric measures, thus providing a complete instrument.

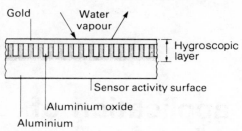

**Figure 7.5.** Aluminium oxide capacitance sensor.

## 7.4. Summary

Although hygrometry is a very old science, dating back to investigations by Leonardo da Vinci, it has until recently been primarily concerned with the measurement of the humidity of ambient air, for the purposes of creature comfort and meteorological records. For these purposes the measure we know as relative humidity has served very well, and the basic instrument technologies, some of which are centuries old, are adequate.

However, many industrial processes now use gases whose moisture content must be maintained accurately within fine limits, often at considerably drier or wetter levels than atmospheric air. To meet this demand a wide range of instruments is now available, working on many different principles and enabling automatic and continuous hygrometric measurements in a wide variety of applications.

The importance of humidity measurement to technologically advanced economies has been recognized in recent times by the standards institutes. The National Physical Laboratory in the UK, the National Bureau of Standards (USA) and other national standards bodies have all established national humidity standards.

# Paper 8

# Development and application of a 0.14 g piezoelectric accelerometer

*Howard C. Epstein*

This paper describes the need for a very low-mass accelerometer, and the factors which affect the design of such a device. The paper is included because it highlights a problem which exists to a greater or lesser extent in all instrumentation systems, that making a measurement can significantly affect the measurement you are trying to make.

The paper also illustrates an initial engineering approach to a problem which, while based firmly on the underlying scientific analysis, allows a rapid 'first-attempt' identification to be made without any precise measurement. The author not only possesses a 'calibrated index finger' but admits its ownership in a scientific paper! (Eds.)

When does the mass and/or size of an accelerometer affect test results?

Ideally an accelerometer has zero size and zero mass, but in actuality the mass of the unit does load the test structure. Simply pushing on a test location with a finger can sometimes help 'get a feel' of the structure's local compliance. This rough estimate of the compliance can be useful in assessing at what frequency test errors result from the weight of the accelerometer.

A piezoelectric accelerometer generates its signal by taking mechanical energy from the test specimen and converting some of that energy into an electrical output. In attempting to attain high sensitivity, high resonance frequency, and a low transducer mass—all desirable qualities—one is limited by the efficiency of the conversion of mechanical energy into electrical energy.

A commercial 0.14 g piezoelectric accelerometer has been developed with a resonance frequency of 54 kHz and a charge sensitivity of 0.5 pC $g^{-1}$. The transducer benefited from increased transduction efficiency at the cost of increased stress levels in the sensor.

Originally published in *ISA Transactions* **12** (1973).

## 8.1. Introduction: accelerometer weight watching

Current applications for accelerometers include uses where the mass of the transducer causes test errors. Scale-model testing, dynamic response analysis of small electromechanical devices, and vibration ruggedness evaluations of integrated circuits are typical of situations that require minimal-weight accelerometers.

A B-1 bomber flutter model is shown being tested in Figure 8.1. Its subsonic spoiler panel, shown in Figure 8.2, is instrumented with an 0.14 g accelerometer. The dynamics engineer in charge of the test rejected using an available 0.5 g accelerometer because it would have affected the resonance frequency of the 1.5 g test specimen.

A vibration test fixture used for environmental testing of integrated circuits is shown in Figure 8.3. The size of the accelerometer allows it to be potted adjacent to the microcomponents that are being evaluated. With a larger accelerometer the vibration could not be monitored at the real point of interest.

Electromechanical devices such as valves, relays, and travelling magnetic recording heads generally have response characteristics that are determined by spring-loaded masses. An accelerometer on such a 'masslike' part will raise the device's response time unless the accelerometer weighs only a small fraction of that mass.

The effective mass at a test location on a panel, shell or other distributed structural member may not be readily known. At such a point the effective mass is a func-

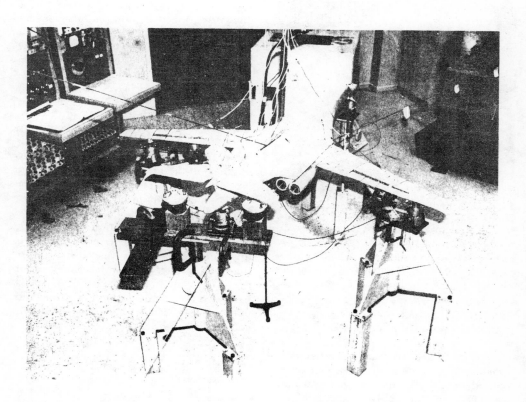

**Figure 8.1.** Testing of a B-1 bomber flutter model, with mounted accelerometers monitoring resonances excited by acoustic drivers.

(Photo courtesy of North American Aviation, Los Angeles Division.)

tion of frequency and will diminish greatly if the location participates in a resonant motion. The added mass of an accelerometer on a resonant position can effect drastic changes in the magnitude of the motion and shift the frequency of the resonance. Also, the stiffness of an accelerometer may produce a node where there ought to be motion.

The compliance of a cantilevered member increases as the third power of its length. Because of this fact, the mass of an accelerometer often induces new bending resonances in a structure. These new resonances not only produce inaccurate information about the test location, but they also distort the motion of other areas.

In short, real accelerometers are not ideal, but rather, can measurably burden the motion of their hosts.

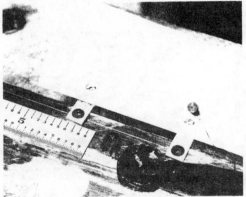

**Figure 8.2.** Subsonic spoiler panel from B-1 bomber flutter model instrumented with a 0.14 g accelerometer. A 0.5 g accelerometer would measurably change the resonance frequencies of the 1.5 g specimen. (Photo courtesy of North American Aviation, Los Angeles Division.)

## 8.2. Rule of thumb — when is an accelerometer too heavy?

On a complicated structure it is often more practical to rely on experience and intuition than on calculation to determine the effect of the added weight of an accelerometer. The rule of thumb outlined below may aid in getting a feel for such a

test situation. The rule assumes that the presence of the accelerometer on the distributed structural compliance produces a new resonance.

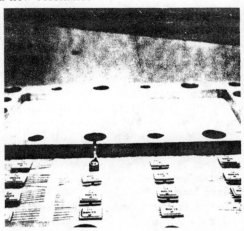

**Figure 8.3.** Accelerometer shown fits in test cavity ensuring that vibration monitored is same as seen by integrated circuits being tested. (Photo courtesy of Motorola Semiconductor Inc., Phoenix.)

For minimal effect, the frequency of this resonance should be more than three times the maximum frequency of interest. This new resonance frequency is determined by the relation $\omega = (1/mc)^{1/2}$, where $\omega$ is the resonance frequency (in rad s$^{-1}$), $m$ is the mass of the accelerometer, and $c$ is the test structure's static compliance at the accelerometer mounting point. The test estimates the magnitude of the compliance by a push on the test location with a finger. The author's 'calibrated index finger' is uncomfortable with a 10 s push of more than about 4 kg (9 lb). The same author can 'feel' motions of $10^{-4}$ m (0.004 inches). Thus, if one pushes firmly on a test point and feels it barely move, then the compliance is at least

$$\frac{10^{-4} \text{ m}}{4 \text{ 'kg-force'}} = 2.5 \times 10^{-6} \frac{\text{m}}{\text{N}}$$

To determine maximum accelerometer weight:

(1) push firmly on the test location with the index finger.
(2) if you can just 'feel' the structure move — then,
(3) the accelerator should weigh less than:   If the maximum frequency of interest is:

| | |
|---|---|
| 10 g | 330 Hz |
| 1 g | 1000 Hz |
| 0.1 g | 3300 Hz |

If it is important to predict the frequency where measurement error results with an accuracy of better than $\pm 1$ octave, then it is recommended that the above rule be refined by making more accurate measurements of applied force and displacement.

## 8.3. Demonstrative test

A demonstrative test was conducted which seemed to verify the validity of the above rule of thumb. An ENDEVCO Model 22 Accelerometer (0.2 g with connector plus cable) was mounted on a violin top plate. The plate was attached to the shaker of a frequency response console. The top trace of Figure 8.4 is a log plot of the response of the accelerometer on the violin plate as compared to a standard accelerometer in the shaker. Three different weight accelerometers were then mounted in turn adjacent to the original accelerometer. The outputs from the new accelerometers were not monitored, but were used to see how the response of the first accelerometer changed in their presence. The bottom three traces of Figure 8.4 show the responses under the influence of a neighbouring 0.2 g accelerometer, of a 1 g accelerometer, and of a 3 g accelerometer. (The above weights include cables.) Figure 8.5 shows the violin plate with the 3 g accelerometer next to the original 0.2 g transducer.

Comparing Trace 1 of Figure 8.4 to Trace 2, one finds that the addition of the 0.2 g accelerometer did not change the response significantly. However, the blocked-off sections of Figure 8.4 indicate there were very significant changes in the response above 1000 Hz when the 1 g accelerometer was present. There were large differences at frequencies higher than 500 Hz when the 3 g accelerometer was present.

By interpolation, our rule of thumb had predicted that the 3 g accelerometer would distort the response beyond 600 Hz, the 1 g accelerometer beyond 1000 Hz, and the 0.2 g accelerometer beyond 2300 Hz. The actual responses were within the $\pm 1$ octave accuracy range of the rule.

## 8.4. Energy-limited tradeoffs of piezoelectric accelerometers

Why not make a lighter accelerometer by just taking a good accelerometer and making it smaller?

The reason is that fabricating diminutive parts is not the only barrier to producing a lightweight accelerometer. If an accelerometer is subjected to a scale size reduction, the sensitivity decreases as the third power of the reduction, while the resonance frequency increases linearly. The relationship between a piezoelectric accelerometer's resonance frequency, charge sensitivity, voltage sensitivity and mass revolves around the transducer's efficiency in converting mechanical energy to electrical energy.

The portion of mechanical potential energy in a piezoelectric material that's available as electrical energy is an intrinsic crystal property defined as $k^2$. $k$ is known as the electromechanical coupling constant. The electrical energy $E_e$ stored in this piezoelectric capacitor is the product of half the charge $Q$ times the voltage $V$.

The potential energy $E_m$ stored in a spring (our piezoelectric crystal) is $1/2\ F^2/y$, where $y$ is the spring constant, $F = ma$ is the force. If we recognize $y/m$ as the square of the resonance frequency (in rad s$^{-1}$), we have

$$E_{\mathrm{m}} = 1/2 \, \frac{m^2 a^2}{y} = 1/2 \, \frac{ma^2}{(2\pi F_{\mathrm{n}})^2}$$

where $F_{\mathrm{n}}$ is the resonance frequency in Hz. Not all of this energy actually gets to the crystal. Let $G$ define the fraction of the transducer's mechanical energy that is in the crystal. The fraction $E_{\mathrm{e}}/GE_{\mathrm{m}}$ is equal to $k^2$, the ratio of electrical energy to mechanical energy in the crystal.

Then,

$$Gk^2 = E_{\mathrm{e}}/E_{\mathrm{m}} = \frac{QV}{\dfrac{ma^2}{(2\pi F_{\mathrm{n}})^2}},$$

$$S_{\mathrm{q}} \quad = \frac{Q}{a} = \text{charge sensitivity,}$$

$$S_{\mathrm{v}} \quad = \frac{V}{a} = \text{voltage sensitivity.}$$

Hence,

$$G \qquad k^2 \qquad = \frac{S_{\mathrm{q}} S_{\mathrm{v}} (2\pi F_{\mathrm{n}})^2}{m}$$

| transducer efficiency | crystal efficiency | accelerometer performance desirables |
| --- | --- | --- |

$$(8.1)$$

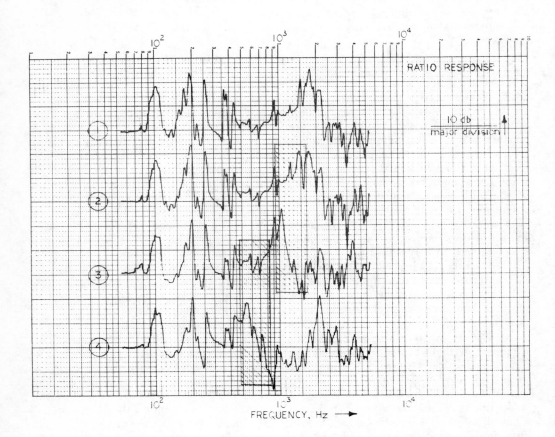

**Figure 8.4.** Effect of mass loading of accelerometers on motion of violin top plate shown in Figure 8.5. Trace 1 is log response (10 dB per major division) of 0.14 g (0.2 g with cable), accelerometers mounted on plate relative to standard in shaker; Traces 2, 3 and 4 show respective responses of same 0.14 g unit under influence of neighbouring 0.2 g, 1 g, and 3 g accelerometers. Boxed areas show onset of major differences between 'true' motion of Trace 1 and Traces 3 and 4.

The right side of expression (8.1) is the mathematical product of desirable accelerometer performance parameters, namely, sensitivity, resonance frequency, and reciprocal mass (i.e., small mass). Given a crystal material and a transducer efficiency, the desired parameters can only be increased at the expense of each other — hardly a surprising result. Since $k^2$ is fixed by nature, the transducer efficiency (always $< 1$) is the only place to make an overall gain.

There are basically two drains on transducer efficiency. The first is due to the accelerometer having weighty components (connector, cover, base, etc.) that do not contribute to signal output. The second is due to compliances other than the sensor being in mechanical series with the 'working mass' and the mounting surface. The amount of energy stored in a spring that is

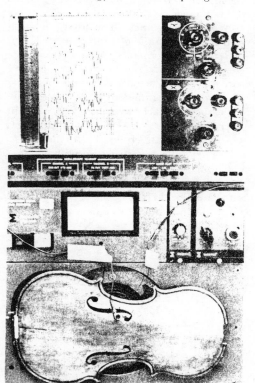

**Figure 8.5.** Shaker activated violin top plate, showing 3 g accelerometer (Trace 4, Figure 8.4) mounted at left of test 0.14 g accelerometer.

subjected to a given force is proportional to its compliance. Hence, any compliance not in the sensor represents an energy loss. Generally, the greatest portion of the total compliance is not in the sensor, because the sensing crystal is stiff and it is purposely isolated from the mounting surface in order to diminish base strain sensitivity.

## 8.5. Design approach for 0.14 g accelerometer

The 0.14 g accelerometer mentioned in the test above resulted from a two-year programme to make a threefold size reduction in accelerometers. In view of the basic tradeoffs we have just discussed, the programme took the approach of increasing the energy efficiency.

We chose a crystal material with one of the highest known electromechanical coupling constants, ENDEVCO PIEZITE P-8, a modified Pb (Zr Ti) $O_3$ ceramic; $k^2 = 0.4$. Sixty per cent of the transducer's mass was made into 'working mass'.

Typically, the portion of an accelerometer's mass that puts stress on the crystal is 30%. The gain in working mass was made largely by reducing the terminating area of the electrical cable. The crystal compliance was reduced relative to other compliances.

Below, we list the key performance parameters of the resultant transducer:

total transducer mass: $m = 0.14$ g
resonance frequency: $F_n = 54$ kHz
charge sensitivity: $S_q = 0.5 \text{ pC g}^{-1}$
voltage sensitivity: $S_v = 2.5 \text{ mV g}^{-1}$

From equation (8.1),

$$\frac{S_q S_v (2\pi F_n)^2}{m} = Gk^2 = 0.01$$

$$k^2 = 0.4$$

so

$G =$ energy coupling efficiency of transducer

$= 2.5\%$.

The accelerometer, which was designed to have high energy efficiency, transmits only 2.5% of the available energy to the sensor. Is this result shockingly low? No. We compare this value with the coupling efficiencies of other accelerometers that are widely used for low mass applications.

ENDEVCO Model 2222B (0.5 g):
 $G = 1\%$

ENDEVCO Model 2226C (2.3 g):
 $G = 0.3\%$

0.14 g accelerometer (ENDEVCO Model 22 'Picomin'): $G = 2.5\%$

The cost of the 0.14 g accelerometer's high efficiency can be surmised from Table 8.1.

High efficiency was bought by increasing sensor stress levels. The consequence was a narrower dynamic range. The 0.14 g transducer has more direct coupling between the mounting surface and the sensor. The result was less isolation from base strains. The Model 22 is 8 times more efficient than the Model 2226C, but the latter has one-third the strain sensitivity, twice the amplitude linearity, and twice the zero shift-free acceleration range.

The 0.14 g accelerometer required many new fabrication techniques. However, the key development was the cable. The 'Picomin' accelerometer necessitated a low-mass, compliant coaxial cable that was 100% shielded and free of triboelectric noise. The resultant cable achieves maximum strength in a small

diameter by utilizing a solid corrosion-resistant steel sheath and centre conductor. The diameter is 0.4 mm (0.017 inches), including a TFE jacket. The cable terminates in a coaxial connector that threads into the accelerometer case.

The operational temperature range of the 'Picomin' accelerometer is $-73°C$ ($-100°F$) to $+204°C$ ($+400°F$). In order to prevent ground loops, the case is electrically isolated from signal ground. The transducer can see 10 000 $g$ acceleration without being damaged.

## 8.6. Conclusions

The mathematical product of a piezoelectric accelerometer's voltage sensitivity, charge sensitivity and the square of its resonance frequency divided by its mass is a direct measure of the transducer's energy conversion efficiency. The above relation outlines basic trade-offs inherent in piezoelectric accelerometers. A high resonance frequency (54 kHz), wide temperature range ($-73°C$ to $204°C$), and low mass (0.14 g) accelerometer was successfully developed by achieving a more efficient energy conversion. The transducer's sensor uses relatively high stress levels.

Pushing a test structure with a finger can sometimes be an aid to estimating the structure's local compliance. The estimated local compliance can help determine the maximum weight accelerometer mass allowable.

### Table 8.1. Accelerometers (same crystal material)

| Property | Model 2226C | Model 22 'Picomin' |
|---|---|---|
| Amplitude linearity (g level of 1% rise in sensitivity) | 500 $g$ | 250 $g$ |
| Zero shift (g level where 2% zero shift is probable) | 4000 $g$ | 2000 $g$ |
| Strain sensitivity (equivalent output at 250 microstrains per ISA RP 37.2 [1964]) | 0.7 $g$ | 2 $g$ |

Paper 9

# The use of ultrasonic techniques for non-invasive instrumentation on chemical and process plant

## R. C. Asher

This paper presents a non-mathematical review of ways in which ultrasound (that is, compression waves at frequencies too high to be detectable by the human ear) can be used to sense measurands such as level, flow and solution concentration. The reasons for using ultrasonic methods in preference to more traditional methods are listed. (Eds.)

Ultrasonic techniques can be used non-invasively on chemical and process plant because it is often possible to penetrate the walls of the plant with ultrasound. Ultrasonic transmitters and receivers typically make use of the piezoelectric effect. Application at high temperatures and in hazardous environments is possible. Ultrasonic techniques are very briefly described for detecting the presence or absence of liquid, or precipitate; determining density or concentration of solutions or dispersions; detecting liquid surfaces and liquid/liquid interfaces; measuring flowrates; providing warning of high or low level or overflow.

### 9.1. Non-invasive techniques

Non-invasive techniques for on-line instrumentation of chemical and process plant are becoming of increasing interest. Some reasons for this are:

they reduce the hazards of operating with poisonous, radioactive, explosive, flammable or corrosive materials
security and accountancy problems with valuable process materials are minimized
contamination of pure or sterile materials (e.g., foods, drugs and pharmaceuticals) is avoided
installation and maintenance may be easier and in many cases may be carried out whilst the plant is on-stream
back-fitting of instruments is facilitated

There is a range of non-invasive techniques available. They make use of, for example, the penetrating powers of electromagnetic radiation (gamma to infrared), particulate radiation (alpha, beta, neutrons, etc.), heat, gravitational fields, magnetic fields, etc. We are particularly

Originally published in *Measurement and Control* 15 (May 1982).

concerned with the use of ultrasound because it also can penetrate the walls of vessels and pipes, thus enabling the resulting instrument to be truly non-invasive.

The ease with which ultrasound can penetrate the walls of a vessel or pipe is influenced by a number of factors (Figure 9.1):

the relationship between the wall thickness ($t$) and the wavelength ($\lambda$) of the ultrasound

the ultrasonic attenuation caused by the material of the wall: this depends on the wall thickness ($t$) and the attenuation coefficient ($\alpha$)

the acoustic mismatch between the material of the wall and the process fluid; this is determined by the acoustic impedance of the wall material ($Z_1$) and the process fluid ($Z_2$).

In clean plant, acoustic mismatch is normally the most serious problem and, because of the enormous acoustic mismatch between solids and gases, there has been relatively little progress in using ultrasound non-invasively in gaseous systems. In the case of liquids, however, where the mismatch is less severe, it is usually possible with intelligent design to get usable ultrasound signals through the walls of pipes and vessels. Deposits on the walls can be a serious problem in 'dirty' plant, but there are methods of alleviating this problem.

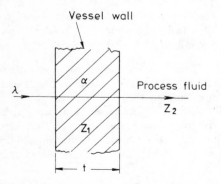

**Figure 9.1.** Transmission of ultrasound through the walls of a vessel.

## 9.2. Generating and detecting the ultrasound

The source of ultrasound — the ultrasonic transducer — is, in the case of non-invasive instruments, sited outside the pipe or vessel. Most commmonly, piezoelectric transducers are used. These can generate either cw (continuous-wave) ultrasound or pulsed ultrasound. The former has the advantage of relative cheapness but does not permit some of the more interesting techniques, e.g., sonar, which are possible pulsed ultrasound.

Piezoelectric transducers of a convenient physical size for incorporating in instruments typically operate in the range 0.5 MHz to 10 MHz. They can either be stuck directly on the outside of the wall or alternatively spaced some distance away (the distance has to be selected intelligently) and the gap filled with a coupling material of suitable acoustic properties (Figure 9.2). Water or greases are often convenient for low-temperature applications, but special high-temperature couplants are available. High-temperature applications may also demand the use of more sophisticated piezoelectric transducers incorporating piezoelectric material which will operate at high temperature and designed to avoid the use of thermally unstable plastics. Alternatively, 'stand-off' transducers, may enable the thermally sensitive part of the transducer to be kept away from the hot area of the plant.

The transmitting transducer, in the case of continuous-wave operation, is energized by means of a tuned oscillator. For *pulsed* ultrasound the simplest energizing technique is to discharge a charged capacitor across the transducer. This gives an exponential rise and decay of the e.m.f. (typically 30–350V). Energizing circuits giving square-wave excitation are available and may give advantages.

The ultrasonic receiver may be one-and-the-same as the transmitter, but clearly only if pulsed ultrasound is used. Alternatively, separate transmitters and receivers can be used for either continuous-wave or pulsed ultrasound.

Although piezoelectric transducers are at present the most popular and convenient way of generating ultrasound for the applications we have in mind, other techniques are worth considering. In particular, electromagnetic acoustic transducers (Whittington 1978) may offer the advantages of being able to operate more readily at high temperature and also of permitting a frequency range to be scanned (this is possible only to a limited extent with piezoelectric transducers) (Figure 9.3).

## 9.3. Using the ultrasound

Having generated the ultrasound and safely transmitted a useful fraction of it through the wall of the vessel or pipe, there are a number of ways in which it can be used to help determine the plant operating conditions.

### 9.3.1. Checking transmission

Checking whether a detectable signal can be transmitted through the contents of a vessel or pipe can indicate whether there is a liquid or a gas at that particular point in the plant. The build up of an ultrasonically opaque sludge or precipitate can also be detected. This simple technique is therefore useful as a high- or low-level warning device (Figure 9.4).

### 9.3.2. Measuring the velocity of ultrasound (v)

This can be done, for example, by pulse echo techniques which give the transit time of ultrasound along a known pathlength of the liquid in the pipe or vessel (Figure 9.5). Under favourable circumstances this enables the liquid to be identified, or its density determined. If the liquid is a solution, its concentration can often be measured. This is a well-developed technique. Certain two-phase dispersions, e.g. emulsions and suspensions, have been studied and found to exhibit a convenient relationship between the velocity of ultrasound and the concentration of the dispersed phase.

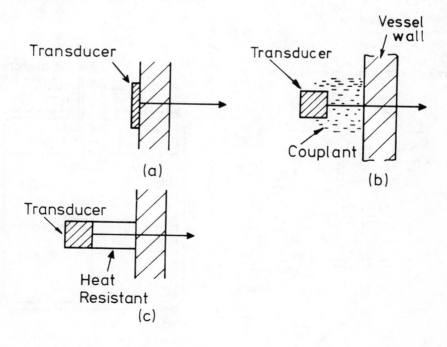

**Figure 9.2.** Arrangement of ultrasonic transducers.

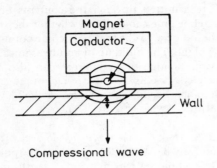

**Figure 9.3.** Electromagnetic acoustic transducers.

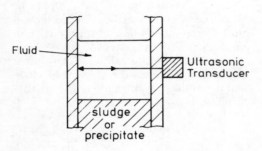

**Figure 9.4.** Ultrasonic presence/absence detector.

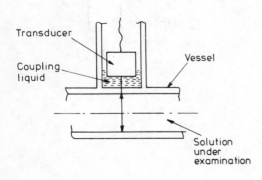

**Figure 9.5.** Determination of concentration of a solution.

The position of the interface between two immiscible liquids can be determined by measuring the apparent velocity of ultrasound over a fixed path passing through the interface at right angles (Figure 9.6); provided the individual velocities in the two liquids are different, then the apparent velocity is obviously dependant on the interface position.

Determination of '$v$' also permits the flow-rates of liquids to be measured; commercial instruments are available which achieve this by comparing $v$ for ultrasound travelling *upstream* with that for ultrasound travelling *downstream* (Figure 9.7).

### 9.3.3. Sonar techniques

Pulsed ultrasound can be used in a way similar to sonar. The pulses can be reflected from liquid/gas interfaces, liquid/liquid interfaces or liquid/ precipitate (or sludge) interfaces. Therefore it is possible to determine liquid levels, the position of the interface between immiscible liquids, and the height of precipitates or sludges (Figures 9.8, 9.9).

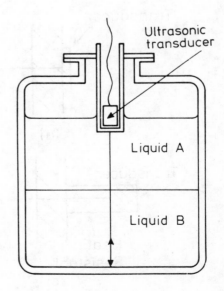

**Figure 9.6.** A method of determining liquid/liquid interface position (schematic).

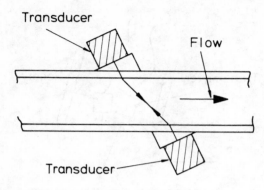

**Figure 9.7.** Flow measurement.

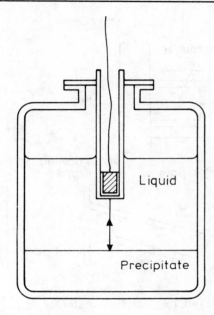

**Figure 9.9.** Determination of precipitate level.

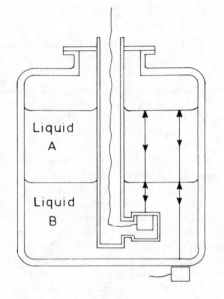

**Figure 9.8.** Determination of liquid level and liquid/liquid interface and position (sonar technique).

### 9.3.4. Near-wall attenuation techniques
These look at the ultrasound reflected back from the *inner* wall of the vessel (i.e., the interface between the vessel and its contents) (Figure 9.10). This technique is more subtle since the amplitude of this reflected pulse depends (but only slightly) on whether the vessel contains liquid or gas. We call this effect near-wall attenuation. A

method of detecting the small difference has been devised and hence is of use to detect the presence of liquids in pipes or vessels, or even to detect the presence of individual drops of liquid trickling down an overflow line. Foams can also be detected. A useful additional feature of this device is that it is applicable to liquids which are opaque to ultrasound.

### 9.3.5. Doppler techniques
The frequency of ultrasound is changed when it is reflected from, for example, particulate matter in the flowing stream of liquid. This 'doppler shift' in frequency is related to the velocity of the particle; this forms the basis of a range of non-invasive doppler flowmeters (Figure 9.11).

### 9.3.6. Cross-correlation techniques
In a flowing liquid the ultrasonic properties vary slightly in a random fashion as a result, for example, of turbulent eddies. Therefore the flowrate can be determined (see Figure 9.12) by means of two ultrasonic transducers spaced along the flow and programmed to compute, by

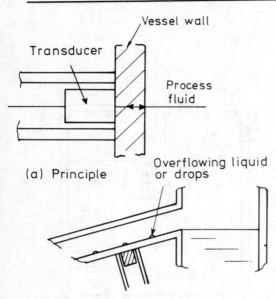

(b) Used for overflow detector
(Schematic)

**Figure 9.10.** Detection of liquid (or foam) by near-wall reflection.

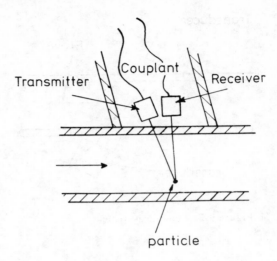

**Figure 9.11.** Doppler flowmeter (schematic).

cross-correlation techniques, the time interval between these 'tagging' random perturbations.

*9.3.7. Surface wave techniques*
Recently the advantages of a new range of techniques have become obvious (Deighton *et al.* 1981). These make use of what may be loosely referred to as 'surface waves' in the walls of the vessel or in components, such as dipsticks, installed in the vessel. These waves have the useful property that, when they are being transmitted along, for example, a dipstick and meet a liquid surface, they are 'mode converted' to the type of ultrasound we have so far been considering (compression waves) (Figure 9.13). The compression wave passes into the liquid at a unique angle (depending on the ultrasonic properties of the liquid and the dipstick material). Consequently a series of reflectors resembling a

venetian blind can be arranged to reflect the compression wave, and ultimately a surface wave, back up the dipstick whatever the liquid level. Therefore, by using pulsed surface waves, an echo can be generated and its timing is directly related to the position of the liquid surface.

The techniques described so far are the basis of a considerable range of commercially available instruments. Other promising techniques need further development. For example the attenuation of ultrasound is difficult to measure, particularly quantitatively, but offers the possibility of giving more information on the identity of liquids, the concentration of solutions, and the fraction of dispersed phase in emulsion or suspensions. The relationship between attenuation and frequency could provide an additional useful clue as to the identity of a liquid. Alternatively the backscatter of ultrasound from two-phase dispersions can, in principle, be used to determine particle (or droplet) size.

In conclusion, a number of non-invasive ultrasonic techniques are already available which enable the determination of plant operating conditions such as liquid levels, liquid composition, solution concentration, liquid/liquid interface positions and flowrates. Other techniques need further research and development before exploitation.

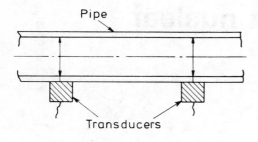

**Figure 9.12.** Cross-correlation flow measurement.

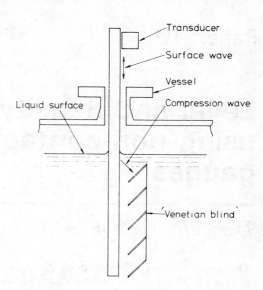

**Figure 9.13.** Typical use of surface wave 'dipstick'.

### References

Deighton, M. O. *et al.* (1981) Mode conversion of Rayleigh to Lamb Waves to Compression Waves at a Metal-Liquid Interface. *Ultrasonics* **19** 249.

Whittington, K. R. (1978) Ultrasonic Inspection of Hot Steel. *Brit. J. NDT*, September, 242.

# Paper 10

# Level and density measurement using non-contact nuclear gauges

## *David R. Carlson*

This paper reviews the basic principles of nuclear gauging, and provides clear definitions of many of the terms in common usage in connection with nuclear radiation sources and detectors. The application of the basic principles to the sensing of liquid density and liquid level is described. For each measurand some practical guidelines are given relating to good installation practice and required precautions. This paper was selected from many on this topic because of its accessibility to non-physicist readers. (Eds.)

Among the many types of level and density gauges available to the instrument and control engineer, the nuclear radiation gauge is unique in that it is completely non-contact with respect to the process material. No probes or interface windows come in contact with the liquid or solid to be measured and maintenance can be performed without shutting down the process.

Thus, the radiation-type gauge can be used on the most difficult of process materials — those that are corrosive, abrasive, very hot, under high pressure, or viscous. This paper will review simple radiation principles, the principle of operation of typical level and density gauges, and then discuss general applications in various process industries.

## 10.1. Basics of radiation

### 10.1.1. Types of radiation
Radioactive isotopes, either natural or manmade, emit radiation in the form of particles or waves as they go through the process of decay. The three most common types of emitted radiation are:

*alpha*: emitted particle consists of two protons and two neutrons — the same as the helium nucleus

*beta*: electrons
*gamma*: electromagnetic waves similar to radio waves and light except of different frequency. Electromagnetic radiation also exhibits particle-like characteristics — these particles are called photons

### 10.1.2. Gamma emitters
Depending upon their relative energies, both beta and gamma radiation are used for thickness measurement in industrial processes such as plastic strip, paper, metal

Originally published in *Measurement and Control* 10 (March 1977).

strip, etc. But for industrial level and density gauging, gamma radiation is used almost exclusively because the energies (penetrating power) of the various gamma-emitting isotopes are found to be suitable for the relatively thick-walled pipes and vessels used in industrial processes. Some commonly used gamma emitters are listed in Table 10.1.

The energy is the energy of each particle (photon) expressed in electron-volts as it is emitted from the isotope. The half-life is a measure of the life of the decaying isotope. Thus, it can be seen that Co 60 has a higher energy than Am 241, and therefore Co 60 would more likely be used on larger vessels with thick walls, but the shorter half-life would limit its usefulness in the long term.

For various reasons (cost, safety, energy, half-life) Cs 137 is the most commonly used isotope for industrial density and level gauging.

### 10.1.3. Activity

Besides the energy per particle, radioactive isotopes are rated as to their 'activity', which is simply the number of particles given off (disintegrations) per unit time. The basic unit of activity is the curie (named after Madame Curie, the French scientist who discovered radium). The practical unit is the millicurie (mCi) and is defiend as:

$$1 \ mCi = 3.70 \times 10^7$$
$$\text{disintegrations/second.}$$

(Note that this definition does not consider the energy of the emitted particle.) For most industrial level applications, practical source sizes (activities) in Cs 137 would range from 5 to 10 000 mCi. It is interesting to note that the activity of an isotope is independent of the effects of temperature and thus it can be considered an extremely predictable source of energy for gauging applications.

### 10.1.4. Half-life and source decay

As the radioactive isotope decays, it loses activity according to a precise exponential

**Table 10.1. Gamma emitters**

| Isotope | Energy (keV) | Half-life (years) |
|---|---|---|
| Americium (Am 241) | 66 | 455 |
| Caesium (Cs 137) | 660 | 33 |
| Cobalt (Co 60) | 1250 | 5.5 |
| Radium (Ra 226) | 1500 | 1620 |

formula. This gives rise to the term half-life, which is defined as the amount of time for a source of stated activity to decay to one-half its original value. For Cs 137 the half-life is 33 years and is graphically shown in Figure 10.1. For example, a 50 mCi source of Cs 137 would have an activity of 25 mCi after 33 years. This means that the number of particles given off per unit time has decreased (but note that the energy per particle remains the same throughout the life of the source).

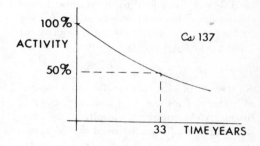

**Figure 10.1.** Source decay — Cs 137.

During the first year, the decrease in activity is approximately 2% and is of no practical concern in level applications where the detector is looking at very large changes in radiation. However, in certain density applications where there is a very narrow measured density span and a small amount of product between source and detector, it is possible that the radiation change caused by source decay over, for example 30 days, would be significant when compared with the radiation change caused by normal density fluctuations in the measured material.

This 'zero shift' due to source decay

can be approximated by

$$N = P \times S$$

where $P$ is the pipe diameter in millimetres, $S$ is the density span and $N$ is the number of days for source decay to cause a zero shift of 1% of span

*Example:* On a typical application where the measuring range is 1.10 to 1.20 $g\,cm^{-3}$ and the pipe diameter is 100 mm,

$$N = (100)(0.1)$$
$$= 10 \text{ days.}$$

This means that in 10 days the zero will shift 0.001 $g\,cm^{-3}$.

Depending upon the required measuring accuracy, the gauge may be manually re-referenced every 10–20 days. If this is not possible, a source decay compensation module may be added to the standard density gauge. This module will automatically compensate the gauge for source decay for a period of three years.

### 10.1.5. Radiation dose
A measure of *absorbed* radiation in a volume of material (not the total radiation passing through). As commonly applied, however, it is considered a measure of the radiation intensity at a given distance from a source and is commonly expressed in milli-roentgens (mR) after Wilhelm Roentgen, the German scientist who discovered X-rays in 1895. By definition, 1 mR produces $2.08 \times 10^6$ ion pairs in 1 cm$^3$ of air at standard conditions.

According to the limits set in many countries, a man can receive a total radiation dose of 100 mR in one week without deleterious effects. This means if he works 40 hours in one week, he can safely work in a radiation field of 100 mR/40 h = 2.5 $mR/h^{-1}$. Most portable radiation monitors (survey meters) are calibrated in mR h$^{-1}$. Conversely, multiplying mR h$^{-1}$ times the total exposure time yields the total radiation dose absorbed in the given period of time.

As will be discussed later, the 'radiation intensity' at any distance from the source is a function of source size (activity), energy of the source, distance from source, and the thickness, density, and type of material in the path of radiation.

### 10.1.6. Half-value
This is a useful concept used in applying radiation gauges and is defined as that thickness of a given material that when placed in the path of radiation, will reduce the incident radiation by 50% or one-half.

For the energy given off by Cs 137 sources, the half-value (HV) of steel is approximately 15 mm. From the diagram it can then be seen that 30 mm of steel (two half-values) will reduce the incident radiation from 20 mR h$^{-1}$ to 5 mR h$^{-1}$ or 75% according to the formula $I_o = I_i/2^{HV}$ (Figure 10.2).

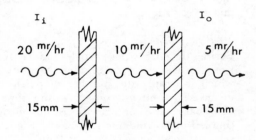

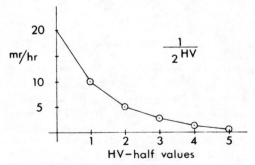

**Figure 10.2.** Half-value concept.

As can be seen from Table 10.2, the half-value is inversely proportional to the density. For materials of density less than 8.0, it is possible to approximate the half-value by dividing the half-value of water

(100 mm) by the density of the material in question:

$$HV_A = \frac{100 \text{ mm}}{\text{density of } A}.$$

Besides its strict mathematical function, the concept of the half-value is useful because if it is desired to maintain a constant radiation level at a detector, then adding or deleting one half-value of material between source and detector will require increasing or decreasing the source size by a factor of 2.

**Table 10.2. Half-values and density**

| Material | Density $(g\ cm^{-3})$ | Half-value $(mm)$ |
|---|---|---|
| Bulk material | 0.50 | 200 |
| Water ($H_2O$) | 1.00 | 100 |
| Refractory ($Al_2O_3$) | 2.25 | 45 |
| Aluminium (Al) | 2.70 | 38 |
| Steel (Fe) | 7.86 | 15 |
| Copper (Cu) | 8.96 | 12 |
| Lead (Pb) | 11.40 | 7 |

## 10.2. Radiation detectors

There are three types of gamma radiation detectors in common usage by suppliers of industrial level and density gauges — the scintillation crystal/photomultiplier tube, the Geiger–Müller tube, and the ion chamber.

### 10.2.1. Scintillation crystal/photomultiplier
Incident radiation (gamma photons) strikes an appropriate crystal, usually sodium iodide (NaI) producing scintillations of light. The photons of light then strike the photocathode of a photomultiplier (PM) tube producing photoelectrons which are attracted to the positively charged dynodes in turn, producing even more electrons by the phenomenon of secondary emission. The resulting current then is a function of the intensity of incident radiation (Figure 10.3).

The crystal/PM detector is capable of very high amplification and sensitivity, but requires a very well-regulated power supply. Also, the PM tube is very temperature sensitive and the photocathode will begin to degrade permanently at temperatures above 50° C. Mechanically, the glass photomultiplier tube is affected by vibration and shock.

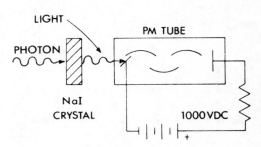

**Figure 10.3.** Scintillation crystal photomultiplier detector.

### 10.2.2. Geiger–Müller tube
This consists of a small diameter (approximately 15–20 mm) thin-walled metal tube filled with an inert gas such as argon at a pressure of about 10 cmHg. The centre electrode is a very thin wire that creates high electric potential stress so that any nuclear particle producing a single ion-pair within the gas can initiate an avalanche of electrons by gas-multiplication. This avalanche of electrons produces a pulse of current which can either be counted or averaged. GM tubes may be paralleled for greater sensitivity (Figure 10.4).

The GM tube has the advantage of relatively small size and low cost, but can be affected by vibration and has a maximum temperature rating of 75 °C.

### 10.2.3. Ion chamber
This is a relatively large-diameter steel chamber with a large-diameter centre electrode. It is filled with a heavy, inert gas such as argon at a specific pressure of several atmospheres. Using a heavy gas under pressure increases the chances of a

collision between a gamma photon and the gas molecule. The resulting current is continuous and proportional to the intensity of the incident radiation (Figure 10.5).

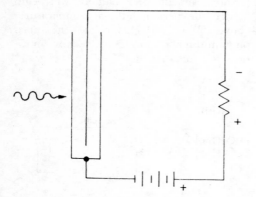

**Figure 10.4.** Geiger–Müller tube detector.

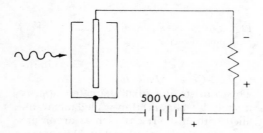

**Figure 10.5.** Ion chamber detector.

The ion chamber is much more rugged in construction than either the crystal/PM or GM tube detectors. It operates at a lower voltage and the output is relatively independent of power supply fluctuations. The ion chamber has a theoretically infinite life. Besides being much less temperature sensitive than the scintillation crystal/PM tube, the ion chamber can withstand ambient temperatures up to 150 °C without permanent damage.

### 10.3. A typical density gauge

Applying some of the previous concepts, it is now possible to construct a typical density gauge. Radiation is absorbed exponen-

tially according to the simple equation:

$$I_o = I_i \exp^{(-\mu \varrho x)}$$

where $I_i$ is the radiation intensity into the process, $I_o$ is the radiation intensity out of the process, $\mu$ is the mass absorption coefficient, $\varrho$ is the density of process material and $x$ is the distance between source and detector (Figure 10.6).

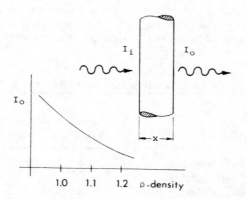

**Figure 10.6.** Characteristic radiation absorption.

The mass absorption coefficient is a characteristic of the process material and can be considered constant for elements with atomic number less than 30. Likewise, the distance $x$ is constant so that the density of the product is the only variable in the equation. The resulting radiation $I_o$ is a function of process density and can be detected using the previous described detector (Figure 10.6).

The simplified diagram of a complete density gauge system is shown in Figure 10.7. The radiation, after passing through the process material of given density, causes a current to flow in the ion chamber circuit. Because the 'hi-meg' resistor is many times greater than the feedback pot $R_1$, essentially all of the resulting voltage appears across the hi-meg. The zero suppression voltage is opposite in polarity to the hi-meg voltage and can be adjusted so that the resultant voltage at the input to the amplifier is 0 V DC. With 0 V DC into the amplifier, there will be 0 V DC out of the amplifier.

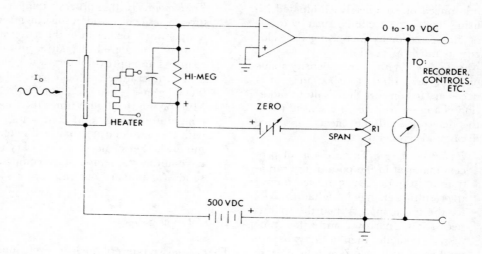

**Figure 10.7.** Basic nuclear density gauge.

If the process density increases, then more radiation will be absorbed and the ion chamber current will decrease. This results in less voltage across the hi-meg so that a net positive voltage appears at the input to the amplifier. The feedback or gain of the amplifier is set by the span pot $R_1$ so that the output voltage as read on the meter corresponds to the desired calibration.

For optimal stability on narrow spans, the ion chamber is heated to a constant 60 °C by a simple, proportional SCR controller and heat blanket. With such a system as described, it is quite possible to practically achieve full-scale spans as narrow as 0.03 g cm$^{-3}$.

## 10.4. Nuclear density gauges: installation and precautions

### 10.4.1. Installation
A few simple guidelines will insure a successful installation:

Mount the source and detector on a vertical pipe with the flow upwards. This assures a full pipe and no settled material between source and detector.
If a horizontal pipe must be used, then mount the source/detector in a horizontal plane so as to avoid settled material. A sample value should be installed near the source/detector for initial calibration and subsequent periodic checking of calibration.
Although a valved by-pass is not necessary from a maintenance standpoint, it is convenient for referencing on an empty pipe or a water-filled pipe.

### 10.4.2. Precautions
As with any instrument, there are several pitfalls that must be recognised if the application is to be successful:

*Air:* Because the nuclear density gauge measures total density in the path of the radiation, it will be sensitive to entrapped air or air bubbles. (This can be an advantage for certain food products where it is desirable to know air content such as pie filling, aerated vegetable oil, candy filling, etc.)
*Build-up:* If the product or deposits settle on the wall of the pipe, then this will represent a density change. If the build-up is slow or eventually reaches a steady-state condition, then the density shift can easily be 'zeroed out' during normal referencing of the gauge.
Build-up or deposits can be

minimized or completely eliminated by using lined pipe sections, such as teflon or glass, and/or by increasing the velocity as much as possible.

*Narrow span:* With narrow density spans on the order of 0.03 to 0.1 $g\,cm^{-3}$, it is necessary to consider the resultant radiation change as regards the sensitivity of the detector, and the effects of source decay and temperature.

*Radiation change:* The amount of process material in the path of the radiation and the density span determines the resulting change in radiation ($\Delta R$) for the given span. When the span is necessarily narrow and the pipe diameter small, then it may be advisable to utilize a Z-section pipe so that the path of the radiation lies along the axis of the pipe. The optimal path length for practical source sizes of 200 to 500 mC is approximately 300 mm (Figure 10.8).

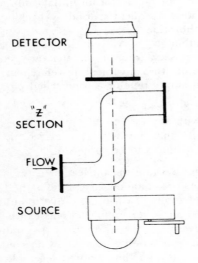

**Figure 10.8.** Use of Z-section pipe.

*Source decay:* This has already been discussed and analysed in section 10.1.4. A simple module will automatically correct for the effects of source decay for a period of three years before resetting.

*Temperature:* If the process temperature varies over a wide range then the temperature coefficient of the material must be considered in relation to the span. As an approximation, water has a temperature coefficient of 0.0005 $g\,cm^{-3}$ per $^\circ$C at 80 $^\circ$C. Multiplying this figure times the total temperature range and comparing it with the total density span will determine whether or not it is necessary to have automatic temperature compensation (a standard option).

## 10.5. Level gauges

There are two types of nuclear level gauge — single-point (simple on-off) and continuous.

*10.5.1. Single-point level gauge*
The simplest of single-point level gauges utilizes one or more GM tube detectors, a simple amplifier and a preset trigger point in order to actuate an output relay that indicates whether or not the process material is between the source and detector. This is simply illustrated in Figure 10.9.

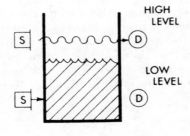

**Figure 10.9.** Single-point level gauge.

More than one single-point gauge may be used to indicate level at discrete points in the vessel. In the usual case, the source is of sufficient strength to produce a field at the detector of between 0.1 to 0.5 $mR\,h^{-1}$ with no material in the path of radiation. When the process material comes between the source and detector, the radiation at the detector is reduced to essentially 0 (except for small-diameter vessels and pipes), thus effecting reliable switching action.

### 10.5.2. Continuous-level gauge

In this case, the radiation beam is wedge shaped (either 30° or 40°) so as to illuminate the detector over the desired measuring range. The rising material 'shades' more of the detector which results in a proportion change in detector current (Figure 10.10).

The amplifier, adjustments, and controls are basically the same as the density gauge described in the previous section. As with the single-point level gauge, multiple sources and detectors can be used to cover greater measuring ranges.

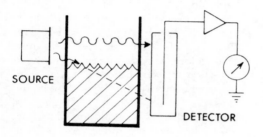

**SOURCE**

**DETECTOR**

**Figure 10.10.** Continuous level gauge.

## 10.6. Level gauges — application and precautions

The following factors must be carefully considered when applying nuclear-type level gauges:

*Source sizing:* The following equation is basic in level gauging and relates all the factors found in a typical level application:

$$mCi = K(d^2)(2^{HV_T})(mR\,h^{-1})$$

where $mCi$ is the required source size in millicuries, $d$ is the distance between source and detector in meters, $HV_T$ is the number of half-values of material between source and detector, $K$ is the

constant of proportionality. (for Cs 137, $K = 3.2$) and $mR\,h^{-1}$ is the minimum radiation intensity required by the detector.

Thus, to accurately calculate the source size, which is the main problem in level gauging, one must know:

*Source-to-detector distance* (*d*): Since gamma radiation intensity, like light, decreases as a function of $1/d^2$, then source-to-detector distance can become an appreciable factor at diameters over 3 m. For continuous-level gauges, the longest radiation path must be considered.

*Total number of half-values of material between source and detector* ($HV_T$): This means all solid material such as steel walls, refractory brick, insulation, water jackets, etc. The HV of the individual materials are summed to make the total $HV_T$.

Usually the density of the process material need not be considered if there are more than two half-values between source and detector (i.e., the inside diameter of the vessel is $2 \times HV$ of process material).

*Sensitivity of the dectector:* The more sensitive the detector, the lower the source strength can be. For GM tubes, it is safe to use a basic sensitivity of $0.5\,mR\,h^{-1}$ although it is possible to go down to $0.1\,mR\,h^{-1}$ with careful application.

Scintillation crystal/PM tube and ion chamber detectors can detect radiation levels almost down to background level — less than $0.04\,mR\,h^{-1}$.

*Ambient temperature:* Nuclear level gauges are commonly applied on very hot processes where contact-type level gauges will not work. All nuclear detectors can operate in ambient temperatures up to approximately 60 °C. Above this temperature water cooling is necessary except in the case of the ion chamber which can operate (with remote electronics) in ambients up to 150 °C.

The source holder can withstand ambient temperatures up to the melting point of lead (Pb) which is approximately 320 °C. For higher-temperature operation, special high density Mallory metal source holders are available that can operate up to 1000 °C.

*Falling material:* If the material coming into a vessel falls in the path of radiation, it is possible to have a false high level indication. For this condition, it would be necessary to increase the source size to compensate for the extra absorption of the falling material (difficult to do in actual practice due to the unknown bulk density) or, in the usual case, offset the source and detector so as to avoid the falling material.

*Build-up:* Material build-up on the wall of the vessel, if large enough, will cause false triggering of the level gauge. In this case, it is necessary to make an estimate of the maximum build-up that may occur and increase the source size accordingly.

*Linearization:* With the continuous level gauge measuring over large ranges, the output is non-linear due to the difference in radiation path length from top to bottom. If a linear output signal is required, a standard six-segment or log linearization module is available as an option.

## 10.7. Level and density applications

Typical applications for nuclear level and density gauges are given below:

*Chemical:* Level in reactors, surge tanks, fluidized beds, storage tanks. Density of polymers, acids, bases, slurries. Interface detection on pipes and vessels. Detection of foam, foam level, foam density.

*Mining:* Level in crushers, storage hoppers, sumps, sand beds. Density of slurries around ball mills, thickeners, cyclones, classifiers. Density of heavy, medium and coal/shale slurries to flotation cells.

*Metals:* Level in continuous casters, cupolas, storage bins, blast furnace. Density of thickener underflow.

*Pollution:* Level in incinerators, fly ash hoppers. Density of $SO_2$ scrubbers, municipal primary sludge, thickener underflow.

*Power:* Level in coal gasifiers, ash lock, coal lock, fly ash hoppers, feed hoppers. Density of coal slurries, filtrates, char/dolomite, scrubbers. Plug-void detection in feed chutes.

*Paper:* Level in chip bin, chip chute, Kamyr, batch digester. Density of strong and weak black liquor, blow line, green liquor, lime mud, coating solutions. Plug-void detection in feed chutes and discharge chutes of TMP and M & D digesters.

*Miscellaneous:* Level of molten glass, food products, powdered metals, bambery mills, sand hoppers. Density of aerated food products, foam rubber, cereals, sugar solutions, sand dredging, ceramic slurries.

Paper 11

# Analytical instruments for process control

## K. G. Carr-Brion

This paper describes some of the chemical analytical techniques which have been used to develop on-line analysis systems for the detection of elements and compounds in chemical process plant. These techniques were originally used in expensive laboratory instruments, used by analytical chemists to make exact determination of the chemical composition of unknown samples.

The need to apply automatic control to chemical plant has led to the requirement for instruments capable of performing these complex analyses on-line; the advent of cheap computing power in the microcomputer has allowed the development of lower cost, faster, rugged versions of the original laboratory instruments which can be used on-line.

To understand fully the descriptions given in the paper requires a substantial knowledge of physics and chemistry. However, the less scientific reader can gain an overall picture of the range of techniques in use and the tasks to which they are being applied. A comprehensive list of references allows the interested reader to pursue the subject in greater depth. (Eds.)

The function and typical applications of on-line analytical instruments used for process control are reviewed. Current limitations and future techniques and requirements are also examined.

## 11.1. Introduction

On-line analytical instruments are becoming of rapidly increasing importance in efficient process control. They merit separate consideration from their laboratory counterparts because the requirements of process control and their operating environment make them very different — in fact they cannot be considered simply as instruments but as systems incorporating sample handling, data processing and instrument control sections.

Where composition is a necessary parameter for efficient control — as it is in a wide range of industry — on-line analysers offer major advantages. Information is available very rapidly, allowing closed-loop control. Where the sensor continuously looks at the whole or part of a process stream, more representative sampling can be achieved, particularly in heterogeneous or rapidly varying streams.

Originally published in *Measurement and Control* **10** (November 1977).

In certain applications they also eliminate the handling of hazardous or unstable samples and allow remote or unmanned operation. There are also limitations. The instruments are generally more expensive than their laboratory counterparts due to the need to include sample-handling systems and to design them to operate reliably in adverse and hazardous environments. Their accuracy on the sample analysed is often less than that achievable in a laboratory, although in many cases this is much more than offset by more representative sampling. They are also of limited flexibility both in terms of the range of components determined and the number of process streams one analyser can reasonably examine.

The plant engineer wishes to treat the process analyser as any other sensor — a 'black-box' source of information which can be bought off the shelf. However, their complexity often prevents this and the design of each system depends on the requirements of the individual process. Only when sampling, sample handling, sensing and data handling are considered relative to the stream to be analysed and the information required, will an effective and reliable system be devised. In the following account of on-line techniques currently in use, an attempt is made to emphasise those factors of especial sensitivity in the application of each technique.

## 11.2. Gas chromatography

This technique (Villalobos 1975) determines volatile compounds by separating them on a column and sequentially measuring each as they emerge in the carrier gas flowing through the column. It is of particular value in the analysis of complex mixtures and low concentrations. Only a small sample — typically 1 $\mu l$ of liquid or equivalent vapour — is analysed and there is of necessity a delay between injecting the sample onto the column and the separated compounds emerging from the column onto the detector — typically between 30 s and 10 min. Sampling is normally carried

out with small precision valves on a side loop from the main stream. These valves require care in design and construction and operate on cleaned samples since entrained solids or gums can rapidly degrade their performance. The chromatograph itself requires controls — for carrier gas flowrate, column temperature and automatic restandardization. Process chromatographs are similar to laboratory units except that they are more ruggedly constructed and designed for work in a plant or hazardous environment. Being complex systems they require an above-average amount of skilled maintenance, a drawback in some applications. As with many on-line instruments the various instrument control requirements and the calculation of concentrations from the detector output are effectively carried out with an on-line microprocessor.

On-line chromatographs (Figure 11.1) have been very widely employed in the petroleum refining and petrochemical industries, where the need to analyse complex mixtures of similar compounds is common and the delay and requirement for clean samples are of limited importance (Verdin 1973). Their field of use is very much wider, ranging from determining trace organics in aqueous effluents (Novak et al. 1973) and hydrogen cyanide and sulphide in coke oven gas (Manka 1975), to control of monomer ratios in copolymerization (Freeguard and Pulford 1972). Recent years have seen the coming of the 'stripped-down' single-component process chromatograph, while future trends seem to be in the direction of wider application and better performance by attention to detailed design and operation, rather than by any revolutionary new procedures.

The detectors used to measure the concentrations of compounds emerging from the column in gas chromatography are also used as continuous analysers. Thermal conductivity (described separately) was one of the earliest techniques of on-line analysis used in the chemical industry. The flame ionization detector can monitor for total hydrocarbons, flame photometric detectors for total sulphur and electron capture analysers for chlorinated compounds and refrigerant gases (Hartmann 1971). They are only semi-selective and care has to be taken to exclude components, such as oxygen, which might affect their response. On-stream techniques using flames also present a potential ignition source, and suitable flame traps and ignition failure devices must be incorporated.

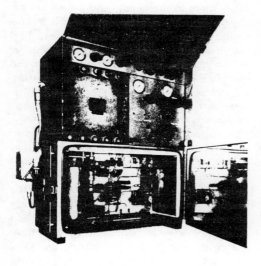

**Figure 11.1.** Series 400 process gas chromatograph (Taylor Servomex Ltd).

## 11.3. Mass spectrometry

This powerful technique (Hill 1972) determines the concentrations of compounds or elements by ionizing the gaseous or vapourized sample material and obtaining a mass spectrum by electrical and magnetic dispersion of the ions. Each component produces a number of ions, differing by mass and charge. These complex mass spectra overlap and computation is generally necessary to separate them and obtain quantitative and qualitative data. Since the spectrometer can only operate under vacuum, the ion source must also be at low pressure. This means material from the process stream — normally at or above

atmospheric pressure — must be transferred into the low-pressure ion source without change in the relative concentrations of its components due to factors such as different diffusion rates. In the majority of cases this problem can be solved by careful design of the sampling system. The analysis itself is very rapid — in the laboratory on-line mass spectrometers are used to identify unknown compounds as they emerge from a gas chromatograph — and is capable of analysing complex mixtures. The potential accuracy is high and limits of detection very low. In the past, because of sampling problems and system complexity, the gas chromatograph has been preferred to the mass spectrometer except where speed of analysis was necessary. With the widespread adoption of quadrupole-type spectrometers, with their speed and ease of installation, the technique has secured a new lease of life for on-line analysis. They are now tackling such difficult problems as blast furnace off gas analysis (Schuy and Reinhold 1971) and multistream multicomponent analysis in monomer production (Dunn 1972).

Computers have been associated with mass spectrometers since their earliest industrial days, due to the need to separate the mass patterns, and the drop in the cost of computers through minicomputers to microprocessors has been a factor in their recent and future development. Here again the future of on-line mass spectrometers appears to be in simplifications and improvements to existing systems, rather than any major leap forward, for the limitation of needing to operate the spectrometer under high vacuum is fundamental.

## 11.4. Infrared absorption

Infrared absorption analysers (Kinsley 1972) determine the concentrations of compounds in gaseous, liquid or solid samples. The technique depends on the characteristic vibration frequencies associated with bonds between the atoms making up the compound. Radiation having these characteristic frequencies is absorbed on traversing the sample, allowing selective quantitative analysis. Because of the wide variety of sample types that can be analysed, a corresponding range of on-line instruments has been developed.

For gas analysis, the most widely used instrument is the Luft-type analyser. Here, the gas being determined is used in a closed cell as the detector. It strongly absorbs its own characteristic frequencies, giving high selectivity and sensitivity. Absorption causes pressure changes in the gas, which are detected capacitively and used to give an electrical output depending on the concentration of the wanted gas flowing between the detector and the infrared source. The main limitations of the Luft analyser are that it is prone to interference from vibration (its detector is in effect a microphone), it can normally determine only one compound for each detector and, in common with all infrared gas analysers, requires a clean sample free of dust and of compounds that would attack the infrared transmitting windows or cause interference due to frequency overlap. Overlap can be reduced with a filter cell containing the interfering gas in the infrared beam, and standard sample preparation systems are available from manufacturers to remove suspended solids and water. These analysers are widely used in process control, mainly for determining carbon monoxide in furnace gases, stacks and the like. They are also used in the chemical industry to determine many gases, one good example of many being ethylene and ethylene oxide in the manufacture of the oxide (Dailey 1973).

With liquid or solid samples, the Luft analyser gives way to instruments using a semiconductor detector with interference filters to isolate the necessary characteristic and reference frequencies. To determine a compound, two filters are moved alternately into the infrared beam. One transmits a narrow band of radiation at the required characteristic frequency, the second at a nearby frequency not absorbed by the sample. The ratio of the two transmitted intensities gives an output dependent on the concentration of the selected compound and largely independent of such factors as variations in source intensity and minor window fouling — an important advantage in liquid analysis. More than one compound can be determined by using three or more filters, the limitations being frequency overlaps, the space required to fit the filters in the rotating wheel which alternates them in the beam, and the reduction in the amount of time each filter is in the beam causing slower response times. The radiation can either traverse the sample in a flow cell or be multiply reflected from the interface between an infrared transmitting crystal and the sample — so called multi-attenuated total reflection. The former is used for infrared-transparent liquids and low concentrations, the latter shown in Figure 11.2, for strongly absorbing liquids, such as water, and at higher concentrations (Carr-Brion and Gadsden 1969). The application range is very wide, the main area being the determination of water in liquids ranging from acetic acid and ethanol to carbon tetrachloride and benzene. Typical non-water applications are the determination of

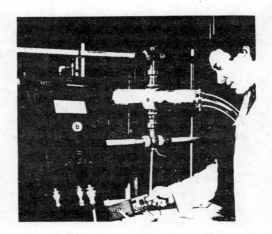

**Figure 11.2.** Immersible attenuated total reflection infrared analyser. (a) measuring head, (b) control unit. (WSL Ref No CON 77/75).

chlorosilanes (Barty 1963) and vinyl cyanide and ethyl benzene in styrene (Barty and Ruhl 1968).

Some solid samples, such as plastic film, can be measured by transmission, but the majority are examined by measuring the infrared radiation diffusely reflected back from the sample surface. The main limitations are the general infrared problem of frequency overlap and that reflection occurs at the surface of the solid, which may not be fully representative of the bulk material. Otherwise it is a simple technique to use with the sensing head, containing both source and detecting system, 10 cm or more from the sample stream. This can be in powder, fibre or pellet form. Manufacturers' application lists for moisture determination are impressive, with materials ranging from tobacco and dried milk to clay and paper (Harbert *et al.* 1974).

Big changes are occurring in the field of infrared analysis. While existing sensors will continue, developments such as tunable solid-state lasers, high-sensitivity detectors and industrialized correlation and interferometric analysers offer a still wider scope for the technique. Tunable infrared lasers (Hinkley and Kelley 1971) can scan across the individual rotation–vibration gas absorption bands of gases, giving very high selectivity and sensitivity. However, at their current state of development, they require cryogenic cooling, which is a limitation for industrial application and need a broader scanning range for liquids and solids. Interferometers (Lephardt and Vilcins 1975) allow spectral regions to be used for measurement rather than the limited frequency bands of filters. This is of especial value in the analysis of multicomponent and heterogeneous samples. Correlation spectrometry (Davies 1970) also has a future in industrial analysis, particularly for low concentrations of gases. One further development which may be of importance is the long-wavelength infrared vidicon, which can be used to obtain two- or even three-dimensional composition information.

## 11.5. Visible and ultraviolet absorption

Colour was one of the earliest physical parameters used for composition control. Instruments for the visible and ultraviolet spectral regions have much in common and are considered together (Kolthoff *et al.* 1964). Measurements are not generally energy-limited as they are in the infrared region, permitting high precision and remote measurement. There are a range of continuous and line emission sources available and room-temperature detectors of very high sensitivity, the photomultiplier and silicon photodiode being the most used. Choice of optical materials is much wider — particularly in the visible region — aiding application to corrosive materials and easing cleaning problems. In some respects instrumentation, while basically similar, is somewhat more advanced than in the infrared field — correlation sensors are used to determine sulphur dioxide in the atmosphere (Davies 1970), immersible probes are available for colorimetric measurements and ultraviolet composition mapping has been used to examine sulphur dioxide in stack plumes. However, these spectral regions are even more prone than the infrared to problems due to scattering and deposition of strongly absorbing compounds on windows and other optical components. Sample streams have to be carefully filtered and windows protected — for 'in-stack' monitors the window is usually isolated from the gas stream by a positive air purge and shutters for automatic isolation in the event of purge failure (Huillet 1975). Applications include the determination of $SO_2$ and NO in stack gases (Huillet 1975), mercury in air (Mayz *et al.* 1971), chlorine in air (Troy 1955) and phenols and aromatics in water (Saltzman 1969).

Tunable dye lasers (Allkins 1975) may have potential for use in the visible region for on-line analysis. Being highly monochromatic and capable of scanning over spectral regions they would be powerful tools especially in the analysis of multicomponent and heterogenous streams. Current systems have been engineered for laboratory use, and problems may occur with dye stability if used for routine monitoring. Further extension of correlation and related techniques also appears likely, and the use of image converters using interference filters and coupled to microprocessors could also have application in on-line analysis.

## 11.6. X-ray fluorescence and absorption

These techniques are almost entirely used on-line for the determination of elemental concentrations. X-ray fluorescence sensors (Owers and Shalgosky 1974) measure the characteristic X-rays emitted by elements when irradiated with primary X-rays from a tube or radioisotope source. The characteristic X-rays generally originate from the surface layer of a sample, hence sample presentation techniques must ensure that this surface is representative of the bulk material. The technique is used both on-line and on-stream. When used on-line, samples and subsamples are taken automatically, pretreated by grinding or fusion, pelleted and presented to the X-ray analyser. This measures the characteristic intensities from the required elements and others that may affect the relationship between characteristic X-ray intensity and concentration. A minicomputer or microprocessor then calculates the necessary concentrations. On-stream the sample is continuously split from the process stream and presented to the X-ray analyser, in a flow cell with an X-ray transparent window, as an open jet (Tily 1974) or, for powders, as a geometrically well-defined strand. A further variation is to build the X-ray analyser as an immersible probe operating in the process stream. With onstream measurements little or no sample pretreatment is possible and more complex mathematical relationships are required to relate accurately the intensities to concentrations. However, more representative sampling and faster response can be achieved on-stream, and it is the preferred way of determining elements in slurries. It is used on-stream in many mineral processing plants — Leppala et al. (1971) describe a typical system and applications — and for determining sulphur in hydrocarbons (Gamage and Topham 1974). Its chief on-line application has been the determination of aluminium, silicon, calcium and iron in kiln feeds for cement manufacture.

X-ray absorption depends on the selective absorption of X-rays by an element — commonly one of relatively high atomic number in a low-atomic-number matrix. It generally uses a radioisotope X-ray source, either directly or with a secondary target, to produce characteristic X-rays. Its application is limited, the main composition use being the determination of sulphur in petroleum products (Cameron and Piper 1972) since it lacks selectivity, except when a two-energy absorption edge method is used, and in most cases has poor limits of detection.

The last five years have seen a revolution in X-ray analysis with the widespread adoption of energy-dispersive spectrometers, which use a cooled silicon detector to simultaneously discriminate between and measure the required characteristic intensities. These are now firmly established in the laboratory and are expanding into the on-stream field (Carr-Brion 1973) due to their mechanical simplicity, increased information and reduced demands on sample presentation when compared with conventional instruments. A typical energy dispersive system is shown in Figure 11.3. Future developments are likely to be the production of integrally cooled detectors requiring no liquid nitrogen supply and the availability of simplified systems using analogue and microprocessing techniques to bring the bottom end of the X-ray instrument range in line with infrared analysers in terms of cost.

X-rays are ionizing radiation and radioisotope sources present potential health hazards from ingestion. Hence they must be used with great care and according to the current legislation for industrial sites. However, acceptance in industry for process control applications does not seem to have been greatly affected by this factor.

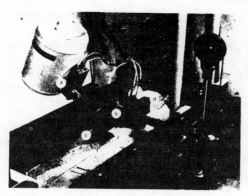

**Figure 11.3.** Energy dispersive on-stream X-ray fluorescence powder analyser. (a) cooled detector, (b) low power X-ray tube, (c) sample presenter, (d) sample strand. (WSL Ref No CON 3/74).

## 11.7. Neutron-based techniques

Neutron absorption, neutron activation and prompt gamma-ray emission during neutron irradiation are all used for on-line analysis (Kliment and Tolgyessy 1972). Neutron absorption depends on certain elements having much larger absorption coefficients for neutrons of selected energy than other elements in the sample. Neutron activation, either with radioisotope or tube source, involves converting the element to be determined into a radioisotope by interaction with neutrons and measuring at a remote location the characteristic gamma radiation it subsequently emits. Prompt gamma ray techniques measure characteristic gamma radiation emitted while the sample is being irradiated. Since the techniques are dependent on nuclear properties, large variations in sensitivity are found from element to element in the periodic table. Large samples can be measured, since both the neutrons and characteristic gamma rays are highly penetrating — markedly reducing heterogeneity effects and easing the problem of representative sampling. They can be used both on-line or on-stream and can examine liquids, slurries, powders and coarse solids.

Their main limitations are high cost,

the need for adequate shielding from the penetrating radiations, potentially toxic sources and the narrow range of elements where their performance excels other techniques. Nevertheless, they present a useful solution to analyses too difficult for other on-line techniques. Neutron activation is used to determine fluorine in the products of the mineral processing of fluorspar ores; nitrogen, potassium and phosphorus in fertilisers; oxygen in steel (Perdijon 1974); silica in taconite ores (Tuttle *et al.* 1972) and aluminium and silicon in cement (Powney 1973). Neutron absorption is largely used to determine boron in water and detergents (Pierce *et al.* 1974), and prompt gamma-ray emission is used to determine the sulphur content of coal on-stream (Parsignault *et al.* 1971). Moderation of fast neutrons by protons is used on-line to determine moisture in materials such as coke and sinter mix (Jones *et al.* 1972).

Neutron techniques have been greatly affected by the availability of the neutron emission isotope Cf 252, which is an attractive alternative for industrial use from the previously used beryllium–plutonium sources and complex neutron generators. Cryogenically-cooled germanium gamma-ray detectors have greatly increased selectivity. Both these developments have increased the potential power of the technique and it remains to be seen if this is sufficient to produce more widespread adoption for on-line analysis.

## 11.8. Electrochemical methods

In liquid samples containing ions or which are easily ionized, the various electrochemical effects associated with these can be used for their on-line determination. The earliest application was the pH electrode, responding to hydrogen ions. The range of ions has been subsequently extended with the development of ion-selective electrodes (Durst 1969). These have an output dependent on the activity of a particular ion, thus response to concentration will be affected by the total ionic strength of the sample. Their logarithmic response, temperature sensitivity and stability present problems in using them for accurate on-stream analysis; an accuracy of ±5 to 10% is typical. Higher precision can be achieved if they are used on-line with chemical and other pretreatment of the sample — in this mode their function overlaps that of automatic chemical analysers, described below. For conditions where fouling of the electrode is a problem manufacturers offer ultrasonic cleaning units which automatically cycle at intervals selected by the user. They are best suited to the determination of lower concentrations of ions — higher concentrations are generally measured after dilution. On-line applications include the determination of cyanide in effluents (Riseman 1972); fluoride in effluents (Sekerka and Lechner 1973) and sodium in boiler water (Diggins *et al.* 1972). A similar electrode, but operating on a galvanic basis, is widely used to measure the dissolved oxygen content of natural and boiler waters (Melbourne *et al.* 1972).

Coulometric analysers, which depend on the electrolytic generation and electrical detection of a reactant, are used for the on-line determination of gases after absorption in a suitable liquid — one established use is the determination of sulphur dioxide in air. Not all electrochemical techniques are used in liquid media — the zirconia solid electrolyte probe used to determine oxygen in flue gases (Church 1973) relies on the potential developed by absorbed oxygen in a cell consisting of platinum electrodes separated by zirconia. This sensor operates at high temperatures directly in the gas stream. Another potential application at high temperatures is the determination of elements in molten metals and slags for control in metal reforming. Given the availability of suitable electrochemical systems, the major problems are in the materials for construction of the probes, which have to stand the solvent effect of the sample and ensure the material in contact with the probe is representative of the furnace load.

## 11.9. Physical parameter measurements

Under this heading are included measurements such as refractive index or thermal conductivity, which can be used to infer composition in sample streams whose composition is binary as far as the technique used is concerned.

Refractive index is extensively used to determine composition on-stream (Clevett 1973). It can show high sensitivity and some degree of selectivity. The ruggedness and simplicity with which in-stream sensing heads can be made permits high reliability and relatively low cost. The use of the differential refractometer increases the power of the technique, for it permits highly accurate final product analysis, using a pure reference material as one optical component, or to determine additions to or purification of a stream of varying composition, using the 'before' and 'after' streams as the differential components. It is used in the hydrogenation of fats and oils (Carlson 1971) and manufacturers indicate a wide range of applications from alcohol in beverages to the analysis of black liquor in the paper pulp industry, where washing techniques have successfully overcome the problem of deposition on the prism face used for measurement.

Thermal conductivity is used on-line in gas analysis. It is normally measured by the filament technique, where a heated filament is immersed in the flowing gas, the filament making one arm of a bridge. As the thermal conductivity of the gas varies, so does the temperature of the filament, causing a resistance change in the filament. It too can be used differentially, with the two filaments in the 'before' and 'after' gas streams forming two arms of a bridge. Where explosion hazards exist the sensor is used on a sample loop with flame arrestors before and after. The main current use for such analysers is for hydrogen determination, although other gases, such as major amounts of sulphur dioxide, are also determined (Verdin 1973).

Measurement of the paramagnetic susceptibility of gases gives a selective method of measuring oxygen (Tipping 1970). NO and $NO_2$ are the only other common paramagnetic gases, and they are less so than oxygen and are normally present in much lower quantities. Various methods are used to apply the technique. Most depend on the loss of magnetism in the gas when heated, the hot gas being displaced in the magnetic field by cooler gas, causing heat loss from the source — often in the form of a filament. Where heating could be hazardous, for instance in monitoring the oxygen content of combustible gases, analysers that directly measure susceptibility are preferred. These use the force exerted by a magnet on a 'dumb bell' using spheres filled with nitrogen. One of the main applications of these analysers is in combustion control and there is also widespread application in cement, metallurgical and chemical processes (Figure 11.4).

Electrical conductivity can also be used to measure composition, either directly on a process liquid or in gases after absorption in a suitable liquid. Examples are the strength of concentrated acids and the determination of lye concentrations in food processing (Legenhausen 1971). It is also widely used to monitor the purity of de-ionized and other high-grade waters. Capacitance has also been used for determining the water content of materials such as paper (Ramaz 1970). Other physical properities such as viscosity, density (Clevett 1973), beta-particle scattering and even temperature are used on-line to infer chemical composition and combinations of these may give a degree of selectivity. However, the division between physical property and composition sensors is ill-defined and selection somewhat arbitrary.

## 11.10. Automated chemical analysers

No account of on-line analytical instruments would be complete without considering automated chemical analysers (Forman and Stockwell 1975) which may function on a continuous–flow or discrete–sample basis. In some respects they can overlap with techniques described previously, for example, ion–selective electrodes being used after chemical pretreatment of the sample and ultraviolet and visible photometers being similarly employed. The analysers are ingeniously engineered equivalents of laboratory chemical analysis equipment, and as such have an extensive range of applications, with gaseous, liquid and readily-dissolved solid samples. Drawbacks are the limitations inherent in a mechanically complex system, such as servicing needs and potential unreliability, the need to supply with reagents, the capacity to analyse only a small sample and generally a rather long response time. Nevertheless, because of their power and flexibility they are extensively used in the process industries — effluent monitoring being a common example.

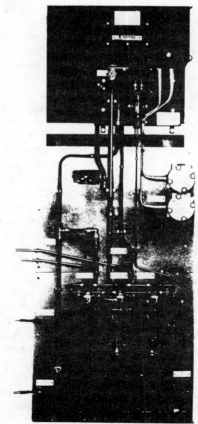

**Figure 11.4.** Two-stream flue gas oxygen analyser (Taylor Servomex Ltd).

## 11.11. Other present and potential techniques

Almost any analytical technique can be used on-line, providing there is a real need and an enthusiast willing to meet that need by developing a system based on laboratory instrumentation. Lesser used techniques include broad-band nuclear magnetic resonance, used for the on-line determination of moisture and fats in the food industry (Simpson 1968), and atomic absorption for zinc and manganese in flotation liquors (Jones and Woodcock 1974).

A number of laboratory composition-measuring techniques may in the future have on-line application depending on industrial need, relevant technical developments and, in some cases, simply the will to go out and sell a technique to industry and back it with efficient service.

Microwave absorption in gases (Lovas 1974) offers very high selectivity and could have application in the analysis of complex gaseous mixtures. Current high-resolution microwave systems are complex, and the need to have the sample at reduced pressure gives the technique few advantages over mass spectrometry. Laser Raman spectroscopy (Gilson and Hendra 1970), now largely used as a laboratory technique for qualitative and structural analysis, could find application in on-line measurement of solids and liquids, particularly since the movement of the sample stream would reduce sample radiation damage. A rugged method of separating the Raman lines from the scattered exciting radiation would have first to be devised.

All materials encountered in process control emit infrared radiation which can in some cases provide information on the composition of the material. For materials near ambient temperatures the intensity is only slightly different from general background emission. The advent of cooled, high-sensitivity infrared detectors (Wright 1973) should make the technique more attractive for determining gas composition or that of thin films — since infrared emission from massive materials approaches that from a 'grey' body and has limited composition information. Massive materials could be made to look like thin films by instantaneous surface heating, just before measurement, with an intense radiation source such as a laser.

In the X-ray field, diffraction has the powerful capability of determining the concentrations of crystalline phases. It is one of the few techniques which has potentially a better quantitative performance on-stream than in the laboratory, since random particle statistics, which can be a major source of error in quantitative work, are much improved by the large sample the sensor would examine in a normal measuring period. It has been successfully used with flowing slurries and powders on belts. Application studies have been made on iron ores, minerals (Barton et al. 1969) and cement (Nutter 1972) but as yet it has not been employed on-stream for control purposes.

## 11.12. Unmet needs

Sticking strictly to analysers and ignoring the need for more effective sampling systems, the main unmet need is undoubtedly reliability (Martin 1975). This is a complex subject and is being attacked on many fronts — better understanding of properties of materials, reduction in mechanical complexity, designing for minimal maintenance and educating users to pay economic prices, to name a few. Many more specific illustrations can be given, such as an attenuated total reflectance analyser for the ultraviolet spectral region, but often the ability to meet the need depends less on technical solutions but on the economic balance between cost of development, market size and payoff to the user — the latter often being very difficult to predict.

## References

Allkins, J. R. (1975) *Analyt. Chem.* **47**, 752A–762A.

Barty, A. M. (1963) *Appl. Optics* **2**, 969–973.

Barty, A. M., and Ruhl, H. D. (1968) *Chem. Eng. Prog.* **64**, 845–849.

Barton, G. H., Carr-Brion, K. G., Hewett Emmett, A., and Johnson, W. (1969) *Trans. Inst. Min. Met.* C **78**, 154–157.

Carlson, D. R. (1971) *Adv. in Instrumentation*, **26**, 731 (1–7).

Cameron, J. F. and Piper, D. G. (1972) *Proc. Technol. Int.* **17**, 915–917.

Carr-Brion, K. G. (1973) *X-Ray Spectrometry* **2**, 63–67.

Carr-Brion, K. G. and Gadsden, J. (1969) *J Phys E* **2**, 155–156.

Church, F. (1973) *Control Inst.* **5** (9), 58–59.

Clevett, K. J. (1973) *Handbook Proc. Stream Analysers* (Chichester: Ellis Horwood), 394–399, 46–65, 274–306.

Dailey, W. V. (1973) *Inst. Technol.* **20** (9), 66–70.

Davies, J. H. (1970) *Analy. Chem.* **42**, 101A–112A.

Diggins, A. A., Parker, K. and Webber, H. M. (1972) *Analyst* **97**, 198–203.

Dunn, B. E. (1972) *Adv. in Instrumentation* **27**, 736 (1–5).

Durst, R. A. (1969) NBS Spec. Pub. 314.

Forman, J. K. and Stockwell, P. B. (1975) *Automatic Chemical Analysis* (Chichester: Ellis Horwood).

Freeguard, G. F. and Pulford, C. I. (1972) *Ind. Eng. Chem. Process Des. Dev.* **11** (1), 78–81.

Gamage, C. F. and Topham, W. H. (1974) *Adv. in X-ray Analysis* **17**, 542–559.

Gilsson, I. R. and Hendra, P. J. (1970) *Laser Raman Spectroscopy* (New York: John Wiley).

Harbert, F. C., Bak, J. and Sjoberg, B. (1974) *Control and Inst.* **6** (1), 36–37.

Hartmann, C. H. (1971) *Analyt. Chem.* **43**, 113A–125A.

Hill, H. C. (1972) *Introduction to Mass Spectrometry* (London: Heydon and Sons).

Hinkley, E. D. and Kelley, P. L. (1971) *Science* **171**, 635–639.

Huillet, F. D. (1975) *Tappi* **58**, (10), 94–97.

Jones, L. M., Pritchard, W. D. N., Williams, K. F. and Williams, R. B. (1972) *Proc. IEE Conf. on Ind. Meas. and Control by Radiation Techniques*, 238–251.

Jones, M. H. and Woodcock, J. T. (1974) *Analyt. Chem Acta* **69**, 275–286.

Kinsley, T. J. N. (1972) *Proc. Inst.* April. 29–32, May 43–96.

Kliment, V. and Tolgyessy, J. (1972) *J. Radioanalyt Chem.* **10**, 273–297.

Kolthoff, I. M., Elving, P. J. and Sandell, E. B. (1964) *Treatise on Analyt. Chem.* Vol 5 (New York: Interscience), 2753–3056.

Legenhausen, R. F. (1971) *Adv. in Inst.* **26**, 735 (1–4).

Lephardt, J. O. and Vilcins, G. (1975) *Appl. Spectroscopy* **29**, 221–225.

Leppala, A., Koskinen, J., Leskinen, R. and Vanninen, P. (1971) *Soc. of Min. Eng. AIME, Trans.* **250**, 261–268.

Lovas, F. J. (1974) *Analysis Inst.* **12**, 103–109.

Manka, D. O. (1975) *Inst. Technol.*, February 45–49.

Martin, D. (1975) *Proc. Eng.*, October, 86–93.

Mayz, E., Corn, M. and Barry, G. (1971) *Am. Ind. Hyg. Assoc. J* **32**, 373–377.

Melbourne, K. V., Robertson, K. G. and Oaten, A. B. (1972) *Water Pollution Control* **71**, 278–288.

Novak, J., Kubelka, V., Zluticky, J. and Mostecky, J. (1973) *J. Chromatog*, **76**, 45–50.

Nutter, J. C. (1972) *Cement Technol.* **3**, 55–57.

Owers, M. J. and Shalgosky, H. I. (1974) *J. Phys. E* **7**, 593–603.

Parsignault, D. R., Wilson, H. H., Mineski, R. and Blatt, S. L. (1971) *Proc Am. Nucl. Soc. Meeting, Neutron Sources and Applications:* NTIS, Rep. 25 50615, IV, 40–46.

Perdijon, J. (1974) *Talanta* **21**, 1047–1064.

Pierce, T. B., Boswell, C. F. and Peck, P. F. (1974) *Analyst* **99**, 774–781.

Powney, C. (1973) *Proc. Eng.*, February, 17.

Ramaz, A. (1970) *Meas. Control* **3**, 720–722.

Riseman, J. M. (1972) *Am. Lab* **4** (12), 63.

Saltzman, R. S. (1969) *Analyt. Inst.* **6**, 79–85.

Schuy, K. D. and Reinhold, B. (1971) *Meas. Control* **4**, T84–89.

Sekerka, I. and Lechner, J. (1973) *Talanta* **20**, 67–72.

Simpson, R. J. (1968) *Meas. Control* **1**, 82–83.

Tily, P. J. (1974) *Control Eng.*, November, 89.

Tipping, F., (1970) *Meas. Control* **3**, 145–152.

Troy, D. J. (1955) *Analyt. Chem*, **27**, 1217–1221.

Tuttle, W. H., Williams, C. J. and Peterson, G. A. (1972) *Min. Cong. J.* **58** (1), 47–54.

Verdin, A., (1973) *Gas Analysis Instrumentation* (London: Macmillan) 298–308 and 17–31.

Villalobos, R. (1975) *Analyt. Chem*, **47**, 983A–1004A.

Wright, H. C. (1973) *Infra-red Techniques* (Oxford: Clarendon Press) 22–55.

# Paper 12

# Silicon sensors meet integrated circuits

*Phillip W. Barth*

This paper introduces some of the techniques used to create a sensing element as an integral part of a piece of silicon which may also contain integrated electronic circuitry for signal processing. The extremely small, relatively cheap sensors which can be made using these techniques have potential applications particularly in the field of medicine, where sensors implanted in the body can detect chemical and physical parameters for diagnosis and control of abnormal body functions. This article places some emphasis on the potential and problems of the technology for these *in vivo* applications.

Some of the terms used will be unfamiliar to readers not having a basic knowledge of semiconductor materials and manufacturing techniques, but lack of a detailed understanding of these terms does not prejudice the comprehension of the trend and pace of development in this growing area of instrumentation. (Eds.)

The combination of sensing and circuit technology in monolithic form is making transducers cheaper and more versatile.

## 12.1 Introduction

Advanaces in silicon sensor technology are now permitting a merger of sensor and integrated circuits into integrated sensors — single silicon chips containing both sensors and circuitry. As a result, the transducer industry stands at a threshold that was crossed by the microelectronics industry in the early 1960s. The next several years will see an explosion of sensors and sensor applications. Silicon sensors will be introduced both in applications where costly hand-assembled devices were previously used and in those where sensors have been too expensive. Used in conjunction with microprocessors, sensors will increasingly contribute to improved system performance with decreased part count and decreased cost.

The sensor revolution will be on a smaller scale than the integrated-circuit and microcomputer revolutions, because a sensor is not as general-purpose an element as an operational amplifier or a microprocessor. But the need for feedback in many measurement and control applications points to a continually increasing role

Originally published in *I.E.E.E. Spectrum* (September 1981).

for inexpensive silicon sensing devices in areas ranging from clinical medicine to automotive engine control and including such exotic uses as spacecraft instrumentation.

The differences between silicon sensors and earlier ones parallel the differences between silicon integrated circuits and discrete vacuum tube circuits: whereas the older technologies required hand assembly on a per-unit basis and on a physical scale compatible with the human eye and hand, the newer technologies permit batch fabrication on a scale limited only by such fundamental concerns as the wavelength of the light used for photolithographic pattern definition. With batch fabrication, small size, and increased experience with fabrication techniques, the per-unit cost of silicon sensors is falling to a point where it will be feasible to use several sensors in a system at small cost. The major factor determining how much sensor costs can be reduced will be the sales volume; greater sales will lower unit cost. This is a chicken-and-egg situation, since use of sensors in high-volume applications is limited at present by their relatively high costs.

A *Spectrum* article (Microprocessors get integrated sensors, February 1980) described several silicon sensors developed in the United States and the Netherlands. The sensors described below were developed in US laboratories, though much sensor work is also being conducted in Europe and Japan. A bias toward biomedical sensor applications is evident, but the technology is also being applied in other fields (automotive, industrial, consumer, aerospace, and military).

Not every silicon sensor is an integrated sensor; indeed, to date most silicon sensors have consisted of the basic sensor element, with little or no on-board signal processing. In addition, not all sensors are amenable to the incorporation of on-board circuitry. For example, the miniature gas chromatography column described below uses an entire silicon wafer (5 cm in diameter) with acceptable yield, but would probably not show acceptable

yield if it included an on-board signal processor. Both dumb and smart sensors will remain important for the foreseeable future.

In contrast to an integrated circuit, which is normally packaged to protect it from the surrounding environment, a sensor must be exposed to its environment in order to function. At the same time, it must transmit an output signal along a path that may be mechanically and electrically unstable. The instabilities can be overcome by including active circuitry on the sensor, but this circuitry must then be protected from the environment.

These three requirements, therefore, define the major areas of work in sensor development: the sensing element itself, the signal transmission technology, and the sensor package. These should be distinguished from other system-oriented considerations, such as signal processing external to the sensor and the use of sensors in control-system feedback loops. Although those concerns will affect the design of individual sensors, they are not as important in determining whether it is possible to fabricate a useful class of sensors.

## 12.2. Micromachining shapes the sensor

A basic requirement in silicon sensor fabrication is the formation of three-dimensional shapes in silicon. The shaping operation, called micromachining, encompasses a three-level hierarchy of techniques: (1) planar technology, typically restricted to structures with a depth less than two micrometers; (2) wet chemical etching technology, suitable for structure depths of from one to several hundred micrometers; and (3) mechanical sawing, a necessary but usually unappreciated aspect of micromachining, useful for kerf depths of several tens of micrometers to several millimeters.

Planar technology comprises the arsenal of techniques developed for integrated-circuit fabrication over the past two decades, including photolithography, oxide etching, thermal diffusion, ion implantation, chemical-vapour deposition, and dry plasma etching. Planar techniques are the beginning point for all micromachining techniques; in particular, photolithography is necessary to define masked areas on the wafer surface for later chemical etching. In addition, planar techniques are sometimes useful in their own right for fabricating micromechanical components. For example, a miniature capillary can be made using a half-micrometer step on the surface of a silicon wafer. If a flat Pyrex plate is bonded over this step using a field-assisted glass-to-silicon technique [see section 12.3], an extremely small capillary is formed because the glass cannot exactly conform to the step.

Planar silicon technology is sufficient for fabrication of two dimensional integrated circuit structures, but fabricating sensing elements in silicon often requires non-planar technology. For example, silicon pressure sensors are fabricated in wafers that are typically 200 $\mu$m thick, but such sensors require pressure-sensing diaphragms only 10 to 20 $\mu$m thick. These diaphragms can be batch-fabricated using silicon dioxide as a mask and a mixture of KOH and water as an etching solution.

Micromachining generally requires attention to both sides of the silicon wafer; for example, features are defined and fabricated on both the front and back surfaces of a pressure sensor and must be aligned with respect to each other. In a well-equipped fabrication facility, such alignment is often accomplished with a two-sided mask aligner that uses an infra-red light source and converter to see through the silicon wafer. Surface steps (which scatter infrared) and heavily doped regions (which block infrared) can be seen on the far side of the wafer aligned with features on the top surface. In less well equipped facilities, a two-sided alignment jig of some type must often be constructed, or alignment holes must be etched completely through the wafer at several points.

The two-sided nature of such processing creates additional headaches, but it also gives an additional measure of freedom in sensor design.

The final fabrication step of a miniature sensor is to saw up the finished silicon wafer. Here again the two-sided aspects of sensor fabrication come into play, requiring judicious use of a high-speed saw (the same type of saw used for die separation in standard integrated-circuit fabrication processes) on both sides of the wafer.

## 12.3. Hermetic seals without glue

The glass-to-silicon field-assisted bonding technique is attractive for sensor fabrication because it creates a hermetic seal without using an intermediate liquid layer such as a glue or a solder. The absence of any liquid permits the bonding area to be precisely defined by the mating areas of the glass and silicon. In a typical application of glass-to-silicon bonding, a piece of optically flat #7740 Corning glass (Pyrex) is placed in contact with a piece of optically flat silicon. The two pieces are heated to a temperature in the range of from $300°$ to $500°$C, and a large, negative DC voltage (in the range of 200 to 1500 V) is applied to that surface of the glass not in contact with the silicon while the silicon is electrically grounded. The Pyrex glass is approximately 3 to 5 % sodium, and the sodium becomes a mobile carrier of electrical current at the temperature used, moving toward the point at which the negative voltage is applied. As it moves, it leaves bound negative charges in the glass near the glass–silicon interface, which create a high electrical field between the glass and silicon and a consequent electrostatic force that pulls the glass and silicon into intimate contact, creating a fusion bond.

The nature of the glass–silicon interface region is not well understood; but when both surfaces are sufficiently clean and flat, a true hermetic seal is created, which is irreversible and is as strong as either the glass or the silicon. This technique has been used in fabricating absolute pressure sensors with electrical feed-throughs under the bonded region (Figure 12.2) and selected sensors have shown good long-term stability in a dry laboratory environment. However, sensors from the same wafer as the stable sensors have been unstable. The bonding technique has been shown to degrade the surface properties of the silicon, making the electrical properties of the silicon and the glass-silicon interface unpredictable. Degradation arises from high electrical fields during the bonding process and from sodium contamination of the silicon.

Despite these problems, glass-to-silicon bonding has several attractive features. Since it uses no glue, both sensor and circuit elements can be placed within several micrometers of the bonding area. The resulting structure is transparent on one side so that the quality of the bond can be optically inspected and pressure sensor diaphragms can be etched and optically inspected after bonding.

Several modifications of glass-to-silicon bonding are currently being investigated. One modification incorporated into an integrated capacitive pressure sensor uses intermediate layers of silicon dioxide and aluminium to screen the underlying silicon from harm. The technique starts with the silicon portion of the integrated sensor. A layer of silicon dioxide is thermally grown on the silicon surface; the layer is known to have good properties at the oxide–silicon interface. Next, a layer of aluminium is deposited on the oxide surface. Finally, a piece of glass is bonded to the aluminium. This technique produces a good hermetic seal, but the soft aluminium layer tends to creep after bonding, producing drift in the sensor's output. In addition, the aluminium is not corrosion-resistant *in vivo*, so the bond area can corrode rapidly.

A second modification of glass-to-silicon bonding now under investigation uses a similar sandwich structure, but a layer of polycrystalline silicon (poly) is deposited on the oxide surface instead of aluminium. The poly is then doped with an impurity (such as boron, phosphorus, or arsenic) to give it the proper electrostatic potential with respect to the silicon beneath the oxide. Finally, a piece of glass is bonded to the poly. The poly layer must be connected to the silicon by a path of low electrical resistance during the bonding operation to prevent high electric fields from damaging the underlying circuit elements during bonding. If the poly layer is doped with phosphorus, some gettering (gathering and binding) of sodium will occur during and after bonding, reducing sodium contamination of the underlying silicon and oxide. After bonding, the poly layer may be used as a MOS field plate to prevent the formation of conducting MOS channels in undesired areas during use.

### 12.4. Sealing the sensor: a tough problem

The world is still searching for the perfect way to protect sensors from their environment. Because the devices must 'sense' and thus be exposed to their surroundings, the problem is much more complicated than it is for a simple integrated circuit, where the naked silicon chip is encased in protective layers of ceramic, metal, and plastic. The problem is least severe for an accelerometer, which can be sealed away from almost all effects except the inertial effects it measures, and is most severe for chemical sensors, which must be exposed directly to an unfriendly world.

With integrated sensors, the protection problem is further complicated. Silicon circuitry is sensitive — to name just a few effects — to temperature, moisture, magnetic field, electromagnetic interference, and visible light. The package must expose the sensing element while protecting the on-board circuitry. Fortunately, the circuit itself can sometimes be used to transmit data about effects that cannot be screened out. For example, integrated pressure sensors can transmit a temperature signal and use external signal processing to remove the temperature dependence of the pressure signal.

A look at the evolution of hermetic seal technology for silicon absolute-pressure sensors gives a sense of the difficulties involved. To fabricate such a device, environmental pressure is applied to the other side of it. Pressure is monitored by measuring displacement of the diaphragm (in a capacitive sensor) or stress in the diaphragm (in a piezoresistive sensor). In either type of sensor, there must be some type of hermetic seal to keep a constant reference pressure in a chamber on board the sensor.

One early way to make the hermetic seal was to put a low-melting-point glass layer between two pieces of silicon, put the resulting sandwich in a clamp, and fuse it together in a vacuum chamber. The result is a sensor with a vacuum inside and a hermetic, low-creep, corrosion-resistant seal around the reference chamber. Because the melting glass forms a sticky glue, the sensing diaphragm and any on-board electronics must be well separated from the sealing area to avoid sticking them to the adjoining silicon piece. Also, because the silicon is opaque it may be necessary to fabricate the sensing diaphragm before the hermetic sealing operation (diaphragm thickness, for a thick diaphragm, is typically monitored by light transmission; silicon transmits red light at about a 20 $\mu$m thickness, and progressively higher frequencies are transmitted with decreasing thickness, until the silicon looks yellow at a 2 $\mu$m thickness). Finally, most glasses available for this sealing operation have thermal expansion coefficients that do not match silicon's thermal expansion coefficient, so high stresses can be frozen into the bonding area. Even given these problems, this sealing technique has been adequate for some commercially manufactured pressure sensors; however, long-term *in vivo* use demands a better sealing technique.

Some sensors have used a gold–germanium preform melted between two pieces of silicon to form a hermetic chamber. This bond has a high mechanical creep rate, however, introducing an uncontrolled drift in the sensor output. In addition, like the melted-glass bond it suffers from the presence of a sticky glue and the opacity of the bonded structure.

The most promising sealing technique for biomedical sensors is field-assisted glass-to-silicon bonding. This technique (known variously as 'anodic bonding', 'electrostatic bonding' and 'the Mallory process') has been used to fabricate several commercial silicon sensors. Its great advantages are that it has no glue or melted layer in the bonded area; glass with a fairly closely matched thermal expansion to silicon's is available; and the glass is transparent, which permits optical inspection of the bond. The technique has its own difficulties, but several modifications now under investigation (see section 12.3) should make it useful for long-term-implantable integrated sensors.

## 12.5. A look at emerging silicon sensors

Sensor prototypes now emerging from research laboratories demonstrate the feasibility of several different types of sensors. A miniature gas chromatography column being developed in the Stanford University Integrated Circuits Laboratory has an integral valve and capillary column and a removable thermal conductivity detector.

In a typical application of gas chromatography for detection of contaminant gases, a pulse of a gas mixture is injected into a stream of carrier gas flowing through a narrow capillary, whose inside walls are lined with a polymer. As the pulse flows past the lining, each component of the mixture is absorbed into the lining, stays in the lining for some time (the time depends on the solubility of that component in the lining), and is then desorbed. This sequence is repeated many times along the length of the capillary, so that some components emerge from the capillary later than others. Each can then be detected and measured by monitoring a characteristic (such as thermal conductivity) of the emergent gas stream.

The miniature silicon gas chromatography column (Figure 12.1) is formed by silicon micromachining techniques: a spiral is defined photolithographically in photoresist and is then etched into an underlying layer of silicon dioxide previously thermally grown on the silicon surface. The oxide layer serves as a mask for a subsequent etch in a mixture of hydrofluoric and nitric acids. This etch creates a long spiral groove in the silicon surface. Next, the oxide is stripped from the surface, and a flat plate of Pyrex glass is bonded to the silicon surface using a field-assisted, glass-to-silicon bonding technique. This forms a tunnel in the silicon under the glass, which is the gas chromatography capillary, visible as a folded spiral.

The thermal conductivity detector consists of a silicon body, a sputtered Pyrex diaphragm, and an evaporated-nickel film-sensing resistor, with a response time constant of 200 $\mu$s to a change in the thermal conductivity of the output gas. The gas chromatography system separates light hydrocarbon gases in less than 10 s, with sensitivities down to the part-per-million level (this compares with separation times of several minutes for typical gas chromatography) The system has been incorporated into a microprocessor-controlled instrument worn by a worker to monitor the gases he or she breathes during a normal working day. The instrument samples the air at 1 min intervals, inspects the output of the column for the presence of up to 10 different gases, sounds a warning if it finds any above a preset concentration and stores data on the cumulative dose of each gas that it 'breathes' during the day. The instrument will greatly expand the ability to monitor and help reduce worker exposure to toxic gases.

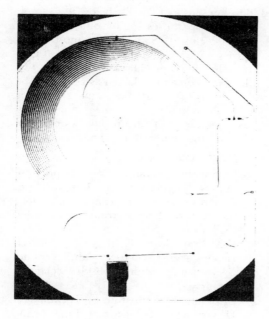

**Figure 12.1.** A silicon wafer is used to fabricate a gas chromatography column. Shallow grooves are etched in the wafer using hydrofluoric and nitric acid mixture, and holes are etched completely through the wafer using a KOH and water mixture. On the gas-chromatography wafer are an integral valve seat formed on the reverse side of the silicon wafer, several side capillaries for injection of test-sample gases, and holes etched through the silicon to connect the side channels and valve to the main capillary (upper left). A window is etched through the silicon wafer at the output of the column (right side) so a removable thermal conductivity detector can be clamped in place on the reverse side of the wafer. Immediately to the left of this window are two feedthrough holes that let the gas stream flow through the wafer and past the thermal conductivity detector.

## 12.6. Silicon pressure sensors in production

Silicon pressure sensors have been produced commercially by several manufacturers; however, the goal of a cheap, stable sensor remains elusive. A major contributor to cost has been the testing and packaging of the sensor element. One approach that has been used is to include on-board calibration data by putting the sensing element on a hybrid substrate and laser-trimming resistors on that substrate to produce a calibrated voltage output. This approach is expensive, using the entire hybrid substrate merely to record a few calibration coefficients in analogue form. The advent of microprocessor-controlled systems means cheaper packages can be produced. Calibration data for each sensor can be stored in system memory at very little cost in system overhead as long as such memory is already present.

Most silicon pressure sensors are suitable only for dry, uncontaminated environments and begin to exhibit calibration drift when placed in more severe situations — under the hood of an automobile, for example. Some work has been done on protecting the sensing element itself from environmental contamination. For example, Honeywell's Microswitch Division's sensors have resistors ion-implanted beneath the silicon surface to protect them from contaminants at the surface, with resulting improved stability. However, the leads attached to the sensor and the hybrid electronics associated with the sensor must still be protected from contamination. The package is still susceptible to drift problems in severe environments.

Perhaps most severe for silicon pressure sensors is the warm, wet, saline *in vivo* environment. Several miniature pressure sensors for biomedical *in vivo* use have been developed at university laboratories. These range from a passive piezoresistive device to a sophisticated capacitive instrument with on-board electronics yielding a pulse-time-coded output. Each new sensor design has revealed new problems with the *in vivo* use of pressure sensors, and no design has yet proven suitable for long-term use — six months or more — with good stability. The problems with these state-of-the-art sensors are instructive for other applications (in automobiles, for instance, where five-year accuracy is desired for manifold absolute-pressure sensors). More important, the previous sensors have served as vehicles for technology development, so that the generation of sensors now under develop-

ment will be stable *in vivo* in the presence of these now-known problems.

## 12.7. Silicon pressure sensors in the USA

Silicon sensors have been developed at several university and industrial centres in the United States. Most of these centres are also developing actuators, often with the same types of technology as those used to develop sensors. The list below is a sampling of these centres and of some of the sensors developed. Some of the most useful theory and development have been at centres that have produced only one or two devices that can be classed as sensors. The following list considers only silicon or silicon-compatible devices — those consisting entirely of silicon, layers on top of silicon, and material attached to silicon:

Case Western Reserve University Engineering Design Centre: pressure sensors (relative and absolute; piezoresistive and capacitive); ion-selective field-effect transistors (called ISFETs or EISFETs).

IBM San Jose (California) Research Laboratory: capacitive cantilever accelerometer with MOS detection circuitry.

Massachusetts Institute of Technology, Department of Electrical Engineering and Computer Science: MOS surface impedance and moisture measurement sensor.

Stanford University Integrated Circuits Laboratory: pressure sensors (relative and absolute; piezoresistive and capacitive); piezoresistive strain gauges; linear thermometer array for cancer hyperthermia; gas chromatography column; thermal conductivity detector for gas chromatography; piezoresistive cantilever accelerometer, piezoelectric-oxide-semiconductor field-effect-transistor (POSFET) arrays for ultrasound detection.

University of Michigan Electron Physics Laboratory: pressure sensors (relative, piezoresistive); thermopile infrared detector.

University of Pennsylvania, Moore School of Engineering: ion-sensitive gate-controlled diode; ion-sensitive electrode.

Commercial manufacturers of silicon pressure sensors include Kulite Semiconductor, National Semiconductor, Honeywell's Microswitch Division, and Foxboro ICT.

Three silicon pressure sensors are especially illustrative of the technology and the problems involved. The simplest, developed in the Stanford University Integrated Circuits Laboratory, is similar in design to several sensors independently fabricated in the Case Western Reserve University Engineering Design Center. This piezoresistive sensor (Figure 12.2) consists of n-type silicon, a thin etched diaphragm, p-type sensing resistors at the edges of the diaphragm, and a sealed reference cavity formed between a shallow well etched in the silicon and a bonded-on Pyrex cap. The sensing resistors are located at the edges of the diaphragm, inside the reference cavity. (Placing the resistors inside the cavity is essential for long-term stability; if they are exposed to the saline environment outside the cavity, their resistance changes with time.) Heavily boron-doped (p + ) regions contact the resistors and are used as feedthroughs under the glass-and-silicon hermetic seal region. Aluminium pads are formed on the glass cap before it is bonded to the silicon, and during the glass-to-silicon bonding operation these pads are sintered to the p + feedthroughs.

In operation, pressure changes cause the diaphragm to flex. The diaphragm acts as a stress amplifier (the stress in the diaphragm is proportional to pressure multiplied by the area of the diaphragm), and stress is monitored by changes in the resistance of the four resistors in the bridge. This sensor, like several similar ones, has shown good long-term stability (less than 1 torr of drift for a full-scale

range of 300 torr) in a dry environment over several months. However, in a wet environment it quickly exhibits drift problems because of lead shunting: very minor changes in the shunt resistance of any attached cable are read as major changes in pressure. In addition, the bonding pads corrode *in vivo*, degrading the output signal amplitude, even when these pads are protected by the best available polymer coatings.

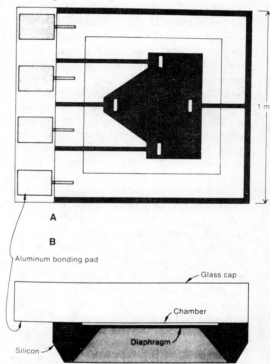

**Figure 12.2.** This piezoresistor absolute pressure sensor has a Pyrex cap. A top view looking through the glass is shown in *A* and a cross section of the sensor is shown in *B*.

The problems of lead shunting and pad corrosion cannot, by themselves, be reduced to the point where a simple passive pressure sensor remains stable *in vivo*. They have therefore become the driving force behind the development of integrated pressure sensors. It has become obvious that a simple analogue voltage amplitude cannot be reliably transmitted from an *in vivo* sensor to the outside world and that signal-conditioning circuitry must be included on the sensor to transmit an amplitude-insensitive signal.

The University of Michigan's Electron Physics Laboratory produced the first integrated pressure sensor. This uses piezoresistors at the edges of a diaphragm for pressure sensing and the 'wasted' area in the centre of the diaphragm for active integrated circuitry (an amplifier and a current-controlled oscillator). Input power and output signal are transmitted over a two-wire cable; the output is an analogue-modulated FM signal with a carrier frequency of 1 to 2 MHz and a frequency deviation of 1.6 KHz torr$^{-1}$. With the relatively high carrier frequency, an oscillator can be built using a very small on-chip capacitor. Several sensors with different carrier frequencies can be assembled into a catheter for short-term use, with their outputs demultiplexed externally by phase-locked-loop circuitry.

Two characteristics keep this sensor unsuitable for long-term *in vivo* implantation. First, it is a relative pressure sensor, with no sealed reference cavity on board, so than an air path to the outside world must be maintained during its use. Second, no provision is made for protecting the on-board circuit from moisture intrusion in the long term; such moisture intrusion can lead to a change in on-board capacitance and resistance and in output frequency during a long-term implant.

The most sophisticated integrated sensor reported to date comes from the Stanford Integrated Circuits Laboratory. It incorporates both on-board circuitry and a sealed reference chamber to form an absolute pressure sensor with a pulse-time modulated output. Instead of piezoresistors, this sensor uses a silicon diaphragm as one plate of a parallel-plate capacitor. The other plate of the capacitor is a metal layer deposited on the surface of the glass cap, bonded to the chip to form the hermetic reference cavity. The capacitor sits in the charging leg of a non-saturating Schmidt-trigger oscillator and is fed by a constant current source. The

capacitance of the two plates increases with increasing pressure, so that the charging time of the capacitor and the period of the oscillator also increase.

The sensor's output is a series of short pulses, with a period proportional to pressure (the period ranges from 73 to 76 $\mu$s over the absolute-pressure range of 760 to 1060 torr). This sensor is specifically designed for low-power telemetry applications and consumes only 54 MW of power during operation. It is the closest approach to date to a stable, long-term implantable sensor, though it has deficient corrosion resistance and hermetic seal technology. In addition, it exhibits a temperature coefficient making it difficult to use in studies that involve raising or lowering the animal's temperature.

The availability of both piezoresistive and capacitive sensing techniques for miniature silicon pressure sensors raises the question of which technique will eventually dominate. Both have advantages, and no clear answer has yet emerged.

Piezoresistive sensing has a lead in terms of the expertise researchers and industry have developed over the past several years. This technique, using a simple, passive resistor bridge and a simple sensor structure, yields a relatively low-impedance output, suitable in many cases for signal transmission over short distances without additional signal conditioning.

Capacitive sensing, in contrast, is inherently a high-impedance method. In a miniature capacitive sensor, the capacitance being sensed is comparable in magnitude to that of the cable attached to the sensor, and signal conditioning is required near the sensing element to prevent signal degradation. For diaphragm structures equivalent in size and thickness, the sensitivity of capacitive sensing (defined as the percentage change of capacitance with pressure, compared with zero pressure capacitance) is higher than that of piezoresistive sensing (defined as the percentage change in resistance with pressure). Capacitive sensing requires a parallel-plate structure with a fluid medium of fixed permittivity between the plates, so that it requires a hermetic seal around the capacitor. A piezoresistive sensor needs only a single diaphragm between two fluid media, so it is simpler to use for sensing the relative pressures of the two media. In a capacitive pressure sensor the temperature coefficient of the output signal is related only to the thermal-expansion coefficient of its structural material. Both capacitive and piezoresistive sensors will have competing applications over the next several years and it may be that neither will ever dominate the industry.

Several research groups are at present working on the next generation of biomedical pressure sensors. The improvements sought are in circuit design, hermetic seal and lead-attachment techniques, and overall packaging. One possibility being investigated is transmitting more than one parameter at once (such as pressure and temperature) and peforming external digital compensation for unavoidable interactions (such as removing a temperature coefficient from a pressure signal) while gaining an additional valuable measurement parameter. Signal transmission options are being evaluated in a communications engineering perspective: it is assumed that the communications channel is poor and that it gets worse with time. Stability and survivability are then determined by how cleverly the designer can incorporate circuitry into a sensor package small enough for implantation *in vivo*.

## 12.8. Accelerometer formed from a cantilever

Researchers at the IBM San Jose Research Laboratory have developed a group of elegant micromachining techniques that combine the orientation-dependent and doping-dependent properties of an ethylenediamine/pyrocatechol/water (EDPW) mixture to form cantilevered beams of silicon dioxide over shallow cavities etched in silicon. These techniques have been used to fabricate several structures such as miniature reed switches and miniature arrays of optical reflectors; one such structure is a capacitive cantilever accelerometer with an on-board MOS amplifier.

The formation of an oxide cantilever begins with a silicon wafer containing a buried, heavily doped boron layer, which can be made by boron diffusion followed by deposition of a lightly doped epitaxial layer. On the exposed silicon surface, a layer of silicon dioxide is formed by thermal oxidation, and a layer of metal is deposited (Figure 12.3). Next, the metal layer is photolithographically formed into fingers, while the oxide layer becomes a pattern of both fingers and a frame around the finger area (Figure 12.3*A*).

The exposed silicon in this area of frame and fingers is then etched with an orientation-dependent etch such as EDPW. [...]

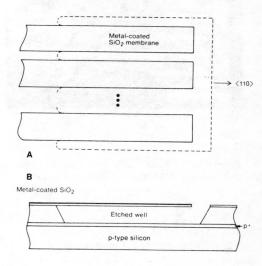

**Figure 12.3.** Etching of silicon to form an oxide cantilever beam. A top view of metal-oxide fingers in an oxide window is shown in *A*, with the SiO$_2$ to be etched away inside the dashed lines. The side view of the etched cantilever in *B* shows the cross section through a silicon wafer with a buried boron layer, epitaxial layer, silicon dioxide, and metal.

This technique for fabricating cantilevers has been incorporated into the fabrication of the world's smallest man-made analogue accelerometer and the first with on-board circuitry (Figure 12.4). At the end of each cantilever beam, a plated-up gold mass is visible; this mass increases the sensitivity of the accelerometer. The cantilever, which is metallized on its top surface, forms one plate of a capacitor, and the silicon substrate forms the other plate of the capacitor. The capacitance of the cantilever changes with acceleration, and the overall sensitivity of the on-board circuit is around $2 \text{ mVg}^{-1}$.

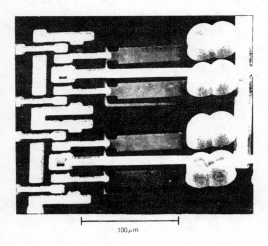

100 μm

**Figure 12.4.** Four adjacent accelerometers with on-board MOS amplifiers are shown.

## 12.9. The outlook for sensors

Sensors like these are now emerging from the laboratory into commercial use [see section 12.7]. Simple piezoresistive silicon sensors have been available for years. At least one vendor (Honeywell's Microswitch division) has recently developed a pressure sensor with an integrated operational amplifier, while the gas chromatography system described above is undergoing further development at a newly founded company. However, much work remains to be done in circuit development, fabrication technology, and packaging for specific environments before sensors will become more widely used. Compared with the integrated-circuit industry, where development has been pushed by military and by large-volume industrial and consumer uses, the sensor industry is limited in the amount of research effort it can invest. Silicon sensor technology has ridden on the coat tails of integrated-circuit technology, and there is no reason to think that this situation will change in the foreseeable future.

An important consideration in integrated-sensor design (as in integrated-circuit design) is the tradeoff between sensor and circuit complexity and yield. It seems that it will not be reasonable to put a high-powered, general-purpose computer on the same chip as a sensor element, though it is obvious from the above that a need for some on-board circuitry exists.

To make the proper compromise between cost and complexity, it may be helpful to consider biological models. For example, the human nervous system uses sensors that do measurand-to-pulse-stream conversion; the pulses are fed to the central nervous system for further signal processing. These biological sensors are frequently recalibrated as the body interacts with its environment, and the calibration data is apparently stored in the central nervous system. In most sensor applications, frequent automatic recalibration of integrated sensors is now a practical possibility, and the correspondence between the central nervous system and a remote microcomputer is obvious. For *in-vivo*-implantable silicon sensors, however, it is often impossible to gain access to the sensor for recalibration, and it may become worthwhile to include on-board recalibration standards of some type.

The next several years will be exciting for engineers involved in sensor development, as work begins to reach fruition and the types of sensors that will become widely used over the next several decades are established.

## 12.10. For further reading

For an overview of sensor technology in both its biomedical and nonbiomedical aspects, see the December 1979 issue of the *IEEE Transactions on Electron Devices*. This is a special issue on solid-state sensors, actuators, and interface electronics. A second special issue on the same topics is scheduled for January 1982. The October 1978 issue of the same journal is devoted to three-dimensional structures in silicon and contains much of the background on sensor fabrication technology.

A look at recent developments in solid-state sensors is given in the technical digest of the International Electron Devices Meeting, IEEE Electron Devices Society, Washington, D.C., December 1980.

A journal on sensor technology, *Sensors and Actuators — An International Journal* edited by Simon Middelhoek, began publication in Switzerland in 1980.

The English translation of *The First Symposium on Sensors — Fundamentals and Applications* sponsored by the IEEE of Japan, Tsukuba City, June 1981, will be available after November 1981.

The proceedings of a conference held at Stanford in 1979 on *Implantable Transducers and Systems: Packaging Methods and Testing Criteria* can be obtained from James B. Angell or Phillip W. Barth of the Stanford University Integrated Circuits Laboratory.

The present state of biomedical sensor technology is considered in some of the papers in *Bioengineering: Proceedings of the Ninth Northeast Conference,* Walter Welkowitz, editor, Pergamon Press, New York, 1981.

The following proceedings of a series of conferences sponsored by the National Institutes of Health give an overview of the biomedical transducer field: *Theory, Design, and Biomedical Applications of Solid State Chemical Sensors,* P. W. Cheung et al., CRC Press, West Palm Beach, Florida, 1978; *Indwelling and Implantable Pressure Transducers,* D. G. Fleming et al., CRC Press, Cleveland, Ohio, 1977; and *Physical Sensors for Biomedical Applications,* Michael Neuman et al., CRC Press, Boca Raton, Florida, 1980.

## Post Script: Unifying the sensor field

In attempting to provide a cross-sectional view of the silicon sensor field in the USA, I made at least one major omission. Work on silicon sensors in the Electronics Research Laboratory at the University of California, Berkeley, has been under way for more than a decade. The research group in that laboratory, headed by Professors R. S. Muller and R. M. White, introduced the idea of the piezoelectric field-effect transistor (PIFET), first embodying the concept in 1971 with a sputtered zinc oxide film on a FET gate. They have since extended the concept to the fabrication of a piezoresistive accelerometer. In addition, the Berkeley group has developed a vapor-sensing system using a surface-acoustic-wave (SAW) device in which propagation time through a SAW delay line is influenced by vapor absorption in a polymer film. In other projects, they are carrying out research on several types of integrated magnetic sensors and are investigating new materials to incorporate into planar processing for integrated sensor fabrication.

The above omission reflects a disunity in the sensor field that is slowly beginning to be remedied. Until recently, there has been no forum in which information on diverse sensor types was regularly gathered. The major route to gathering new sensor information has been attendance at diverse conferences, which are not devoted to sensors, and with proceedings which are not always widely published. The journal *Sensors and Actuators,* now entering its third year of publication, is a partial remedy for this situation, but is not the whole answer. There is growing sentiment within the sensor research community (expressed in discussions at the recent Materials Research Society meeting in Boston) for a regularly scheduled workshop devoted to solid-state sensors. With a little effort in the next year or so, such a workshop may become a reality.

# Paper 13

# Optical measurement methods

*Robert Jones*

This and Paper 14 form a pair which should be read in order by readers unfamiliar with recent developments in optical sensing. The first part of the paper introduces the laser as a light source, and a range of light detectors. The technical detail of this first section is of interest, but a full understanding of the physical principles is not necessary for later sections of the paper. This paper deals with some of the 'traditional' optical sensing methods using reflection and refraction of light, as well as newer devices using holography and interferometry. The paper ends with the introduction of optical fibres into sensing methods, a subject which is enlarged upon in the next paper. (Eds.)

Despite their high sensitivity, optical measurement techniques can often be used in hostile engineering environments. This paper describes the basic physical principles of a wide range of optical measurement methods and demonstrates how they may be used to solve specific engineering problems.

Two inventions fundamental to the diversity of modern optical measurement techniques are the laser, and solid-state light-sensitive detectors and arrays.

It is important, therefore, that the principles and significance of these devices be appreciated before specific methods of measurement are discussed. (Readers who are familiar with these areas may prefer to consult briefly Tables 13.1 and 13.2 and proceed to the following section).

## 13.1. The laser

The characteristics of the laser are best understood by comparing the features of laser light with those of the light emitted by a conventional source. Consider, for example, the light field produced by a tungsten-filament light-bulb. This consists of a continuous spectrum of radiant energy distributed over the wavelength range 0.3 to 15 $\mu$m. (The visible spectrum extends from roughly 0.4 to 0.7 $\mu$m). Emission of this form is referred to as black-body radiation and is the result of the energy released when thermally excited electrons pass from a broad band of discrete high-to-low energy states. In accordance with Planck's law (Transition energy ($E$) = Planck's constant ($h$) $\times$ frequency of emitted radiation ($\nu$)) the resultant light field may be regarded as the superposition of a large number of individual sinusoidal waves of differing frequency, each with random relative phase and hence mutually incoherent. In certain cases, for example gas discharge,

Originally published in *Engineering* (February 1983).

electron transitions occur between a small number of energy levels and consequently narrow wavelength bands or spectral lines are emitted. Such a phenomenon is called selective emission and can be used to produce approximately single-wavelength, or monochromatic, light of relatively narrow spectral width. The latter is achieved at the expense of considerable power reduction by passing the light through a narrow bandpass optical filter.

Laser light is generated by stimulated emission, a process not subject to any of these limitations. It enables high-intensity light of both narrow spectral width and spatial divergence to be obtained. The basic energy-level distribution within the light-emitting or lasing medium, which enables this process to take place, is shown in Figure 13.1. Electrons are excited from the ground state $E_0$ to an upper energy level $E_1$. They then spontaneously decay to a metastable level $E_2$. This is so called because the electrons have a sufficiently long lifetime in the state for a growth of the electron population to occur. If the build-up of electrons is sufficient for greater than 50% of electrons to exist in the metastable state then a state of population inversion is said to have been established. Under this condition the process of stimulated emission and light amplification may occur. This happens in the following way: a photon generated by a transition from the metastable state to ground level stimulates a second electron transition which in turn liberates a photon in phase with the first. These photons each stimulate further transitions and a burst of in-phase *coherent* photons in the form of laser illumination is generated provided that the emitting medium is contained within a suitable cavity. If the lasing material is kept in a state of population inversion by continual excitation of the lasing medium a continuous wave of coherent light is emitted. The most convenient source of visible continuous laser radiation is the helium–neon (He–Ne) laser which generates light at wavelength ($\lambda$) of 0.633 $\mu$m. The properties of this and other types of commonly used laser are summarized in Table 13.1.

It is clear from all of the foregoing discussion that three basic factors distinguishing laser light from incoherent and conventional monochromatic sources are: are:

high coherence and hence narrow spectral bandwidth, (typically less than $\pm 10^{-3}$ $\mu$m);
high intensity;
small beam diameters accompanied by low angular divergence.

The importance of these characteristics in measurement systems will become apparent from later discussions.

In optical measurement systems one is faced with the problem of measuring the spatial or temporal variation of the light intensity. The characteristics of some commonly used forms of light-sensitive detector are summarized in Table 13.2. All of the devices outlined in Table 13.2 will generate an analogue electrical signal proportional to the incident light intensity. The scan-operated devices, i.e. TV vidicons and solid-state self-scanned arrays, transform the spatial distribution of the light field into a video signal. The latter consists of a standard-format voltage waveform in which the voltage level at a given time is proportional to the intensity at a given coordinate in the scanned light field. In the case of TV vidicons the scan geometry is subject to errors induced by thermal drift, random tube voltage fluctuations etc. They cannot, therefore be used in the quantitative measurement of the coordinate of light intensity variations. This

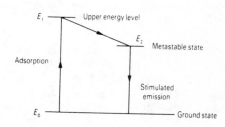

**Figure 13.1.** An energy level diagram for laser operation.

limitation does not apply in the case of self-scanned arrays. In these devices the spatial intensity distribution is derived from the output of each discrete detecting element in the array. (For photo detector arrays the output is read from a corresponding number of integrated diodes. For CCD arrays a charge packet proportional to the intensity of each element is generated and shift read in sequence via a single output diode). The position of each element in a self-scanned array is controlled precisely during production. Consequently their output suffers from negligible scan distortion and can be used in the precision measurement of light intensity variations.

The characteristics of laser illumination, just outlined, have led to many of the major innovations in optical measurement over the last 20 years but the principles of some of these techniques, such as interferometry and diffraction phenomena, were established well before the invention of the laser using conventional monochromatic sources. However, the limitations imposed by these light sources meant that these methods were never considered seriously as tools for routine optical measurement until the emergence of the laser.

## 13.2. Laser-based position sensors and alignment devices

The techniques described in this section rely primarily on the fact that a laser is capable of emitting a narrow low-divergence pencil of intense illumination. (See Table 13.1).

Consider the arrangement shown in

**Table 13.1.  Summary of laser characteristics**

| Lasing medium | Wavelength ($\mu m$) | Beam characteristics | Typical power range | Comments |
|---|---|---|---|---|
| He–Ne | 0.633 | Divergence $10^{-4}$ to $10^{-3}$ rad, source beam diameter $\simeq 1$ to 3 mm, circular Gaussian intensity profile | 1 to 50 mW continuous | General-purpose, low-cost laser useful for many optical systems. Laser cavity lengths, 0.3 to 1 m, typically |
| He–Cd | 0.442 or 0.325 | Divergence $10^{-4}$ to $10^{-3}$ rad, source beam diameter $\simeq 1$ to 3 mm, circular Gaussian intensity profile | 1 to 10 mW continuous | Useful as a coherent, near UV source. Cavity length, 0.3 m typically |
| Argon | Tuneable over the range 0.476 to 0.514 | Divergence $10^{-4}$ to $10^{-3}$ rad, source beam diameter $\simeq 1$ to 3 mm, circular Gaussian intensity profile | 100 mW to 15 W continuous | High powered tunable source. Water cooling and three phase power generally required. Cavity lengths 1 to 5 m |
| Ruby | 0.654 | As above with beam diameters 3 to 4 mm at source | 1 to 10 J (Single or dual pulsed) | Useful in study of transient phenomena |
| Gallium arsenide | 0.905 (typical) | Beam divergence $10^{-2}$ to $10^{-1}$ rad, elliptical Gaussian intensity profile | Continuous (low power) 10 mW or repetitively pulsed (up to 20 kHz) for higher power (max typically 100 W) | Broader spectral width than above devices combined with higher beam divergences. Have dimensions typical of a power transistor package |

Figure 13.2. Light from the laser source is transmitted across the region $R$. (The transmission process may be optimized using various forms of beam-shaping optics).

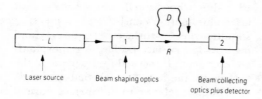

**Figure 13.2.** Optical sensor arrangement for the detection of a moving object $D$.

The transmitted light is then detected by a receiver, 2, which may consist simply of a lens designed to focus the collected light onto a suitable photosensitive detector (Table 13.2). If the passage of the beam is interrupted by an opaque object ($D$) the event is registered as a change in detector output level. Figure 13.3 shows how this method may be extended and used to measure the coordinate of the edge of a moving strip of material. This method is used frequently in strip mill quality control. Light from the source is scanned along a line $pr$ by means of the rotating mirror $M$. An image of $pr$ is formed in the plane of a self-scanned array by the lens $L$. The length of the self-scanned array $q'r'$ over

### Table 13.2. Light-sensitive detectors

| Detector type[*] | Principle of operation | Spectral range[**], $\mu m$ | Structure | General comments |
|---|---|---|---|---|
| 1 Photo-multiplier (PM) tube | Acceleration of photo-electrons by high electric field | 0.3 to 0.9 | Vacuum tube | Ultra-low-noise, high-sensitivity performance. Special applications where size and voltage supply not a problem |
| 2 Silicon photo-diode | Light induced electron/hole liberation at silicon p–n junction | 0.4 to 1.1 | Solid-state transistor package | Utility use |
| 3 PIN photo-diode | As 2 but with enhanced high frequency performance | 0.4 to 1.1 | Solid-state transistor package | High-frequency, fast-rise-time applications |
| 4 Silicon avalanche photo-diode | Silicon p–n junction operated at reverse bias close to breakdown | 0.4 to 1.1 | Solid-state transistor package | High-sensitivity device suitable for low light level applications. Nearest solid-state equivalent to a PM tube |
| 5 TV vidicon | Two-dimensional electron beam scan of charge analogue of image | 0.3 to 0.9 (standard) 0.4 to 1.1 (silicon target tube) | Vacuum tube | Standard TV camera applications, scan rate 25 frames s[-1] |
| 6 Photo-diode and CCD, self-scanned arrays (SSAs) | High-precision one- or two-dimensional solid-state detector arrays | 0.4 to 1.1 | Solid state | Used for precision measurement of spatial variations of light intensity. Minimum resolution 12 $\mu m$ at scan rate of 60 frames s[-1] |

[*]Detectors 5 and 6 are image forming devices. [**]Typical values.

which a signal is detected is proportional to the cooordinate $qr$ of the strip edge (clearly $qr = q'r'$ for $1:1$ image formation). The width of the strip can be monitored automatically when a similar sensor is used to measure the coordinate of the other edge of the strip, provided that the separation of the two detector arrays is known. In some applications it is better to use common optics to transmit and receive the laser beam. Such an arrangement is shown diagrammatically in Figure 13.4. Light from the source passes through a partially-reflecting mirror (or beam splitter) $B$ into the beam-shaping optics. Light back-scattered from the object $D$ then passes back through the beam-shaping optics and is focused onto the detector via $B$. Under these conditions an increase in output signal at the detector indicates the presence of the object. Various forms of sensor based on the above principles have been used to measure the

position of objects to a resolution of better than 1 mm over distances of 10 m in such environments as furnaces, tunnels and pipe lines.

The arrangement shown in Figure 13.5 shows how a self-scanned array and laser illumination can be used to measure simultaneously surface roughness and macroscopic surface profile. An unexpanded laser beam is focused in the plane of the surface by the lens $L_1$ such that the angle of incidence relative to the surface normal is $45°$. The surface is viewed by the lens $L_2$ so that the included angle between the illumination and viewing direction is $90°$. It can be seen from the figure that for this geometry, the $x$ coordinate of the image spot is a function of the surface height, $h$, with respect to a defined mean level. The output from the self-scanned area in the general form of a high-frequency ripple (surface roughness term) and slowly varying DC level (surface profile) can therefore be used to monitor the surface geometry. In one application a sensor of this design was mounted in a towed vehicle (towing speed typically $100 \text{ km h}^{-1}$) and used to measure the characteristics of the road surface. Surface height resolutions of about $5 \mu m$ can be attained using this method.

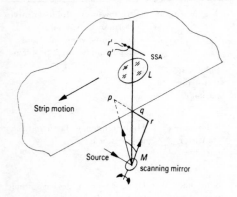

**Figure 13.3.** Strip gauging using an extension of the system shown in Figure 13.2.

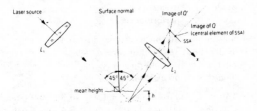

**Figure 13.5.** A surface finish gauge based on laser illumination and a self-scanned array.

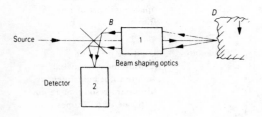

**Figure 13.4.** An object sensor in which the transmitting and receiving optics are combined in one unit.

An unexpanded laser beam provides an ideal light source for an optical level which can be used in precision mechanical and surveying alignment applications. The principle of the technique is illustrated in Figure 13.6. Here the laser is mounted on a suitable reference surface using, for example, a tripod mount and the beam directed at a small mirror attached to the structure under investigation. It can be seen that the coordinate of the return beam relative to that of the source can be used to calculate the angular misalignment of the structure. Relatively short ($\simeq 1$ m) optical level lengths ($L$), will result in return beam offsets of the order 10 mm for angular misalignments of typically $1^\circ$.

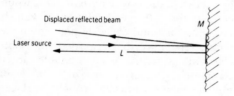

**Figure 13.6.** The use of an optical level as an alignment device.

## 13.3. Classical interferometry

Interferometric techniques depend upon the coherent nature of laser light and provide the basis for the most sensitive optical measurement methods. The phenomenon of interference may be observed when waveforms are superimposed. The resultant wave motion may then be expressed as the sum of the individual waveforms. This is known as the principle of superposition. Consider the case where two sinusoidal waves with the same wavelength ($\lambda$) have a relative path difference $d$, which is the difference between the total distance travelled by each wave. The relative phase of the waveforms is then $\phi$ where

$$\phi = 2\pi \frac{d}{\lambda}. \tag{13.1}$$

When

$$\phi = 2n\pi, \; n = 0, 1, 2, 3 \ldots \tag{13.2}$$

the waveforms superimpose to generate a state of maximum reinforcement referred to as a state of 'constructive interference'.

Conversely when

$$\phi = n\pi, \; n = 1, 3, 5 \ldots \tag{13.3}$$

the waveforms cancel one another out and a state of destructive interference is said to exist. These pheonomena are summarized in Figure 13.7. A variation between a state of constructive and destructive interference results in the formation of an interference pattern. Light consists of an electromagnetic waveform, and when light fields are superimposed under suitable conditions an interference pattern in the form of sinusoidal variation in light intensity may be observed. This is often referred to as an interferogram or fringe field. (See, for example, Figure 13.10).

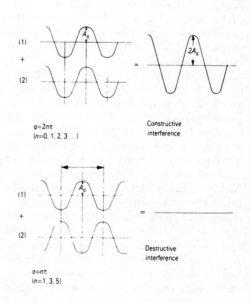

**Figure 13.7.** Condition for constructive and destructive inteference.

Various arrangements exist which enable optical interference phenomena to be observed. Many of the most useful systems are based on the interferometer devised by Michelson; this is shown in Figure 13.8. The input light beam, ideally derived from a laser, is split into two components 1 and 2 at the beam-splitter $B$. It is clear from the diagram that light entering the lens $L$ consists of the superposition of the fields reflected by $M_1$ and $M_2$ with the latter 'seen' in the dashed position. As a result a fringe field, $F$, may be observed in the resultant light field. If, as is shown, $M_1$ and $M_2$ are arranged to have a small relative angular displacement, $\theta$, then the relative path difference at a coordinate $x$ across the illuminated portion of the mirror is $x\theta$. In accordance with equation (13.1) a maximum of intensity in the fringe field will be seen when

$$2x\theta = n\lambda, \quad n = 0, 1, 2, 3. \tag{13.4}$$

Fringes of spacing $\simeq 1$ mm will be observed for value of $\theta \simeq 10^{-4}$ rad. This is the principle of methods by which the relative shape of optical components are compared interferometrically.

Consider now the case where the mirror $M_1$ is held static and the mirror $M_2$ is attached to a moving object. It follows from equation (13.4) that every time $M_2$ moves a distance $d = \lambda/2$ ( $\simeq 0.3$ $\mu$m) the intensity at a given point in the fringe field will pass through one cycle. If $N$ cycles are counted the total displacement is simply $N\lambda/2$. $N$ may be determined automatically using a light-sensitive detector placed in a fringe field coupled to a digital counter. In this way displacements may be measured to high levels of accuracy and resolution. It is important to note, however, that the lack of coherence of conventional monochromatic sources means that relative displacements of the order 1 mm will usually result in a complete loss of fringe contrast. By comparison displacements of order 10 m may be measured when a stabilized He–Ne laser is used as a light source. This feature enables the technique to be used in such applications as precision machine-tool calibration and length measurement.

## 13.4. Holographic and speckle interferometry

A Michelson interferometer of the type just described may be classed as an instrument of the 'classical' type in that the optical components must have high-quality specular surfaces. The introduction of a non-specular or optically rough surface, i.e. one with random, high-frequency surface height fluctuations of amplitude greater than the wavelength of light, would cause the fringe field to vanish. Holographic and speckle-pattern interferometers may be regarded as 'non-classical' systems which enable interferometric measurements to be made on non-specular, optically rough surfaces. This means tht all the advantages of interferometry, i.e.:

high resolution and sensitivity ( $\simeq 0.3$ $\mu$m displacement resolutions)
non-contacting measurement
field of view observation

may be extended to a wide range of engineering problems.

In holographic interferometry a hologram of the object under test must first be recorded, processed and relocated in the

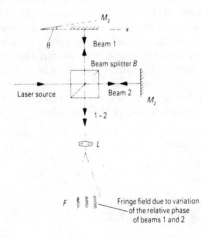

**Figure 13.8.** The layout of a Michelson interferometer.

original recording plane. For the purpose of this discussion the reader may regard the hologram simply as a device which will reconstruct the light field corresponding to the original object when it is illuminated by a reference wavefront. The basic arrangement is shown in Figure 13.9. An observer at $V$ sees the superposition of light from the virtual, holographically reconstructed image and that scattered from the real object. When the latter is displaced, for example, by the application of a load, a fringe field representing the displacement of the object relative to the virtual image is observed. This is referred to as a holographic interferogram.

Speckle interferometry is based on the observation of the speckle pattern effect. The latter consists of a granular variation in light intensity and is observed when any optically rough surface is illuminated by coherent light. It is due to the interference of a large number of wavefronts of random phase scattered from the surface. In speckle-pattern interferometry the image-plane speckle pattern of an object is made to interfere with a reference wavefront analogous to that used in holographic interferometry. The relative phase of the individual speckles is thereby recorded, hence their intensity changes when the

object is displaced with respect to the static reference wavefront. This causes fringes representing the displacement of the object to be formed when the image-plane speckle pattern of the object in its displaced and undisplaced positions are correlated.

The strength of holographic and speckle interferometry lies in their ability to generate high sensitivity field view strain and vibration patterns of complex objects typically over the size range $1 \, \text{mm}^2$ to $1 \, \text{m}^2$. One can therefore identify immediately regions of strain concentration, component fault, vibration mode, etc. This method has been applied to a large number of engineering design problems. For example, Figure 13.10 shows a speckle-pattern interferogram in which the fringe field indicates a region of material fault in a composite material.

## 13.5. Diffraction phenomena and measurement techniques

When a light beam is incident on an aperture, edge or surface which is rough compared with the wavelength of light it behaves in a way which is not consistent with geometrical ray optics, in that the

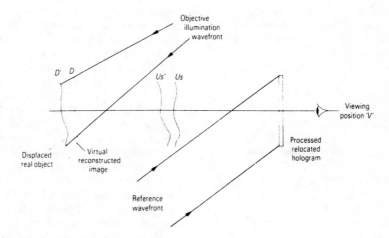

**Figure 13.9.** The arrangement for the observation of holographic interference fringes.

line of propagation departs from the geometrical ray path. This phenomenon is known as diffraction. The basic effect is shown in Figure 13.11. The diffracted wavefront is seen to contain an interference pattern the form of which is a function of the aperture dimension and geometry together with the position of the observation plane. This is due to the interference of secondary wavefronts generated within the aperture. When the distance between the aperture and observation plane is small ($\leqslant 1$ m) the pattern is called the Fresnel diffraction pattern, at larger distances the Fraunhoffer or far-field pattern.

Diffraction effects may be used in a number of measurement systems, a novel example of which is shown in Figure 13.12. A precision knife-edge is placed close to a rotating shaft ($d \simeq 0.2$ mm). The $x$ distance between the maxima and minima in the resultant diffraction pattern is proportional to $1/d$. Hence the intensity detected at a given point in the diffraction pattern changes ($l_1 - l_2$) as the shaft rotates and varies due to, for example, run-out and bearing faults. In this way shaft errors of the order $0.1$ $\mu$m may be detected.

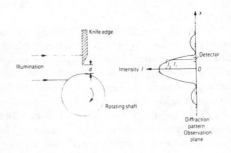

**Figure 13.12.** Use of the diffraction effect to measure errors in shaft rotation.

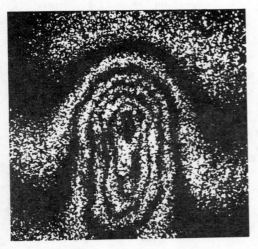

**Figure 13.10.** A speckle interferogram in which the area of fringe discontinuity corresponds to a region of material fault. (Notice the superimposed speckle effect.)

Component gauging and inspection is an area in which the diffraction related technique of Fourier filtering can be used to considerable advantage. The principle is illustrated in Figure 13.13. The optically rough surface $D$ under inspection is illuminated by a coherent light. For simplicity only the light scattered from a small element of the surface is considered. It can be seen that the scattered light field consists of a specular or undiffracted component plus diffracted components. The angle of diffraction $\alpha'$ is inversely proportional to the spatial frequency of the surface feature. When the scattered light field is viewed in the plane a focal distnce $f$ from the lens $L_1$ (the Fourier plane) it can be seen that these components are separated. A filter in the form of an aperture may be placed in this plane so that only specific components are transmitted. As the simple

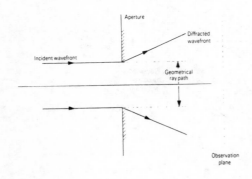

**Figure 13.11.** The diffraction effect.

example of the square grid object shows, the information content of the final image formed by $L_2$ may then be modified.

In this way the visibility of various features such as edges, defects, cracks, etc. can be enhanced as required. This makes the task of automatic image analysis by, for example, a self-scanned array, considerably less complex. Automatic inspection procedures are thereby simplified.

discovered that the light transmission characteristics of the fibre were sensitive to external stimuli such as acoustic vibration and temperature. Various workers then recognized that fibre optics could be used as sensing devices. These fall into two basic categories.

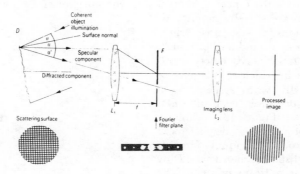

**Figure 13.13.** Image processing using the principle of Fourier-plane filtering.

### 13.6. Fibre-optic sensors

This area of optical technology is still at a relatively early stage of development. Yet it already promises to lead to a new generation of high-sensitivity, non-contacting transducers having a wide range of potential applications.

In its simplest form a fibre optic consists of a thin, drawn strip of glass of circular cross-section. Light propagates along such a fibre by means of a process of repeated internal reflections. In practice the propagation process is optimized by minimizing impurity contents and the introduction of a refractive index variation across the fibre diameter. The latter is achieved by the controlled addition of dopants during manufacture. Attenuation levels of 3 dB km$^{-1}$ are now readily obtainable. Such fibres are used to carry analogue or digital information in the form of a modulated light signal. During the development of these systems it was

*Active:* Here the fibre is exposed directly to the phenomenon under investigation in such a way that the resultant change in the mode of light propagation may be detected.

*Passive:* These systems rely on the fibre to transmit the light to the region of interest where it is used to operate a conventional optical sensor. The intensity-modulated light field derived from the local sensor is fed out, via a second optical fibre, for subsequent detection.

An example of an active sensor based on the principles of interferometry is shown in Figure 13.14. Light from the laser source is split into two components at the

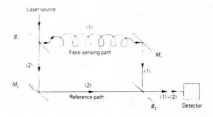

**Figure 13.14.** A fibre-optic interferometer.

beam-splitter $B_1$. Component 1 is made to propagate along a fibre which is exposed to the stimulus under investigation whilst component 2 propagates along a reference path which is insulated from the stimulus. The beams recombine at $B_2$ after reflection from $M_1$ and $M_2$ respectively. The variation in intensity in the resultant interference pattern is measured at the detector. Variations in phase of the light propagating along the fibre with respect to that in the reference arm are thereby monitored. In many applications the system is designed so that the former are induced primarily by pressure fluctuations. Extremely sensitive hydrophones in which sensitivities of better than 20 dB with respect to a reference sound-pressure level of 1 $\mu$Pa have been reported. Figure 13.15 shows a simple form of passive displacement sensor. Light from the input fibre is focused to a small spot at the surface $D$ by the lens $L_1$ ($B$ is a beam-splitter). The power of the scattered light collected by $L_1$ is proportional to $\tan^2 \alpha$ and hence $d$, the distance between the lens and the surface. Under these conditions the level of light intensity propagating along the output fibre via $L_2$ varies with respect to $d$. This variation may be used to monitor the displacement of $D$, d($t$), with respect to time. Displacements of order 1 $\mu$m can be detected in this way.

Major advantages of fibre-optic sensors are their immunity to external noise (e.g. electric fields), non-contacting operation, and high sensitivity. They do not carry electric currents, therefore introduce no dangers of sparking in any potentially hazardous environments.

## 13.7. Conclusions

Optical methods will play an increasingly important role in engineering measurement. The techniques described in this brief survey serve to demonstrate the scope and flexibility of these methods. Each of the applications described is relevant to an important area in engineering measurement and has provided a successful and economic solution to a difficult problem.

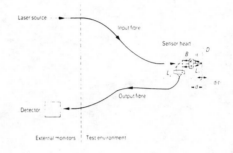

**Figure 13.15.** A fibre-optic coupled displacement sensor.

# Paper 14

# Sensing with optical fibres: an emerging technology

*Albert R. Tebo*

This paper assumes the reader has some knowledge of what an optical fibre is and how it works. For readers unfamiliar with optical fibres, Paper 16 has a section which describes their basic properties.

The use of optical fibres for sensing extremely small changes in measurands is a rapidly developing area of transducer technology. Since optical fibres can also be used to provide a very efficient communication path, sensing and transmission can be accomplished entirely optically, and there are suggestions that in future the signal processing functions of instrumentation systems may be achieved by optical devices. This paper, however, concentrates only on the sensing applications. (Eds.)

Developments in fibre optics have been so rapid that already the term 'fibre-optic sensing' is well established. Here is a look at fibre-optic sensors for the measurement of liquid level, fluid flow, temperature, acoustics, magnetic field, electric current, and rotation.

## 14.1. Introduction

Often 'fibre-optic sensing' includes detection applications in which the fibres only transmit optical signals from detectors at remote or inaccessible spots (such as in medical probing) and when the fibres are basically passive to the process. However, I shall consider only cases in which the parameter to be measured causes a change in the transmission properties of the fibre itself. In such cases the fibre takes an active part in the detection process.

In the past three years the technical journals have published many reports on the development of fibre-optic sensors. Now articles are beginning to appear in other technical publications. In one publication Kessler Marketing Intelligence covers the basic types of sensors in a market survey that includes brief descriptions (KMI 1981). In another, an adaptation of a Gnostic Concepts report discusses the market prospects for fibre-optic sensors (Van Meter 1981). A good technical summary of work on fibre-optic sensors

Originally published in *Electro-Optical Systems Design* (February 1982).

was written by scientists of the Naval Research Laboratory (Cole *et al.* 1981). A Bell Laboratories scientist points out that features in optical fibres that are undesirable for optical communications can be very useful for detection of environmental parameters (Kaminow 1980). And finally, Sperry Research Center scientists discuss several types and applications of sensors, and include references to market forecasts (McMahon *et al.* 1981).

Interest in fibre-optic sensors has been generated by the need for sensors that are immune to electromagnetic interference, that create no electrical hazard, and that can be used in spots devoid of electrical power. Fibre-optic sensing is possible when changes in length of refractive index or polarization of transmitted light are made by temperature changes or external forces that cause deformation or movement.

Let's look at examples of several types of fibre-optic sensors in some promising applications.

## 14.2. Liquid-level sensors

The simplest concept in fibre-optic sensing, a liquid-level sensor, has become an established product on the market. When the unclad fibre sensor is submerged in a liquid, the increased refractive index at the interface allows light to escape from the fibre, enabling a go/no-go type of detection. Lewis Engineering produces these sensors for petroleum tank cars (called TABS for 'top and bottom sensor'), and TRW has developed the same type of item for cryogenic propellant tanks.

## 14.3. Flowmeters

University College London, England, developed a fluid flowmeter that uses a single fibre mounted transversely to the flow within a pipe (Lyle and Pitt 1981). (See Figure 14.1.) The fibre is vibrated by the natural phenomenon of vortex shed-

ding, causing phase modulation of the optical carrier within. The modulation is detected by the 'fibredyne' technique, and the flow rate is determined from the vibrational frequency (Bell *et al.* 1980). The performance of this flowmeter, using multimode fibres of 200 and 300 $\mu$m core diameters, compared favourably with other flowmeters at flow rates $0.3-3.0$ m s$^{-1}$.

Optronics Ltd of Cambridge, England, recently announced the availability of a fibre-optic flowmeter, but did not give details of its operation, saying only that it 'uses the modulation of an optical signal carried by a loop of fibre optic cable to measure flow'.

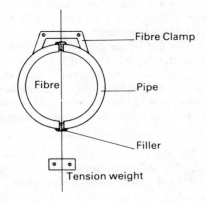

**Figure 14.1.** An optical fibre flowmeter, developed by University College London, in which the transverse fibre is vibrated by vortex shedding. The flexible filler allows tension to be adjusted. (Adapted from Lyle and Pitt 1981).

## 14.4. Temperature sensors

Several techniques have been used to measure the temperature of a medium with optical fibres. Two of these have been developed by Westinghouse for detecting and monitoring hot spots in electrical machinery. One method senses the thermal radition generated within a quartz fibre by the hot spot (Gottlieb and Brandt 1981a). The portion of the fibre exposed to the heat from the hot spot becomes, in effect, a blackbody cavity whose temperature is to be determined. No light source is required, and the temperature of a hot spot located anywhere along the fibre can be determined. For a fibre in thermal equilibrium with its surroundings, the temperature of the surroundings is also determined.

Using a quartz fibre 1 mm in diameter, a hot spot can be determined in the temperature range from room temperature to 1000 °C. By using a two-colour detector (germanium/silicon) at each end of a 1 m commercially available fibre, it is possible to determine the temperature, the position, and the length of a hot spot in the temperature range 135–725 °C.

The second Westinghouse technique, using the loss effects in the cladding and jackets of fibres, is still in the laboratory stage of development (Gottlieb and Brandt 1981b). The underlying mechanism involves temperature-dependent coupling of the evanescent field through a relatively thin cladding layer onto a lossy jacket. The temperature dependence comes about through the refractive indices of the core and cladding materials, which are chosen along with the fibre dimensions to give the desired response.

The sensor must be designed so that the loss takes place in the jacket rather than in the cladding. The technique now uses two fibres that differ in cladding thickness but cover the same path through the detection region. Dependence of sensor response on the wavelength of light is still a problem.

The Naval Research Laboratory has made use of the optical phase shifts induced by temperature changes in a monomode fibre to determine the temperature of the medium surrounding the fibre (Lagakos *et al.* 1981a). The phase shifts are caused mostly by axial strain (length change) from thermally induced stresses.

In a test, the fibre was exposed to a furnace that produced known temperatures, and the phase shifts from a 632.8 nm He–Ne laser were measured by a Mach–Zehnder interferometer, producing results that were in good agreement with theoretical calculations. Proceeding with a study of high-frequency temperature changes, NRL hopes to develop sensors that can respond reliably to a varying temperature in the kilohertz range.

## 14.5. Acoustic sensors (Hydrophones)

Acoustic fibre-optic sensors have been developed for hydrophone applications, based on several different physical principles. One of these, making use of frustrated total internal reflection, has been successfully built and tested by Sperry Research Center (Spilman and McMahon) 1980). A multimode fibre is cleaved and polished at a large enough angle to the fibre axis to cause total internal reflection for all modes propagating in the fibre (using 633 nm wavelength light from a He–Ne laser) when the medium in the gap is air. See Figure 14.2. By bringing the fibre ends sufficiently close together, a large fraction of light power can be coupled between the two fibres.

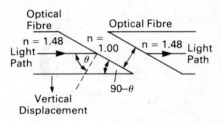

**Figure 14.2.** In Sperry Research Center's fibre-optic hydrophone an acoustic signal displaces one fibre vertically, changing the gap spacing and thus modulating the light coupled between the two fibres. (Adapted from Spillman and McMahon 1981.)

A pressure wave in the air displaces one fibre relative to the other, varying the gap thickness and thus modulating the light power coupled between the two fibres. With this fibre arrangement mounted behind an acoustic diaphragm in a sealed box that is submerged in water, the device responds in good agreement with predictions. It provides a sensitivity higher than needed to detect the deep-sea noise level from 100 Hz to 10 kHz. The minimum detectable pressure is 62 dB relative to 1 μPa at 500 Hz.

A second type of acoustic sensor, developed at the University of Central Florida, consists of a single-mode fibre whose end is cut at an angle close to the critical angle (Phillips 1980). See Figure 14.3. The amount of light reflected by the fibre changes when the critical angle changes as the acoustic pressure alters the refractive indexes of the fibre and its surrounding medium differently. The acoustic wave thus modulates the transmitted light.

A third type of acoustic sensor is a fibre-optic linear accelerometer based on intensity modulation produced by lateral displacement of a cantilevered fibre (Rines 1981). (See Figure 14.4.) Developed by Sanders Associates with a multimode fibre of 63 μm core diameter, the model reported on has a displacement sensitivity of $6.4 \times 10^{-3}$ m.

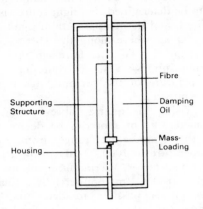

**Figure 14.4.** This acoustic sensor of Sanders Associates, Inc., is a linear accelerometer based on intensity modulation produced by lateral displacement of a cantilevered fibre. The loading mass sets the resonance of the beam to a desired frequency. (Adapted from Rines 1981.)

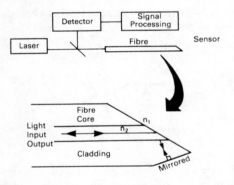

**Figure 14.3.** In this acoustic sensor, designed by the University of Central Florida, acoustic pressure alters the refractive index of the fibre differently from that of the medium, causing a change in the critical angle and thus a change in internal reflectivity. (Adapted from KMI 1981.)

A fourth type of acoustic sensor, developed both by Catholic University of America and Hughes Research Laboratories and now being pursued by the Naval Research laboratory, functions on the principle of displacement-induced microbending (Lagakos et al. 1980, 1981b, Fields et al. 1980). The microbending is introduced to the fibre by a set of two corrugated plates (sometimes called the deformer) which sandwich the fibre. (See Figure 14.5.)

A displacement of one plate against the other by acoustic pressure causes loss of light from the core modes of a multimode fibre into the clad modes, thus effecting a detectable intensity modulation of the light.

## 14.6. Spectrophone

The Naval Research Laboratory recently built what is claimed to be the first fibre-optic spectrophone, an instrument used to study a sample of gas contained in an optical absorption cell (Leslie et al. 1981). A

In a hydrophone application it has a sensitivity limit corresponding to 75 dB relative to 1 μPa at 1 kHz. The device is still in the experimental stage, but it shows promise.

modulated He–Ne laser beam at 3.39 $\mu$m wavelength irradiates the gas, producing a temperature change and hence a pressure modulation in the container.

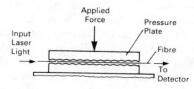

**Figure 14.5.** In this hydrophone sensor, a development of Catholic University of America, of Hughes Research Labs., and of the Naval Research Lab., the microbending of the multimode fibre by acoustic pressure applied to the corrugated plates causes loss of light from the core modes, effecting a modulation of the light. (Adapted from Fields *et al.* (1980.)

The pressure fluctuations were detected using a 9.2 m coil of single-mode optical fibre wound with a 2.54 cm diameter inside the cell cylinder. (See Figure 14.6.) The coil was fused into one arm of an all-fibre Mach–Zehnder interferometer. The pressure modulation in the cell induced a phase modulation in the light (from a 0.6328 $\mu$m He–Ne laser) propagating in the fibre coil. A spectrum analyser at the interferometer output provided the analysis of the gas properties.

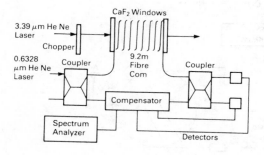

**Figure 14.6.** The characteristics of gases are analysed in this Naval Research Lab. spectrophone whose fibre coil detects thermally-induced pressure changes in the test cell. (Adapted from Leslie *et al.* 1981.)

## 14.7. Magnetic field sensors

The Naval Research Laboratory has been active in the development of fibre-optic sensors for measuring magnetic fields. One such sensor was constructed by winding, under tension, a single-mode fibre around a nickel cylinder in the magnetic field under study (Rashleigh 1981). The magnetic field, by means of the magnetostrictive effect changes the state of polarization of the light from a He–Ne laser (0.633 $\mu$m) propagating through the fibre. (See Figure 14.7.) With this method a sensitivity of $1.76 \times 10^{-2}$ rad(mOe)$^{-1}$ was demonstrated, permitting detection of fields as small as $4.4 \times 10^{-6}$ Oe per meter of fibre.

In another version of a magnetic field sensor, NRL used a GeO$_2$-doped silica fibre coated with either a commercially available nickel jacket or a metallic-glass coating (Siegel 1981). The magnetostrictive stress created by a change in the magnetic field caused a change in the fibre length, which was detected as a phase shift in the output of an interferometer. A change of $10^{-7}$ gauss in a magnetic field was measured with the metallic-glass jacketed fibre.

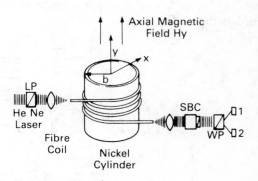

**Figure 14.7.** The fibre coil under tension detects changes in the surrounding magnetic field by means of the magnetostrictive effect that changes the state of polarization of the laser light, in this Naval Research Lab. magnetic field sensor. LP: laser polariser. SBC: Saleil–Babinet compensator. WP: Wollaston prism. (Adapted from Rashleigh 1981.)

## 14.8. Current sensors

Optical fibres were used as alternating-current sensors by the John Carroll University and NRL in single-mode fibre Mach–Zehnder interferometers (Dandridge *et al.* 1981). Two separate approaches were used to transform the current to a phase shift detectable in the interferometer system. The first depends on the magnetic field produced by the current acting on a magnetostrictive material bonded to the fibre.

A fibre was bonded to a piece of thin-walled (0.1 mm) nickel tubing in a sensor length of about 10 cm. The sensor was placed within a 5 $\Omega$ coil through which passed the current to be detected. The interferometer, using a ruggedized 'bottle' coupler (NRL design) and a single-mode Ga-Al-As laser, was capable of detecting phase shifts as small as $10^{-6}$ rad.

In the second method a single-mode fibre was coated with a 2 $\mu$m layer of aluminium in a length of 10 cm with a resistance of 3 $\Omega$ through which passed the current to be detected. The same interferometer was used to detect the phase shifts.

These two methods provided experimental sensitivities of about $7 \times 10^{-9}$ A m$^{-1}$ with an input impedance of less than 5 $\Omega$.

In another electrical-current sensor Thomson-CSF Central Research Laboratory used the Faraday effect in a closed-loop fibre-optic interferometer (Arditty et al. 1981). This Sagnac interferometer — a discussion of which is given in the next section — was implemented in a strictly reciprocal configuration and used a reciprocal phase modulation scheme. (See Figure 14.8.) A 160 m fibre loop was used to reduce the modulation frequency to an acceptable 620 kHz. A 30 m segment of the loop fibre was twisted to ensure some polarization stability.

This current sensor offers total insensitivity to the environment, long-term zero stability that allows DC operation, and a very large dynamic range ($>10^5$). Its capability extends from mA to MA.

## 14.9. Fibre-optic gyroscopes

An application of optical-fibre sensors that is receiving a lot of attention is the gyroscope, using an adaptation of the Sagnac interferometer. If two coherent beams of light (derived by a beam-splitter from a single laser source) are sent in opposite directions around a fibre loop of area $A$ that rotates with an angular velocity $\Omega$ about an axis perpendicular to the plane of

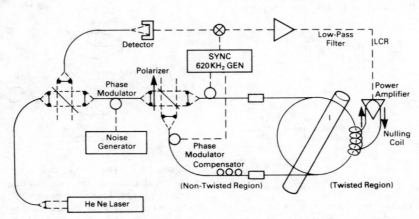

**Figure 14.8.** The fibre loop in this Sagnac interferometer, developed by Thomson-CSF, senses electrical current in the conductor that is aligned perpendicular to the plane of the loop. (Adapted from Arditty *et al.* 1981.)

the loop, the beam rotating with the loop will travel a longer distance. In the first-order analysis this results in a phase difference of $8\pi A\Omega/\lambda c$ between the two existing beams, where $\lambda$ is the monochromatic wavelength and $c$ is the vacuum speed of light.

To achieve the sensitivity necessary for navigation ($10^{-3}$ of the earth's rotation rate) the gyroscope must be able to detect, in terms of phase difference, a frequency difference of less than one part in $10^{17}$ — an extremely delicate physical measurement. The approach in fibre-optic systems is to cause the light beams to traverse many turns of the fibre, thus multiplying the phase difference.

A group at Stanford University has developed an all-fibre gyroscope that consists of 580 m of single-mode fibre (wound around a 14 cm diameter spool), a polarizer, polarization rotators, a piezo-electric phase modulator, and fibre-optic couplers (Schwarzchild 1981). (See Figure 14.9.) A Ga-As laser light source and a photodetector complete the integrated fibre system. The polarizers ensure the identities of the clockwise and counterclockwise optical paths. The system achieved a rotational sensitivity of 0.1°/hour.

A group at Massachusetts Institute of Technology has constructed a fibre-optic gyroscope of comparable noise level and rotational sensitivity without using full

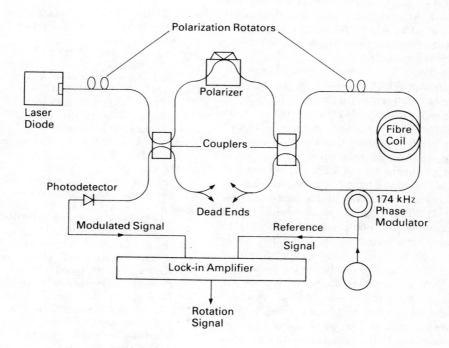

**Figure 14.9.** The 580 m coil of single-mode optical fibre in this splice-free all-fibre Sagnac interferometer, developed by Stanford University, detects gyroscopically-induced phase shifts between counter-rotating beams of light through the coil. (Adapted from Schwarzchild 1981.).

integration of fibre components. They used 200 m of single-mode fibre wound around a spool of 19 cm diameter, and achieved the same $0.1° h^{-1}$ sensitivity.

Their system uses a He–Ne laser, half-silvered mirror beamsplitters, polarizers, an electro-optic phase modulator, and an acousto-optic isolator. Acousto-optic Bragg cells shift the frequencies of the counter-rotating beams to restore a null interference signal when the gyroscope is rotating.

In another development of optical fibre gyroscopes, a group at the Naval Research Laboratory achieved equivalent sensitivity with a non-reciprocal $\pi/2$ phase shift between the counter-rotating beams that was electro-optically induced with a pulsed voltage applied to a $LiNbO_3$ channel waveguide device (Bulmer and Moeller 1981). At zero rotation the operating point is shifted to where the two outputs are in quadrature and show maximum sensitivity to small rotations.

Present interest in fibre-optic gyroscopes has climaxed in an international conference on fibre-optic rotation sensors, held at MIT in November 1981 (Klass 1981). At this meeting Stanford University revealed that they are now using a re-entrant rotation sensor, in which the multiturn optical fibre coil is closed so that the pulsed light can rotate as many as 10 times around the coil (Pavloth and Shaw 1981). The Sagnac phase shift is then accumulative. The prospects for this technique are encouraging, but the whole technology is still a long way from replacing laser gyros with fibre-optic gyros.

A scientist at Stanford says we are approaching the photon limit of measurement techniques, and he is excited because every month new physical phenomena are encountered. Participants in the meeting feel that the technology developed for the gyroscope is certain to have a large payoff later for more demanding applications in telecommunications.

## 14.10. Future of optical-fibre sensors

Of all the sensors I have discussed — liquid-level, flow, temperature, acoustic, magnetic field, current, and rotation (gyro) — only the liquid-level sensor has become an established product. However, this is not because of lack of effort or interest. The whole technology is so new that we cannot expect finished products on the market until the usual research and development period is completed. Judging from the activity and progress evident now, that period is likely to be shorter than for any previous technology.

## References

KMI Market Report (1981) *Fiberoptic sensors* (Kessler Marketing Intelligence, Newport, RI).

Meter, L. Van (1981) A bright future for fiber optic sensors systems. *International Fiber Optics and Communications* March, 19.

Cole, J. H., Giallorenzi, T. G. and Bucaro, J. A. (1981) Research update on fiber optic sensors. *International Fiber Optics and Communications* March, 23.

Kaminow, I. P. (1980) Polarization in fibers. *Laser Focus* June, 80.

McMahon, D. H., Nelson, A. R. and Spillman, W. B. Jr. (1981) Fiber-optic transducers. *IEEE Spectrum* Dec., 24.

Lyle, J. H. and Pitt, C. W. (1981) Vortex shedding fluid flowmeter using optical fiber sensor. *Electronics Letters* 17, 244.

Ball, P. R., Culshaw, B. and S. A. Kingsley (1980) Recovery of phase modulated signals in multimode optical fibres. *Proceedings of International Optical Computing Conference.* Washington. DC. April.

Gottlieb, M. and Brandt, G. B. (1981) Fiber-optic temperature sensor based on internally generated thermal radiation. *Applied Optics* 20, 3408.

Gottlieb, M. and Brandt, G. B. (1981) Temperature sensing in optical fibers using cladding and jacket less effects. *Applied Optics* 20, 3867.

Lagakos, N., Bucaro, J. A. and Jarzynski, J. (1981) Temperature-induced optical phase shifts in fibers. *Applied Optics* 20, 2305

Spillman, W. B. Jr. and McMahon, D. H. (1980) Frustated-total-internal-reflection multi-mode fiber-optic hydrophone. *Applied Optic* **19**, 113.

Phillips, R. L. (1980) Proposed fiber-optic acoustical probe *Optic Letters* **5**, 318.

Rines, G. A. (1981) Fiber-optic accelerometer with hydrophone applications. *Applied Optics* **20**, 3453.

Lagakos, N., Macedo, P., Litovitz, T., Mohr, R. and Meister, R. (1980) Fiber optic displacement sensor. *Proceedings American Ceramic Society Conference on the Physics of Fiber Optics.* p. 128. Chicago.

Fields, J. N., Asawa, N. K., Smith, C.P. and Morrison, R. J. (1980) Fiber optic hydrophone. *Proceedings American Ceramic Society Conference on the Physics of Fiber Optics*, p. 125. Chicago.

Lagakos, N., Trott, W. J. and Bucaro, J. A. (1981) Microbending fiber-optic sensor design optimization, paper THE2 at the *Conference on Lasers and Electro-Optocs*, Washington, DC.

Leslie, D. H., Trusty, G. L., Dandridge, A. and Giallorenzi, T. G. (1981) Fibre-optic spectrophone. *Electronics Letters* **17**, 581.

Rashleigh, S. C. (1981) Magnetic-field sensing with a single-mode fiber. *Optics Letters* **6**, 19.

Sigel, G. H. Jr. (1981) Recent progress in fiber optic magnetic sensors. *Proceedings, FOC' 81, International Fiber Optics and Communications Exposition in the 4th Year*, p. 174. Cambridge, MA.

Dandridge, A., Tveten, A. B. and Giallorenzi, T. G. (1981) Interferometric current sensors using optical fibers. *Electronics Letters* **17**, 523.

Arditty, H. J., Bourbin, Y., Mapuchon, M. and Puech, C. (1981) Current sensor using state-of-the-art fiber-optic interferometric techniques. *Technical Digest of 100C'81, 3rd International Conference of Integrated Optics and Optical Fiber Communmications*, page 128. San Francisco.

Schwardzchild, B. M. (1981) Sensitive fiber-optic gyroscopes, *Physics Today*, Oct. p. 20.

Bulmer, C. H. and Moeller, R. P. (1981) Fiber gyroscope with nonreciprocally operated fiber-coupled LiNbO3 phase shifter, *Optics Letters* **6**, 572.

Klass, P. J. (1981) Work progresses on fiber optic gyro. *Aviation Week and Space Technology*, Dec. 7, p. 68.

Pavloth, G. A. and Shaw, H. J. (1981) Re-entrant fiber-optic rotation sensors, paper presented at the *International Conference on Fiber-Optic Rotation Sensors and Related Technologies*, Massachusetts Institute of Technology.

Paper 15

# Feedback in instruments and its applications

*B. E. Jones*

Many transducers incorporate feedback in order to improve accuracy, linearity and speed of response of a measurement. This paper outlines the advantages of feedback, and shows many examples of its use. (Eds.)

Instruments can be considered as systems, and there are only a few structural schemes employed in the construction of such systems; one such scheme is the use of feedback. The general mode of operation of feedback-measuring systems is ascertained, and the reasons for applying feedback to a measurement situation are established. Using feedback it is possible to improve accuracy and speed of measurement, reduce the effect of interfering and modifying inputs, and allow remote indication and non-contact measurement. The property of inversion assists instrument design, and can provide for digital indication. The common balance variables are listed, and a wide range of instruments is discussed, making explicit the general properties of feedback.

Measuring devices having a frequency output maintained by feedback, and the use of feedback for control purposes in instruments, are considered. The pedagogic implications are referred to, and an extensive list of references is provided.

## 15.1. Introduction

Application of the feedback principle had its beginnings in simple machines and instruments, some of them going back 2000 years or more (Mayr 1970). In fact, the ancient water clock is the earliest known device for feedback control. It was invented in the third century BC by a Greek mechanic called Ktesibios working in Alexandria, and he employed a float regulator. Records also exist of water clocks using float valves in the ninth century. The invention of the feedback amplifier (Black 1934) was considered by Greig (1950) to be probably the most important single influence on measurement technique from a conceptual and practical point of view in the period that followed. Many instruments incorporating feedback techniques for the measurement of a wide range of variables are available today, and the future extensive development and use of feedback-measuring systems seems certain.

In an introductory article to instrument science, Finkelstein (1977) has

Originally published in *J. Phys. E Scientific Instruments* Vol. 12 1979.

pointed out that instruments can be considered as systems, that is assemblies of interconnected simpler components organized to perform a specific function. There are only a few structural schemes employed in the construction of instruments and instrument systems, and one such scheme is the use of feedback. To aid instrument science it is desirable to ascertain the general mode of operation of such instruments and the reasons for applying feedback to a measurement situation.

One can think of a feedback system as a system which tends to maintain automatically a prescribed relationship of one system variable to another by comparing functions of these variables and using the difference as a means of control. The main characteristic of a feedback system is its closed-loop structure. The author has chosen to call a measuring system in which feedback is the basic structural arrangement, a feedback-measuring system (Jones 1974), and has defined a feedback-measuring system as one that measures a variable by using error sensing through a closed loop (Jones 1977).

A general block diagram of such a system is shown in Figure 15.1. Here the output signal (usually electrical) is converted to a form (usually non-electrical, for example force) suitable for comparison with the quantity to be measured (for example force). The resultant error is usually transduced into electrical form and amplified to give the output indication. Normally a transducer and associated circuit has a non-electrical input and an electrical output, for example a thermistor, strain gauge, and photodiode, whereas a so called 'inverse transducer' or precision actuator has an electrical input and a low-power non-electrical output; for example, a piezoelectric crystal, translational and angular moving-coil elements can be used as inverse transducers. The transducer, inverse transducer and usually the amplifier must be close to the point of measurement, whereas the indicator may be some distance away. Of course the feedback loop must have sufficient negative gain and the system must be stable. The system is driv-

ing fairly low-power devices at its output, and the inverse transducer essentially determines the characteristics of the system, although noise connected with the transducer and amplifier input stage may well be important. In practice the measurand may not be directly connected to the 'null' or 'balance' point, and there may well be one or more primary sensing elements inserted between points $X$ and $Y$ in Figure 15.1.

When feedback is used for control purposes, the block diagram of a simple control system looks similar to that of Figure 15.1, but in this case the actual quantity to be controlled (usually a non-electrical quantity) is measured and compared with a demanded quantity (usually this quantity is electrical) to produce an error which is amplified to drive an actuator producing power to drive the controlled quantity. The operation of the feedback is similar for both the control and measuring systems and design for accuracy and stability is basically the same.

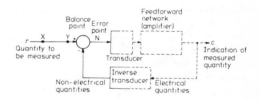

**Figure 15.1.** Diagram of a feedback-measuring system. Broken line, electrical units or signals; full line, mechanical units or signals.

## 15.2. General properties of feedback systems

A measuring system may have many components or elements and it is convenient to consider each such element as a block with its own input–output relationship, that is its own transfer function. The blocks representing the various elements of a system are connected to use their functional relationship within the system, thus producing a block diagram for the system.

A generalized block diagram of a simple feedback system is shown in Figure 15.2. The diagram illustrates negative feedback in that the gain around the loop has negative polarity. In this case the measurand $r$ and the output indication $c$ are in phase but if in the practical system they are in antiphase, then the amplification $A$ will have a negative sign associated with it and both inputs to the comparator will have positive signs. In Figure 15.2, a fraction $\alpha$ of the quantity to be measured ($r$) is fed in, and a fraction $\beta$ of the output quantity ($c$) fed back to a summing point which takes the difference and amplifies it by gain $A$ to produce the output quantity; $\alpha$, $\beta$ and $A$ are transfer functions, while $n_1$ and $n_2$ are noise sources at the input and output of the amplifier, respectively. Assuming $n_1 = n_2 = 0$, the following equation expresses the situation of Figure 15.2:

$$\alpha r - \beta c = c/A$$

or

$$\frac{c}{r} = \frac{a}{\beta} \frac{A\beta}{1 + A\beta}. \tag{15.1}$$

The ratio $c/r$ is called the closed-loop gain, while $A\beta$ is the open-loop gain and both are transfer functions. Provided that $A\beta \gg 1$, $c/r \approx \alpha/\beta$, and $c/r$ is largely independent of $A$ and is simply determined by transfer functions $\alpha$ and $\beta$. Thus the sensitivity of such a system can be changed by altering $\alpha$ or $\beta$. In some practical feedback-measuring systems, the feedback also makes $\alpha$ and $\beta$ well defined. Actually, $\alpha$ and $\beta$ may alter, but so long as their ratio remains fixed, $c/r$ is well defined.

We shall see that the main properties of this general system can be summarized as well-defined gain (giving accuracy), wide bandwidth (giving fast response), low or high output impedance (allowing remote indication), reduction in noise effects and extraneous disturbances, and the property of inversion. Negative feedback can be employed to produce instrument high-input impedance, and unity-gain feedback often allows non-contact measurement.

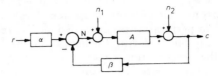

**Figure 15.2.** Generalized block diagram of a simple feedback system.

Measurement devices providing a frequency output usually employ feedback, and feedback can be used to enhance the characteristics of frequency-selective networks.

To evaluate the effect of changes in the amplifier gain, let $A$ increase to $A + \Delta A$, so that $c$ increases to $c + \Delta c$; $A$ may alter because of temperature, non-linearity and ageing effects. From equation (15.1) we have

$$\frac{c + \Delta c}{r} = \frac{\alpha}{\beta} \frac{(A + \Delta A)\beta}{1 + (A + \Delta A)\beta},$$

which becomes

$$\frac{\Delta c}{c} = \frac{\Delta A}{A} \frac{1}{1 + A\beta},$$

and a change in amplifier gain is reduced in its effect on the output when the open-loop gain $A\beta$ is much greater than unity.

In practice the amplifier transfer function might be of the form $A/(1 + s\tau)$, and so equation (15.1) becomes

$$\frac{c}{r} = \frac{\alpha}{\beta} \frac{A\beta}{(1 + A\beta)[1 + s\tau/(1 + A\beta)]},$$

indicating that the effective time constant is $\tau/(1 + A\beta)$. Thus feedback around the amplifier reduces gain and increases bandwidth, so that in this case the gain–bandwidth product remains fairly constant.

The effect of loading at the output of the feedback system can be best understood by considering the amplifier to be a voltage amplifier with an output impedance $Z_0$. This impedance is effectively in series with the output of the amplifier shown in Figure 15.3($a$). The system output is now a voltage $v_0$ and if this changes by, say, $\Delta v_0$ due to change in the system electrical load (load current change $\Delta i_0$), the amplifier output changes by $-A\beta\Delta v_0$ due to the feedback around the loop. If it is assumed that the load change of current occurs in the amplifier output, then

$$\Delta i_0 = \frac{\Delta v_0 + A\beta\Delta v_0}{z_0},$$

and the effective output impedance of the feedback system

$$Z_{oe} = \frac{\Delta v_0}{\Delta i_0} = \frac{Z_0}{1 + A\beta} \approx \frac{Z_0}{A\beta}.$$

Thus the effective output impedance in this case is that of the amplifier divided by the open-loop gain. If the current (or a component of it) is fed back, then the feedback operates to maintain a fixed current output, and the effective output impedance in this case is high. It is this property of feedback which allows accurate remote indication to be achieved.

The effect of noise sources $n_1$ and $n_2$ (Figure 15.2) can be evaluated by writing down the expression for the system when they are present,

$$\alpha r - \beta c = \frac{(c - n_2)}{A} - n_1,$$

which for $A\beta \gg 1$ becomes

$$c \approx \frac{\alpha}{\beta} r + \frac{n_2}{A\beta} + \frac{n_1}{\beta}.$$

The effects of disturbances such as noise, hum, and varying loads at the output on the system output are considerably reduced by the open-loop gain, while the effects of disturbances such as noise and drift at the amplifier input on the system output are not reduced.

The feedback fraction (transfer function $\beta$) may be a linear inverse transducer or a non-linear element. When $A\beta \gg 1$ and $\alpha = 1$, $c/r \approx 1/\beta$, and the inverse transducer characteristics completely determine the

closed-loop gain characteristics of a feedback-measuring system. The property of inversion can also be employed to make better use of available equipment; for example, if $\beta$ has a square-law characteristic, and this is relatively easy to produce, then the closed-loop gain has a square-root characteristic. Another example is the production of an instrument with digital output, by using a well-defined digital-to-analogue converter in the feedback path of a feedback system. Another application of the property of inversion is for transducer linearization purposes, and an example of this will be given in section 15.7.

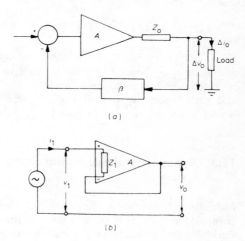

**Figure 15.3.** Application of feedback to produce ($a$) low output impedance, and ($b$) high input impedance (amplifier output with respect to ground).

Unity-gain feedback can be used to increase input impedance levels, and this is particularly necessary when high-impedance transducers are used or non-contact measurement is desirable. The basic technique can be understood by considering the voltage amplifier shown in Figure 15.3($b$); the amplifier has a voltage gain $A$ and input impedance $Z_1$. The voltage to be measured, $v_1$, is connected in series with the amplifier differential input and the amplifier output, which provides a voltage $v_0$ with respect to ground. The effective input impedance $Z_{1e} = v_1/i_1$, and the following equations express the situation of Figure 15.3($b$):

$$i_1 Z_1 A = V_0$$

$$v_1 = i_1 Z_1 + v_0.$$

Using these expressions we find

$$Z_{1e} = Z_1 (1 + A)$$

and

$$v_0 = m v_1,$$

where

$$m = 1/(1 + 1/A).$$

Transistor emitter and source followers use this technique. Some transducers need to operate into a low impedance, and the 'virtual earth' point of a feedback amplifier provides this (section 15.7). [ . . . ]

## 15.3. Types of feedback-measuring system

Systems incorporating feedback involve comparison of two physically similar variables and production of a minimum or null (at point $N$ in Figures 15.1 and 15.2). In feedback-measuring systems, per-variable (or 'through' variable) balance occurs with force, torque, current and heat flow, while in the case of transvariables (or 'across' variables), voltage, temperature and displacement balances are common. Examples of some force and torque balance systems have been described by Rohrbach (1967), Oliver (1972), Neubert (1975), Doebelin (1975), Welkowitz and Deutsch (1976) and Jones (1977), while Jones (1977) has described the other forms of balance: the list below shows the very wide application of these systems. Many of these instruments and circuits are commercially available, and some are extensively used in industrial plants. Others have been developed to solve specific measurement problems in research laboratories.

### 15.3.1. Examples of feedback-measuring systems (the balancing variable heads each list).

*Force and/or torque*
precision microweight chemical balance;
instruments for measurement of liquid surface tension, liquid level, density of solids and liquids, fluid specific gravity, absolute and differential pressure, vacuum, fluid flow, momentum and torque;
translational and angular accelerometers, jerkmeter;
electrical current balance;
attracted-disc voltmeter;
two- and three-wire DC output transmitters of alternating voltage, current, active and reactive power, frequency and phase;
pneumatic pressure transmitter;
electrical-current-to-air-pressure converter;
paramagnetic analyser of oxygen–gas mixtures;
moving-coil microphone;
angle-of-attack transducer;
vibrating-contact transducers;
thread-tension monitor.

*Electrical current (and flux)*
transformer ratio-arm autobalance bridge;
direct current comparator;
fluxgate magnetometer;
clip-on DC ammeters;
contact DC ammeters.

*Heat flow*
hot-wire anemometer;
thermistor fuel flowmeter.

*Voltage*
self-balancing bridges and potentiometers;
automatic Kelvin–Varley divider digital voltmeter;
feedback AC voltmeter;
feedback amplifier circuits to provide resistance-to-voltage, voltage-to-current and current-to-voltage converters, and improved meter sensitivity;

Wheatstone bridge resistance-to-current converter;
high-fidelity galvanometer;
AC–DC converters;
magnetic flowmeter servosystem.

*Pressure*
pressure balance flowmeter.

*Temperature*
automatic cooled-mirror dewpoint hygrometer;
instrument for measurement of internal human body temperature.

*Displacement*
instrument for creep and micro-movement measurement;
extensometer;
optical position follower used in precision pressure measurement;
light-spot-follower recorder;
automatic refractometer;
Hall displacement transducer;
pneumatic servo-follower;
servomanometer;
non-contact displacement followers.

*Phase (and frequency)*
phase-lock loop;
automatic vector voltmeter.

*Radiation*
automatic-emissivity-compensated radiation pyrometer.

## 15.4. Force and torque balance

Force and torque balance systems are very common and are employed, for example, in precision weighing, for the measurement of acceleration, pressure, flow, level, electrical power and high voltage. These quantities can be converted to a force or torque; for example, acceleration of a fixed seismic mass produces a force on the mass proportional to acceleration, and a dynamometer movement produces torque on a shaft proportional to electrical power in the movement. The relationship between the quantity to be measured and force may be non-linear.

The main methods of force and torque balance are shown in Figure 15.4, where $F$

is the force to be measured and $F_f$ is the balancing or feedback force generated in the particular instrument. For the beam and shaft, $F$ and $F_f$ are converted to torques $T$ and $T_f$, respectively. In each case the element being used (seismic mass, diaphragm, beam or shaft) is displaced a small amount in the direction in which $F$ acts before the equilibrium balance $F = F_1$ is achieved. This displacement is the information required by the instrument to generate $F_f$. [ . . . ]

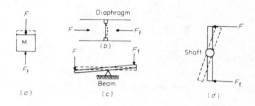

**Figure 15.4.** Force balance with ($a$) a seismic mass $M$. ($b$) a diaphragm: torque balance with ($c$) a beam, ($d$) a shaft. Broken line, position before application of $F$, when $F = F_f = 0$.

Various force- and torque-balance instruments are shown schematically in Figure 15.5, and illustrate the methods of balance, some typical devices used, and the wide application of this form of balance. The arrangement of Figure 15.5($a$) can be used to transmit force information over a long distance. An input movement on a force spring, due to a measurand, creates a torque on a beam with a flexure pivot. Any movement of the beam is detected by the position detector and a current is created in the moving-coil actuator (or force motor) to produce the balancing torque. The output DC is linearly related to the input movement. This two-wire transmitter has a live-zero current output, and its sensitivity can be altered by changing the spring stiffness. A pneumatic pressure transmitter is shown in Figure 15.5($b$). The pressure unit is completely sealed and $P_2$ could be the pressure at the bottom of a petrol tank, enabling the level of petrol in the tank to be

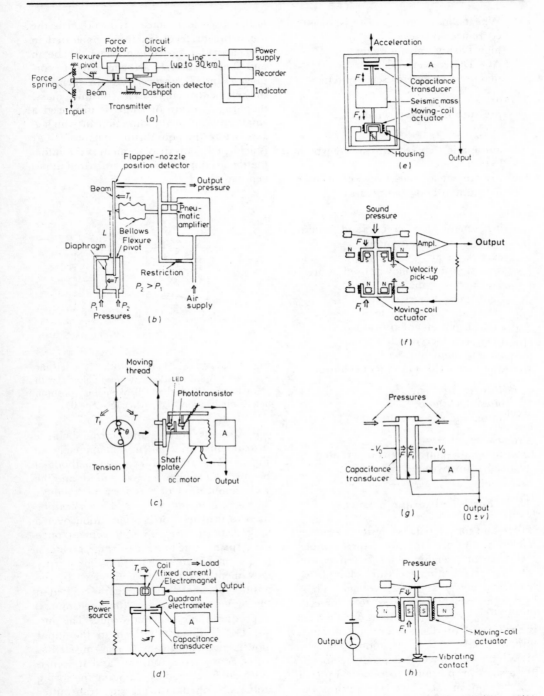

**Figure 15.5.** Various force- and torque-balance instruments: (*a*) two-wire transmitter, (*b*) pneumatic pressure transmitter, (*c*) thread-tension monitor, (*d*) automatic electrostatic wattmeter, (*e*) seismic-mass accelerometer, (*f*) force-balance microphone, (*g*) electrostatic pressure sensor, (*h*) vibrating-contact pressure monitor (A, amplifier and controller).

monitored. The output of this transmitter is air pressure, and the sensitivity of the transmitter is inversely proportional to length $L$. Various forms of feedback transmitter are widely used in industry.

The monitor of Figure 15.5($c$) employed in the textile industry uses the torque of a DC motor to balance torque created by tension in the thread; an optical transducer is used to determine the angular position of the motor shaft (Auckland and Hawke 1978). The monitor sensitivity is proportional to $\sin \theta$. The wattmeter of Figure 15.5($d$) uses the torque of a non-contact moving-coil actuator to balance torque created by the load power acting on the quadrant electrometer; here the quadrant pairs act as a push–pull capacitance displacement transducer (Jones and Mirza 1977). It is also possible to use an optical position detector, and voltage feedback to the quadrants so that both torques are due to electrostatic forces (Jones and Gharakhanian 1977). The torques are very small, typically about 10 nN m.

In the accelerometer of Figure 15.5($e$), the inertial force derived from the acceleration is balanced by the feedback force created by current in the moving-coil actuator (MacDonald 1961). The force-balance microphone of Figure 15.5($f$) (Neubert 1975, Welkowitz and Deutsch 1976) employs a velocity pick-up as the transducer and a moving-coil actuator to create the balancing force. In contrast to other microphones there is no primary resonance and the microphone frequency response is adjustable by electrical means. Of course this arrangement does not respond to static pressure. The pressure sensor of Figure 15.5($g$) has been used for mesurement of small gas differential pressures (Neubert 1975). It employs a push–pull capacitance construction which acts both as the displacement transducer and as the means of providing an electrostatic balancing force. The final force-balance instrument shown (Figure 15.5($h$)) is a simple device using a vibrating contact (Neubert and Price 1969). The contact is broken by an upward force generated by

the moving-coil actuator, and then the de-energized coil allows the contact to be closed again. The on–off ratio of the circuit current is controlled by the pressure.

It should be noted that the moving-coil actuators (both translational and angular) make use of the electromagnetic force created when a current-carrying conductor is at right angles to a magnetic field. The force is at right angles to both the conductor and the field, and is proportional to both the current $i_t$ and field flux density; as a consequence, force-balance instruments employing such actuators usually have linear characteristics with a current as the output. The ratio $T_t/i_f$ may vary from about 0.1 to 50 mN mA$^{-1}$, while the ratio $F_t/i_t$ may vary from about 0.1 to 20 N A$^{-1}$. In comparison electrostatic forces are much smaller in magnitude, and the forces are inherently non-linear functions of voltage. Nonetheless, with proper design the arrangement of Figure 15.5($g$) can provide an output voltge $v$ linearly related to the pressure differential (Neubert 1975). The feedback attracted-disc voltmeter proposed by Broadbent *et al.* (1965) has a square-law characteristic.

## 15.5. Current (and flux) balance

A useful instrument for the measurement of direct current without breaking a printed circuit track is shown in Figure 15.6($a$) (*Electronics Industry* 1976). The central two connections of the probe sense current in the track, and the amplifier feeds back a current $I_f$ through the track in the opposite direction to the current $I$ of the circuit. Provided that amplification is high, at balance $I_f \approx I$ and there is negligible voltage drop along this portion of the track.

The alternating-current transformer is an inherent current- or flux-balance system, and the transformer ratio-arm bridge uses this current-balance technique. An autobalance arrangement is shown in Figure 15.6($c$) (Wayne Kerr 1975). The multi-winding current transformer acts as a current-balance detector. The output of the search coil is amplified to provide a bridge-balancing voltage which is applied through a resistance to the feedback winding. The initial current $i_1$ derived from the voltage transformer and unknown impedance is opposed by this feedback current $i_f$. The amplifier output is usually connected to two phase-sensitive detectors, to provide resistive and reactive indicators.

The principle of minimum flux change can be used for the non-contact measure-

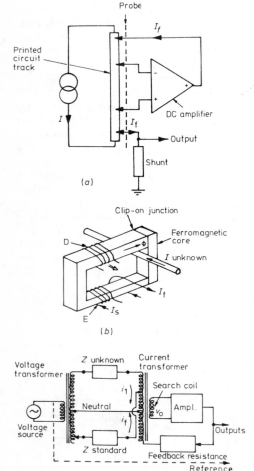

(a)

(b)

(c)

**Figure 15.6.** Current- and magnetic flux-balance instruments: ($a$) contact DC ammeter, ($b$) clip-on DC ammeter (E, excitation winding; D, detector winding), ($c$) autobalance transformer ratio-arm bridge.

ment of direct current (Jones 1977). The measuring head of a clip-on DC ammeter is shown in figure 15.6($b$). The unknown direct current $I$ and a feedback direct current $I_f$ link with the ferrite magnetic circuit as shown. The magnetic circuit is driven into saturation by the AC excitation winding having current $I_s$ of frequency $f$, and flux changes at $2f$ are detected with another winding. The amplitude of the $2f$ component is related to $I - I_f$ and at balance, with high loop gain, $I_f \approx I$. Because at balance there is almost zero flux in the measuring head, measurement is largely independent of the type of cable or wire and its position in the magnetic circuit. Methods of sensing flux changes and nulling the changes by another flux are discussed by Geyger (1964). A discussion of Hall-effect feedback transducers and their applications has been given by Warsza (1976).

## 15.6. Heat-flow balance

The self-balancing Wheatstone bridge shown in Figure 15.7($a$) is a DC substitution RF power meter (Larsen 1976). The bridge voltage is automatically adjusted so as to heat the bolometer to a point which closely approximates bridge balance. The output voltage is measured before and after the RF power is applied, and the unknown RF power is related to the change in the computed DC power in the bolometer. Heat loss from the bolometer and its temperature (and resistance) are maintained constant.

The feedback hot-wire anemometer is also a heat-balance instrument, and the instrument shown in Figure 15.7($b$) employs a simple self-oscillating technique (Somerville and Turnbull 1963). It should be pointed out that the conventional oscillator can be explained in terms of a simple feedback system, and considering equation (15.1), it can be arranged that at some frequency the open-loop gain $A\beta = -1$ so that the system has infinite closed-loop gain and oscillations commence at this frequency. In order to control the magnitude of this oscillation some form of negative feedback is needed. In Figure 15.7(*b*) a thin heated wire ($R_w$) is connected to a comparison resistance ($R_0$) in a bridge driven by the centre-tapped secondary of the transformer. The bridge output is amplified to complete the feedback loop as shown. A steady state oscillation occurs with $R_w = R_0$. With $R_w < R_0$, positive feedback occurs and oscillation amplitude increases, with $R_w > R_0$ negative feedback occurs and oscillation amplitude decreases, and when $R_w \approx R_0$ the bridge is balanced and steady-state oscillation is reached. The loss of heat from the wire is related to fluid flow velocity, and so the amplitude of the carrier gives a measure of the flow velocity.

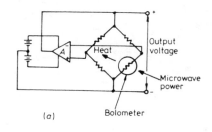

(a)                              Bolometer

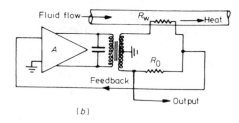

(b)

**Figure 15.7.** Heat-flow balance instruments: (*a*) self-balancing DC substitution RF power meter, (*b*) self-oscillating anemometer.

## 15.7. Voltage balance

Voltage balance is extensively used in measurement practice, in particular in bridge and potentiometer circuits. A useful DC self-balancing Wheatstone bridge which acts as a resistance-to-current converter is shown in Figure 15.8(*a*) (Korobcv et al. 1965). A fixed compensating resistance $R_k$

is placed in series with the unknown $R_x$, and if causes a compensating voltage across $R_k$ which is opposed to the error voltage generated by a change of resistance $R_x$. Note that calibration is not affected by changes in $V_s$. Details of AC self-balancing bridges for the measurement of resistance have been given by Aalto and Ehnholm (1973), Giffard (1973) and Pollock (1975). The differential voltmeter of Figure 15.8(*b*) is a special form of potentiometer whereby an unknown voltage $V$ is applied against one that is accurately known ($V_1$). The feedback activates the switches in the Kelvin–Varley divider such as to make $V_1 \approx V$ and provide a digital output reading. Many AC-to-DC converters are based on the feedback circuit of Figure 15.8(*c*) (Hermach 1976). When $I_{AC}$ is applied, $I_{DC}$ increases until the outputs of the two matched thermocouples are nearly equal.

Negative feedback around an amplifier is to achieve voltage balance, as shown in the circuit and block diagrams of Figure 15.8(*d*). If the open-loop gain $A\beta \gg 1$, then the voltage at point $N$ is virtually zero ('virtual earth') and $v_0/v_1 = -Z_2/Z_1$. If $Z_2$ is a resistance transducer, with $v_1$ a constant, the circuit is a resistance-to-voltage converter. The current in $Z_2$ is solely dependent upon $v_1$ and $Z_1$ and so feedback AC voltmeters can measure small AC voltages, largely independent of meter and diode resistances. Current fed into the 'virtual earth' point $N$ from a transducer is converted into a voltage $v_0$. Various frequency characteristics can be generated by correct choice of $Z_1$ and $Z_2$ values. Almost all instrumentation amplifiers use some form of feedback to define their characteristics (Morrison 1970). Particular configurations with high-impedance transducers are shown in Figures 15.8(*e*) and (*f*) (Jones 1977). In the former case a piezoelectric transducer is coupled to an amplifier with a feedback capacitor, this arrangement being known as a charge amplifier. The amplifier effectively grounds its input terminal thereby substantially reducing the loading effect of signal cable capacitance and leakage on the charge signal generated by the transducer. In Figure 15.8($f$) feedback is used to linearize the signal obtained from a displacement capacitance transducer where capacitance $C$ is varied by separation of the plates, $d$. Another method of achieving this with feedback and using a charge carrier amplifier has been given by Walton (1975).

The high-input-impedance clip-on AC voltmeter is an example of unity-gain voltage feedback (Turnbull and Jones 1966). This device shown in Figure 15.8($g$)

enables the voltage $v_s$ of an insulated cable to be indicated on a conventional AC voltmeter ($v_0$) which has one of its terminals connected to ground. The operation relies on the capacitance $C_1$ between a conducting cylinder, which is clipped around the cable, and the inner conductor of the cable. For high open-loop gain, at balance $v_0 \approx v_s$, and the calibration is sensibly independent of cable size and its position in the cylinder. This circuit has been combined with a clip-on ammeter to give a clip-on AC multimeter (Jones 1971), and

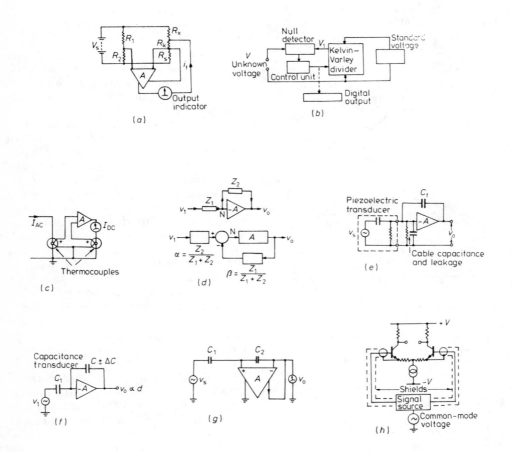

**Figure 15.8.** Voltage balance in instrumentation: ($a$) self-balancing Wheatstone bridge ($R_k \ll R_x$; $R_1$, $R_2$, $R_s$, fixed resistances), ($b$) automatic differential voltmeter, ($c$) automatic thermocouple AC-to-DC converter, ($d$) negative feedback around an amplifier, ($e$) charge amplifier, ($f$) capacitance transducer signal linearization, ($g$) clip-on AC voltmeter, ($h$) common-mode rejection improvement with driven shields.

has been extended to DC voltage measurement (Jones 1970). Common-mode rejection can be improved by unity-gain voltage feedback to electrostatic shields (Oliver 1972). This technique is shown in Figure 15.8(*h*); the transistor stages act as emitter followers driving the shields, thus eliminating common-mode currents from the leads, as well as reducing the effective lead-to-shield capacitance.

## 15.8. Pressure balance

A flowmeter using pressure balance is shown in Figure 15.9 (Maurer 1977). In this instrument the pressure drop across the orifice plate or venturi is balanced by the pressure rise generated by a centrifugal-type rotor or impeller. The null is sensed by the diaphragm and displacement transducer, and at balance the presure rise across the rotor equals the pressure drop across the orifice plate, such that the rotor speed $n$ equals the fluid velocity $v$; it is easy to present the ouput speed in a digital form. This is a feedback system in which both the feedforward and feedback fractions have square-law characteristics.

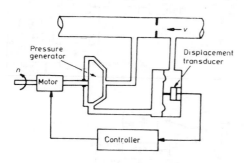

**Figure 15.9.** A pressure-balance flowmeter.

## 15.9. Temperature balance

Temperature balance is usually employed when non-contact measurement of temperature is required. A compensation pyrometer has used this method (Euser

1963), as has the deep-body thermometer (Togawa et al. 1976). The principle of this latter instrument is shown in Figure 15.10(*a*). The heat flow from the surface is nulled by measuring the temperature across a thermal insulator and feeding back heat. At balance the temperature $T_1$ on the front face of the sensor approximates to the internal body temperature $T_B$.

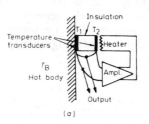

(*a*)

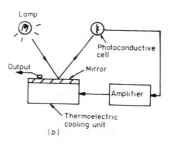

(*b*)

**Figure 15.10.** Temperature-balance instruments: (*a*) deep-body thermometer, (*b*) dewpoint hygrometer.

An optical dewpoint hygrometer is shown in Figure 15.10(*b*) (Goldsmid 1960, Allgeier and Reger 1968). In this instrument, a beam of light is reflected from a mirror on to a photoconductive cell, and this in turn drives a thermoelectric cooling unit. As the mirror surface is cooled, condensation occurs at the dewpoint. This condensation is detected by the reduction in reflected light level, and the feedback is arranged to maintain the mirror surface at the dewpoint, which can be measured with a temperature transducer.

## 15.10. Displacement balance

Two displacement followers are shown in Figure 15.11. The non-contact analogue tachometer (Jones and McNaughton 1977) employs a cathode-ray tube as a voltage-to-light-spot-position actuator, a black–white contrast edge on a diameter of the shaft end, and a photomultiplier light detector. The closed-loop system operates to maintain the light spot on the contrast edge and the input to the circle generator is the tachometer output. The pneumatic non-contact follower of Figure 15.11(*b*).

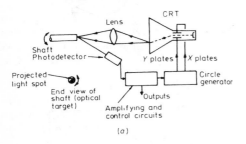

(*a*)

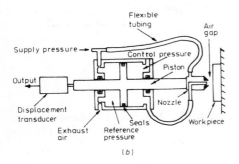

(*b*)

**Figure 15.11.** Displacement-balance instruments: (*a*) non-contact analogue tachometer, (*b*) pneumatic follower.

(Herceg 1976) consists of a double-acting air cylinder in which the reference pressure is about half the supply pressure. The nozzle back-pressure is dependent upon the nozzle-to-workpiece air gap, and this pressure is fed back to the control pressure chamber. At a particular air gap, there is zero pressure drop across the piston, and the piston will move to maintain this

balance and the particular air gap. The transducer provides the output representing the workpiece profile. Since both reference and control pressures are obtained from the same supply pressure, variation in this pressure has little effect on the follower operation.

## 15.11. Phase balance

The automatic vector voltmeter of Figure 15.12(*a*) is a phase-insensitive detection system (Jones 1977). The lower phase-sensitive detector (PSD) is used in a feedback loop to control the phase shifter which locks with the signal phase. The output of the lower PSD is the error point of the loop. The integrator sums the phase errors and provides the required DC voltage to drive the phase shifter and give a phase output. The DC output from the top PSD is proportional to the amplitude of the signal. The full signal recovery capability of phase-sensitive detection is maintained (Blair and Sydenham 1975).

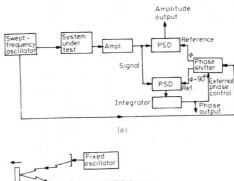

(*a*)

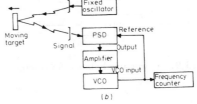

(*b*)

**Figure 15.12.** Phase-balance instruments: (*a*) automatic vector voltmeter, (*b*) phase-locked loop for measurement of Doppler shift. PSD, phase-sensitive detector; VCO, voltage-controlled oscillator.

A phase-locked-loop system can be used to automatically track a frequency of interest, and such a system for measurement of Doppler shift is shown in Figure 15.12($b$). In lock the signal and reference inputs to the PSD are approximately in phase quadrature, and the output of the PSD is the error point of the feedback loop. The frequency of the VCO is measured on a counter, its difference from that of the fixed oscillator being a measure of Doppler shift. A useful reference on phase-lock techniques is that by Gardner (1966).

## 15.12. Radiation balance

A pyrometer using radiation balance is shown in Figure 15.13 (Kelsall 1963). The radiation in beam $a$ (due to the hot surface and the heater), and that in beam $b$ (due to the heater) are applied to the detector alternately via the mechanical chopper, and the PSD produces a signal if they are unbalanced. This signal is used to adjust automatically the heater temperature until balance is achieved, when the heater temperature is equal to the surface temperature, regardless of the emissivity of the surface.

## 15.13. Instruments with digital outputs

Digital feedback instruments can be considered to range from those that provide a frequency output to those that employ analogue-to-digital conversion. The automatic differential voltmeter of Figure 15.8($b$) provides a digital output. The null detector feeds the logic circuits which control the steps on the voltage divider network. The measurement sequence usually selects the largest steps of the internal voltage first, the magnitude of the steps decreasing until the null point is reached. This analogue-to-digital voltmeter basically operates by comparing the input voltage with the output of a digital-to-analogue converter. Thus a well-defined digital-to-analogue converter can be used in the feedback path of a feedback system to make the system into a well-defined analogue-to-digital converter.

It is possible in force- and torque-balance systems which use moving-coil actuators (section 15.4) to produce the feedback current in a digital manner, and an example of this for liquid density measurement is given by Wightman (1972). Force-balance digital transducers for pressure measurement have been described by Serra (1966), Sherwood (1969) and Johnson (1977). The pressure-balance flowmeter (section 15.8) can provide an output pulse suitable for a counter with a digital display.

In the discussion of the self-oscillating anemometer (section 15.6), an oscillator was explained in terms of a simple feedback system. Clearly capacitance and inductance transducers can be used to determine the frequency of oscillators. Lövborg (1965) has described a method of converting temperature to a frequency signal by means of a thermistor which is part of a frequency-determining network of an $RC$ oscillator. Mechanical elements have a natural frequency of vibration, which is determined by the phsyical properties of the body and its environment, and so a number of transducers have been built (Wightman 1972, Woolvet 1977) that

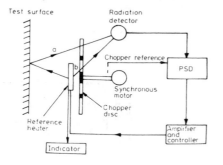

**Figure 15.13.** Radiation balance in an emissivity-compensated pyrometer.

depend on the mechanical vibration properties of stretched wire (Wyman 1973),
beams (Voutsas 1963) and diaphragms
(Woolvet 1977); displacement, force and
pressure have been measured with this
range of transducers. Vibrating-cylinder
transducers are used for measurement of
gas pressure (Halford 1964), gas density
(Woolvet 1977) and liquid density (Agar
1969).

## 15.14. Instrument control systems

Feedback can be used for control purposes
in an instrument application. The potentiometric recorder (Figure 15.14($a$)) is used to provide high-accuracy data recording
at low frequency. The output of the potentiometer is the feedback signal $v_p$ for comparison with the recorder input $v_i$, and if
there is a difference or error $v_e$, this is
amplified and a motor is arranged to drive
the pen until there is negligible difference.
A modification of the recorder whereby the
moving chart is replaced by a stationary
piece of paper and the pen is free to move
in two orthogonal axes across the paper
results in an $X - Y$ recorder. In this case the
pen is controlled independently in both
directions by separate closed-loop servomechanisms.

Piezoelectric actuators have been used
in instruments (Jones and McNaughton
1974, N-nagy and Joyce 1972), and to
reduce their hysteretic and creep effects,
they have been incorporated in the feed
forward path of feedback systems, as for
example shown in Figure 15.14($b$). Strain
gauges are mounted on the two faces of the
piezoelectric cantilever construction called
a bimorph; the gauges are used in a bridge
network to provide a feedback voltage
which is proportional to the actual deflection of the bimorph. This voltage is compared with the input voltage, representing
the required position, by an integrator.
Any error is integrated and forces the drive
amplifier output to change until the error
is corrected. The overall accuracy of this
position control system depends solely
upon the performance of the strain-gauge

bridge and its associated amplifier, and
such a system (with a mirror surface on the
cantilever) has been used for precise laser
deflection.

Feedback can be used to enhance a
filter selectivity, as for example, in the
constant-$Q$ wave analyser of Figure
15.14($c$). Here the input and output of a
Wien bridge filter network are compared,
and the difference is subtracted from the
input to the system. Thus a frequency
component at $f_0$ is unattenuated, while
frequency components on either side of $f_0$
are attenuated because of the feedback.
Feedback is often used to maintain constant the output amplitude and frequency of
instrument oscillators, and automatically
adjust the gain of parts of a measuring
system.

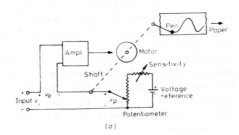

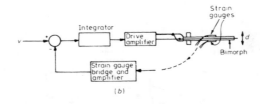

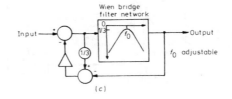

**Figure 15.14.** Instrument control systems:
($a$) potentiometric recorder ($v_e = v_i - v_p$), ($b$)
bimorph position-control system, ($c$) constant-$Q$
wave analyser.

## 15.15. Conclusions

Feedback is widely used in instruments, and may well be the main structural element of a measuring system, linking the output indication with the measurand. Using feedback it is possible to improve accuracy and speed of measurement, reduce the effect of interfering and modifying inputs, and allow remote indication and non-contact measurement. The property of inversion assists instrument design, and can assist in provision of digital indication. In general the main disadvantages are increased complexity, size and cost, but the advent of small cheap integrated circuits has reduced these. The further development of transducers and accurate precision actuators seems certain to extend the development and use of feedback-measuring systems.

Measuring devices where the output is in frequency form usually employ feedback to maintain oscillation. Feedback is used in instruments for control purposes, and the operation of the feedback is similar in both the control and measuring applications; the design for accuracy and stability is basically the same.

The developments outlined in this paper have pedagogic implications (Jones 1974, Jones 1976). At present, ideas about feedback are usually considered solely in the context of servomechanisms or control engineering. However, instrument designers have equal claim to feedback, and principles of measurement and instrumentation, and general concepts of feedback can be combined into a unified course of study by relating them to practical feedback-measuring systems. To control one must be able to measure; now, to measure one must often be able to control.

## References

Aalto M. I. and Ehnholm G. J. (1973) A self-balancing resistance bridge. *J. Phys. E.: Sci. Instrum.* **6**, 614–8.

Agar J. (1969) Frequency-modulating transducers *Radio Electron. Engng* **38**, 89–97.

Allgeier R. L. and Reger J. A. (1968) New automatic dewpoint atmosphere control. *Proc. 18th ISA Iron and Steel Instrumentation Symp.* (Pittsburgh: ISA) pp. 105–121.

Auckland D. W. and Hawke J. R. (1978) A thread-tension transducer using torque-balance about the axis of a motor. *Proc. Transducer '78 Conf.* (Tavistock: Trident International Exhibitions).

Black H. S. (1934) Stabilised feedback amplifier. *Bell Syst. Tech. J.* **13**, 1.

Blair D. P. and Sydenham P. H. (1975) Instrument science: phase-sensitive detection as a means to recover signals buried in noise. *J. Phys. E: Sci. Instrum.* **8**, 621–7.

Broadbent T. E., Cooper R. and Elliott C. T. (1965) A direct-reading, attracted-disc, absolute voltmeter *J. Sci. Instrum.* **42**, 618–20.

Doebelin E. O. (1975) *Measurement Systems — Applications and Design* (New York: McGraw-Hill).

*Electronics Industry* (1976) In-circuit DC current measurement. **2**, No. 1115.

Euser P. (1963) Compensation pyrometer. *Instrum. Pract.* **17**, 487–9.

Finkelstein L. (1977) Instrument science: introductory article. *J. Phys. E: Sci. Instrum.* **10**, 566–72.

Gardner F. M. (1966) *Phase-lock Techniques* (New York: Wiley).

Geyger W. A. (1964) *Nonlinear Magnetic Control Devices* (New York: McGraw-Hill).

Giffard R. P. (1973) A simple low-power self-balancing resistance bridge. *J. Phys. E: Sci. Instrum.* **6**, 719–23.

Goldsmid H. J. (1960) Principles of thermoelectric devices. *Br. J. Appl. Phys.* **11**, 209–17.

Greig J. (1950) Developments in the teaching of electrical measurements. *Proc. IEE* **97**, 55 1–4.

Halford R. J. (1964) Pressure measurement using vibrating-cylinder pressure transducer. *Instrum. Pract.* **18**, 823–9.

Herceg E. E. (1976) *Handbook of Measurement and Control* (Pennsauken: Schaevitz Engineering).

Hermach F. L. (1976) AC-DC comparators for audio-frequency current and voltage measurements of high frequency. *IEEE Trans. Instrum. Meas.* **25**, 489–94.

Johnson A. W. (1977) Digital indicator brings DVM accuracy to pressure measurement. *Control Instrum.* **9**, 63–5.

Jones B. E. (1970) Clip-on DC voltmeter. *J. Phys. E: Sci. Instrum.* **3**, 101–4.

Jones B. E. (1971) Portable clip-on AC

multimeter to measure voltage, current, active and reactive power, and power factor. *Control Instrum.* **3**, 37–41.

Jones B. E. (1974) Feedback-measuring systems. *Electron. Power* **20**, 566–9.

Jones B. E. (1976) Unified course in measurement and control. *Control Engineering in Undergraduate Courses, Proc. Conf., Teesside Polytechnic UK* B2/1–B2/5.

Jones B. E. (1977) *Instrumentation, Measurement and Feedback* (Maidenhead: McGraw-Hill).

Jones B. E. and Gharakhanian A. (1977) An electrostatic wattmeter employing automatic voltage feedback to the quadrants to achieve torque balance. *Euromeas 77: Precise Electrical Measurement: IEE Conf. Publ.* **152**, 129–31.

Jones B. E. and McNaughton H. M. (1974) An investigation of open- and closed-loop tilt monitors employing electrolytic spirit levels. *J. Phys. E: Sci. Instrum.* **7**, 582–5.

Jones B. E. and McNaughton H. M. (1977) A cathode-ray tube actuator in a non-contact analogue tachometer. *Proc. IEE* **124**, 669–72.

Jones B. E. and Mirza N. A. (1977) Feedback electrostatic wattmeter using a non-contact moving-coil actuator. *Proc. IEE* **124**, 1259–62.

Kelsall D. (1963) An automatic emissivity-compensated radiation pyrometer. *J. Sci. Instrum.* **40**, 1–4.

Korobcv J., Piguet G. and Pun L. (1965) High-precision temperature-measuring device. *Control* **9**, 168.

Larsen N. T. (1976) A new self-balancing DC-substitution RF power meter. *IEEE Trans. Instrum. Meas.* **25**, 343–7.

Lövborg L. (1965) A linear temperature-to-frequency converter. *J. Sci. Instrum.* **42**, 611–4.

MacDonald W. R. (1961) *Flight Test Instrumentation*, ed. M. A. Perry (Oxford: Pergamon) pp. 15–23.

Maurer R. (1977) V-Delta-P digital fluid flowmeter *Proc. Flow-Con 77* (London: Institute of Measurement and Control) pp. 139–48.

Mayr O. (1970) The origins of feedback control. *Sci. Am.* **223**, 110–8.

Morrison R. (1970) *DC Amplifiers in Instrumentation* (New York: Wiley).

Neubert H. K. P. (1975) *Instrument Transducers* (London: Oxford University Press).

Neubert H. K. P. and Price E. F. (1969) Vibrating-contact pressure monitor. *Instrum. Control Syst.* **42**, 81–4.

N-nagy F. L. and Joyce G. C. (1972) Micro-movement control systems *Proc. 5th World Congr. IFAC* Part 2b No. 23.6 pp 1–9.

Oliver F. K. (1972) *Practical Instrumentation Transducers* (London: Pitman).

Pollock N. (1975) A simple high-performance device for measuring strain-gauge transducer outputs. *J. Phys. E: Sci. Instrum.* **8**, 1049–52.

Rohrbach C. (1967) *Handbuch fur elektrisches Messen mechanischer Grössen* (Düsseldorf: VDI Verlag).

Serra G. F. (1966) Force-balance principle raises accuracy of digital pressure transducer. *J. Instrum. Soc. Am.* **13**, 51–54.

Sherwood W. M. (1969) The search for a true digital transducer. *Control Engng*, 95–8.

Somerville M. J. and Turnbull G. F. (1963) Self-generating HF carrier feedback anemometer. *Proc. IEE* **110**, 1905–14.

Togawa T., Nemoto T., Yamazaki T. and Kobayashi T. (1976) A modified internal temperature measurement device. *Med. Biol. Engng* **14**, 361–4.

Turnbull G. F. and Jones B. E. (1966) High-input-impedance clip-on AC voltmeter. *Proc. IEE* **133**, 908–14, 1695.

Voutsas A. M. (1963) Twisted beam transducer. *AIAAJ.* **1**, 911–3.

Walton H. (1975) Developments in accurate non-contact measuring techniques. *J. Br. Nucl. Energy Soc.* **14**, 341–5.

Warsza Z. L. (1976) Hall effect feedback transducers and their application. *Proc. IMEKO VII* BTS/274 1–8 (London: Institute of Measurement and Control).

Wayne Kerr (1975) *Some Notes on Bridge Measurement* (Bognor Regis: Wayne Kerr Co.).

Welkowitz W. and Deutsch S. (1976) *Biomedical Instruments: Theory and Design* (London: Academic Press).

Wightman E. J. (1972) *Instrumentation in Process Control* (London: Butterworths).

Woolvet G. A. (1977) *Transducers in Digital Systems* (Stevenage: Peregrinus).

Wyman P. R. (1973) A new force-to-frequency transducer. *Digital Instrumentation: IEE Conf. Publ.* **106**, 117–23.

Paper 16

# Communications in process control

*P. R. Matthews*

This paper covers, in a comprehensive way, the various methods used for transmitting signals in instrumentation systems. Only radio telemetry is omitted.

Readers without a knowledge of electronics may find the details of some parts of the paper difficult to follow. However, it is still worthy of thorough study as it contains much that a student of instrumentation needs to know about signal transmission.

Note that there is a glossary of terms in the appendix. (Eds.)

The last decade has seen many new measurement and control techniques developed. This has been principally due to the development of low-cost electronic circuits and, more recently, the microprocessor. As a result of this rapid development the instrument and process control engineer is faced with the arduous task of interfacing new and old equipment with differing transmission and communication requirements. The situation is further complicated by the numerous transmission and communication techniques that are available and the usual dogma of finding differing communication interfaces at either end of the connecting cable.

The paper provides an insight to the array of transmission and communication techniques currently in use and looks at new techniques which are being developed. The application of the transmission and communication techniques to process control systems is a subject which is now under careful scrutiny as the establishment of an effective transmission and communication configuration can result in an appreciable cost advantage and increased transmission efficiency. A number of these configurations (networks) are discussed.

## 16.1. Introduction

For the human race, communication is a fundamental requirement of life. Without the ability to pass information it is unlikely that we could survive. This same basis ap-plies in industry. There must be the ability to transmit information around a process control system, be it a simple measurement or a plant automation system.

The development of control systems from simple mechanical indicators and

Originally published in *Measurement and Control* (November and December 1982, January 1983).

distributed microprocessor-based systems has required equal developments in the methods of transmitting and communicating information between these devices. Although the development of control and associated communication systems has occurred over many years, the principles that were first discovered, e.g., mechanical transmission, are still very much a part of modern-day systems. Consequently today's instrument or process control engineers are faced with a mixture of old and new technology equipment, with the need to be able to interconnect it. As a result of manufacturers standardizing on instrument interfaces this has become a relatively simple task. The interconnection of instrumentation to computer systems and intercomputer communication is a different matter. Although various standards do exist, there are still systems which require customized interfaces. Whilst the requirements of a communication or transmission system can be expressed in very simple terms, the means of satisfying that requirement can be very complicated in comparision.

The situation is now being further complicated by a new fashion which is likely to grow, known as 'networking'. Companies are having to monitor and control production more closely than ever, and as a result more and more computer intercommunication is required to enable the required management information to be collected and analysed. As well as these changing system requirements, new communication media such as fibre optics are coming on to the market and are likely, in time, to cause little less than a revolution in new instrumentation and in data transmission systems.

In order that the merits of the differing communication and transmission systems can be appreciated it is worth reviewing some of the transmission techniques. There are basically three transmission carriers: pneumatic, electrical, and light.

## 16.2. Pneumatic transmission

The use of a compressive fluid to transmit analogue signals is a well-tested and time-served communication medium. In this day of microprocessor technology one might think that there is a weighted argument for using an electrical transmission method, but there are also arguments for preferring to use pneumatic-type transmission methods for particular services such as applications in hazardous areas.

The internationally accepted standard for pneumatic transmission pressures of $3-15 \, \text{lbf in}^{-2}$, which approximates to $20-100 \, \text{kPa}$ if described by the SI system of units, means that practically any manufacturer's pneumatic instrument can communicate with any other manufacturer's instrument by simply providing a regulated pneumatic supply and the piping to connect the two. So simple, but is it really as straightforward as that? How does the quality and type of the pneumatic fluid affect the transmission? What type and size of piping should be used? Is the transmission time fast enough to ensure that a control loop can adequately respond to a process change?

The quality of the pneumatic fluid, typically compressed air, natural gas or nitrogen, is of paramount importance. No instrument will operate reliably if the signal line is being supplied with a gas contaminated with oil, carbon, rust, water or any other foreign matter. It is a necessity that pneumatic supplies are properly dried and filtered. This usually amounts to drying to a dewpoint of 20°C below the lowest expected ambient temperature, filtering down to 5 $\mu$m and ensuring that the air is oil free.

The choice of piping to connect the various instruments is usually dependent upon tradition, environment, cost, length and response time.

Tradition is fine, providing it is used as only one of the reasons for selecting a particular manufacturer's equipment and not for selecting either the piping material or size. The selection of the piping material is normally only dependent upon the environment through which the piping is to be run and the cost of the piping. The materials available include copper, stainless steel, carbon steel, nylon, high-density polyethylene and plastic-coated aluminium. Of these copper and the polymer-based materials are probably the most commonly used. The criteria which will affect the choice of material include temperature, particular environmental corrosion problems, and mechanical strength. Temperature can have an eventual effect on some of the polymer pipes, but in the past the main problem associated with plastic piping has been its degradation due to ultraviolet radiation. However, piping is now available with a protective coating, or a black pigment can be incorporated into the material which prevents this degradation. Copper tends to age-harden and will slowly oxidize if left open to the elements. The oxidization problem may be overcome by using plastic-coated copper. The coating is available in a range of colours which facilitates colour coding of supply and signal pipes. If mechanical strength is the all-important factor then either carbon steel or stainless steel is the answer, even though the latter is rather expensive.

The specification of pipe diameter is dependent upon the time response required; this is, in itself, dependent upon pipe length and the volume at the receiving end. The transmission pipe is analogous to an electrical circuit in that it includes a capacitive component which is proportional to the volume of the tube, and is distributed along the length of tube. A resistive component is similarly distributed along the length of the tube. The calculation of pipe size for a particular time response, although not difficult, is rather tedious; therefore sizing is invariably done by experience or by referring to graphs similar to that shown in Figure 16.1 which are based on approximately 63% response to a step change, e.g. $100 (1-1/e)$. The important point to remember in deciding if a particular response is acceptable is that the time response for a control loop is the time

for the signal to arrive from a transmitter, be acted upon by a controller and for the output signal to reach and have an effect on the final element.

For many applications pneumatic transmission still provides the best solution; this is especially true for field-mounted control loops, where a compressed-air supply is locally available. Pneumatic transmission is also ideally suited for instrumentation located in hazardous areas. Where computer control is involved, pneumatic transmission may not usually be regarded as suitable due to the pneumatic/electrical interface. However studies (Wilson 1978) have shown that there is not always a clear-cut cost advantage in using an electrical transmission technique over pneumatics. As a result of this, the inherent immunity to cross-talk, and the ability of pneumatics to transmit more power than most other transmission techniques, pneumatic transmission is likely to continue to be very much a part of modern-day control systems for many years to come.

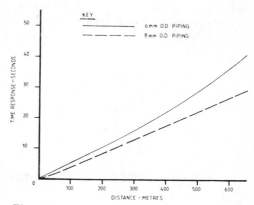

**Figure 16.1.** Pneumatic piping time responses.

## 16.3. Electrical analogue

Industrial process control is being increasingly dominated by electronic instrumentation. This is mainly due to the majority of sensor development being supported by electronic detection circuits. The use of an electrical method for transmission is therefore the obvious method of communication between instruments. Over the years it has been fashionable to use particular signals and ranges. The trend has changed from voltage to current and from true-zero to live-zero signals. The preference for using current signals over voltage is principally due to the low level of noise interference associated with current signals as a result of the somewhat lower input impedance. The preference for using live zero over true zero can be explained with reference to Figure 16.2.

The preferred current ranges for true and live zero are specified in BS 3586 and IEC 381 as 0–10, 4–20 mA, and 0–20, 4–20 mA respectively. Figure 16.2a shows a typical true-zero transmission where a three-wire cable powers the transmitter and carries the output signal, the earth return being shared by both. The live-zero configuration shown in Figure 16.2b only requires a two-wire cable, the power for the transmitter being provided by the 4 mA live zero. Therefore, the choice of live zero results in lower cable and installation costs; live zero also provides loop-break detection and due to the very low power consumption of the transmitter is ideally suited for

intrinsically safe applications.

The loop power supply is provided either from within the receiving instrument or from a separate power source which would typically be mounted in an interface room or could be field mounted in the case of non-intrinsically-safe installations. The voltage required by a transmitter can be as little as 5 V, but may be considerably more. In specifying transmitter power supplies the main considerations should be loop impedance and the transmitter voltage requirement. The detailed analysis of power supply selection is beyond the scope of this paper and has been well documented by others (Cook 1972). However, an insight to some of the considerations required is given in the following example.

If we consider a control loop with a live-zero transmitter which requires a voltage of 10 V to operate and the combined series impedance of the loop due to other instruments and cable losses amounts to 1000 $\Omega$, then the power supply voltage requires to be 10 V + (20 mA × 1000) = 30 V. However, if further instruments are likely to be added to the loop at a later date, allowance for the increased loop

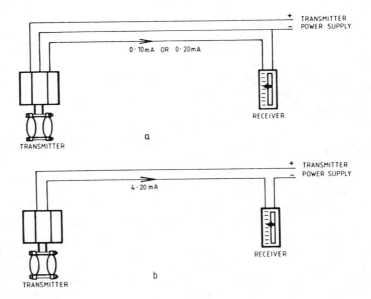

**Figure 16.2.** True- and live-zero transmission: (a) three-wire true-zero loop; (b) two-wire live-zero loop.

impedance should be included. Another consideration worthy of mention is the fact that as the signal current increases from 4 mA the load on the power supply increases. This can result in a reduction of the supply voltage which may be a problem, particularly with intrinsically safe circuits. This can be overcome by the use of a negative-resistance power supply where the supply voltage will actually increase as the load increases, thus ensuring the correct supply voltage at the transmitter. The situation is also complicated if Zener barriers are included for intrinsically safe systems as the loop impedance is then increased considerably, the amount depends on the barrier used.

Only those analogue signals specified by BS 3586 and IEC 381 have so far been considered; there are, of course, others. On the current side there is 10–50 mA; this live-zero signal obviously requires more power to drive each loop and as a result is not really suitable for hazardous-area applications. On the voltage side the most common voltage signal is 0–10 V DC. This is fast becoming a control room standard for interconnecting instrumentation, especially chart recorders. The main reasoning behind this is that when a voltage is used to drive a number of instruments the failure of any one instrument has no effect on the operation of others.

Other voltage signals which are quite commonly used are 0.25–1.25 and 1–5 V DC. These signals are derived from 4–20 mA signals across either a 62.6 or 250 resistor. The 1–5 V DC signal is, in fact, an ISA specified signal covered by ANSI MC 12.1 (1975). The combination of using a current signal with its high noise immunity to transmit from the field to the control room and then to interface with the receiving instruments using either 0.25–1.25 or 1–5 V DC is particularly advantageous if there is a substantial number of receivers in the loop making the loop impedance prohibitively high.

Once the signal type has been selected the next step is to specify the cabling which will carry the signal. This is the area where the operational success of a control loop is usually determined. The specification of cable type and connection is so often inadequately researched and can result either in an over-specified system with the associated cost penalty or a poorly specified one which results in an unsatisfactory loop operation. Additional problems are the considerable extra time required during commissioning periods to diagnose the problem, and also the resulting cost outlay to either replace the cables concerned or redesign the loop configuration to produce an operational system. To explain all of the cable criteria that should be considered requires more than a short digression, therefore, the following comments are intended solely as a guide to the steps that must be taken.

There are four criteria which, if satisfied, will go a long way to ensuring an efficient and safe system: cable length, electromagnetic interference, electrostatic interference, cable routing and instrinsically safe circuit considerations.

The distance between instruments can often be considerably longer than initial estimates as the distance is determined by the cable routing. The distance is of particular importance as it has a direct effect on the transmission line resistance. When calculating the cable resistance the cable lengths must be doubled to account for the two wires which form the loop. Once the size of cable has been determined the next consideration is how to prevent electrical noise in the form of induced electromagnetic or electrostatic interference affecting the transmission signal. Electromagnetic noise reduction is the simplest of the two and is achieved by ensuring that the area between the cores that form the loop is minimized. This is best achieved with the use of twisted-pair cables with a lay of 25–50 mm. Twisted-pair cables also have the advantage of aiding the cable installer to identify the paired cores; this is especially useful when terminating multipair cables.

Electrostatically induced noise, more commonly known as capacitive coupling, is more difficult and usually more expensive to cure. The solution here in terms of cable parameters is insulated screening. The use of an aluminium foil is preferred to braided screens as the foil can be twice as effective in reducing the noise effect. However, if the screen is incorrectly terminated it can be the cause of induced noise. The rules to follow when terminating screens are:

Terminate the screen to the loop zero-volt reference and, if possible, connect this termination point to mains earth.
Terminate each screen to the loop zero-volt reference at only one point.
At field junction boxes, ensure that the continuity of screens is maintained but effectively insulated from earth.

Cable routes are very often decided simply as a matter of convenience; this can result in expensive re-routing if induced electrical noise levels are too high. The following rules should therefore, where possible, be implemented:

Where signal cables run parallel with power cables, the cables should be separated by at least 250 mm for 100 V AC power cables and 600 mm for 440 V AC power cables. Current-carrying cables should also be separated by the same amounts for 10 and 100 A respectively.
Signal cables should cross power cables at right angles.
Signal and power cables should not be run on the same cable tray;

Where a plant includes hazardous areas and intrinsically safe circuits are used, it is imperative that in addition to the above, the cable capacitance and inductance/resistance ratio are considered together with the certified requirements. This is particularly important where it is intended to use multicore cables as the problems of interference between cores could have disastrous consequences.

## 16.4. Digital communication

The 1970s saw an explosion in the development and application of digital computers in process control. Initially, this consisted of centralized systems which incorporated the interfaces to connect the computers via analogue transmission to the production process. It was not until the advent of distributed control, telemetry systems and the need for computer intercommunication that digital communication really found a place in process control systems. The microprocessor has reinforced this trend by providing relatively cheap data collection, logging and display equipment which can be located at remote sites. Digital communication — or, as it is more correctly called, digital data communication — is the transmission of encoded data from one point to another by means of switched electrical signals. The individual digital signals received from field-mounted pressure, level and temperature switches are not included in this category as these are straightforward, uncoded signals. The digitally encoded data can represent both status and analogue values, a two-condition status value can be directly represented by a binary signal. Analogue values are digitized using an analogue to digital converter and a technique known as pulse code modulation (PCM). This involves periodically sampling the analogue value and converting the measured magnitude into a stream of binary signals which can very accurately represent the true analogue value, e.g. an eight-binary-bit stream can represent an analogue signal to within 0.5%.

The distance between the various items of equipment which now need to communicate can vary from as little as a few metres up to a number of kilometres. Consequently there are many differing techniques using various media to transmit over these distances. This paper does not intend to cover all of the techniques in detail but concentrates on those which are in use, or will be used in the near future, for process control and industrial management information systems.

Digital data communication can be divided into two categories, parallel and serial. A description of the techniques and equipment used for parallel and serial communication and the way in which data are arranged in these communications is given in the following sections. A glossary of the technical terms used is given in the Appendix. [ ... ]

### 16.4.1. Parallel communication

Parallel communication is a method of transmitting data from one point to another using a cable system which has the same number of data-transmitting cores as there are data signals. Therefore, if there are eight data bits (a byte) to be transmitted, e.g. 10100110, eight data cores are required; the cable will, in fact, contain more than eight as various control and address cores are also required. The parallel communication concept is basically an extension of the internal computer bus. The central processing unit (CPU) requires fast access and communication to the system memory and peripheral boards. Parallel communication satisfies this requirement as whole bytes of data can be transferred in literally the same time as a single data bit. The connection of the computer system to the outside world is not a simple matter of extending these buses. This is particularly true in the case of process control where equipment can be of various manufacture using differing communication techniques; also the internal computer system buses use very low-power devices and are particularly sensitive to changes in peripheral loads. These problems can be overcome by the use of a buffered interface which separates the internal system buses from the outside connection, or as it is sometimes called the 'data highway'.

The implementation of this concept was initially hampered by the multitude of computer system buses which were available and the lack of an interface to connect the bus to a suitable parallel data highway. The rapid development of computer system architecture resulted in manufacturers each developing a computer system bus suitable for their own needs. However, the need for standardization was recognized and has resulted in a small number of computer system buses being

produced which are being supported by the main, peripheral and small-system manufacturers. One of the most common of the computer bus standards is the S100 bus developed by MITS. Others include Multibus (Intel), STD Bus (Prolog) and Unibus (DEC).

Parallel data highways have been subject to much thought and argument which has resulted in the development of two main standards, CAMAC and the IEEE 488/IEC 625 series. The bit-parallel, byte-serial transmission systems specified by these standards are generally intended for laboratory-type environments; this restriction is due to the limits placed on transmission distance and the requirement for a relatively electrically noise-free environment. Parallel communication, being a high-speed transmission technique, requires that each transmitted bit of a byte of data arrives at the receiver simultaneously. For distances exceeding a few metres, imperfections in the transmission line parameters result in some byte bits arriving out of phase, therefore synchronization becomes unmanageable unless the transmission speed is reduced. The requirement for an electrically clean environment has similar grounding; any induced noise into the transmission circuit would also result in synchronization being lost or data corrupted.

CAMAC (computer automated measurement and control), although often referred to as a communication standard, is in fact a specification for a modular instrument system which was developed primarily to provide a specialized instrumentation bus for the nuclear power industry.

CAMAC is based around devices known as crates. A crate consists of slots for up to 25 slide-in boards. The boards will either be crate controllers or functional and interfacing modules. These slide-in boards plug into an 86-line crate bus which is known as the dataway. The crate controllers handle all communications on the dataway and also act as an intermediary between the crate and any external computer or other crates. A range of modules has been developed to provide functional and interface facilities such as multichannel analogue-to-digital conversion, parallel and serial interfaces and also data gathering and control facilities.

CAMAC has the facility to connect up to seven crates into a number of configurations; the most common of these uses a daisy-chain technique in which the crates are interconnected one after the other using an interconnecting cable known as the branch highway. The total length of the branch highway should be restricted to below 30 m if data rates of up to 5 Mbit s$^{-1}$ are to be achieved.

CAMAC was accepted and has been used for many years in the industry for which it was designed; there are in fact over 70 companies manufacturing CAMAC-compatible equipment. However, it has not received the same support from the other process-control users, the main reason for this being cost. CAMAC has for many years been regarded by the process-control industry as an over-specified standard. For most small applications it is considerably cheaper to provide a custom-built parallel bus using proven microprocessor techniques rather than to use CAMAC. CAMAC in its specified form is only really viable for the larger control systems, but for these systems the need for such high data rates is not often a requirement.

CAMAC is specified by both American and European standards. BS 5554 (1978) and BS 5836 (1980) detail the functional, mechanical and electrical requirements but the full specification is more easily referred to using the American IEEE series standards 583, 596, 675, 683, 726 and 758. Although specified primarily as a parallel communication technique, CAMAC can be configured as a serial branch highway, which enables communication over distances up to 2 km (IEEE 595).

The IEEE 488/IEC 625 data highways are both derivatives of the Hewlett Packard interface bus. The IEEE adopted the interface as a standard within

two years of it being launched by HP. The IEC, however, waited until 1979 before publishing IEC 625 Part 1 which covers the electrical, mechanical and functional requirements of the bus. IEEE 488 did not originally include any description of how the transmitted data should be coded or formatted; this is, however, now covered in the IEC 625 Part 2 standard which was published in 1980. The IEC 625 bus is intended for use as an interface for programmable measuring instruments. The bus consists of a data highway that utilizes a 24-core screened cable which is normally terminated with a 25-way plug/socket. The IEEE specify a 24-way plug/socket to protect circuits against accidental connection to other communication ports which use a 25-pin connector (RS232). The connectors fitted are a back-to-back plug/socket arrangement which facilitates device daisy-chaining. Although a 24-core cable is specified, only 16 of the cores are used for signal lines, i.e. eight data bus cores, three control bus cores, five interface management cores. The remaining eight cores are ground returns for the three control bus and five interface management cores.

The maximum distance that this bus is specified as being capable of transmitting over at its top speed of 1 Mbyte s$^{-1}$ is 15 m; however, this sort of distance can only be achieved if the signal lines are correctly loaded. Longer transmission lengths are possible with bus extenders but this also requires a reduction of transmission speed; too great a reduction would, of course, defeat the object of having a parallel bus. With a capacity to handle up to 15 devices this bus is ideally suited to monitoring and controlling small processes where a very fast response is required. However, although this bus is aimed at supporting instrument systems it must be restricted to either laboratory-type duties or areas where electrical noise is at a minimum.

In addition to the two standard buses mentioned there are other parallel communication systems. These are generally customized multiplexed systems. The simplest example of these systems is shown in Figure 16.3. This system could be used to transmit the position of a remote selector switch. The example shows that the status of any one of the 15 possible combinations can be transmitted over 4 + 1 wires where the extra core provides the power supply. Similarly, 31 states can be multiplexed over 5 + 1 wires. This system is obviously rather restrictive in that only one state can be transmitted at any time.

### 16.4.2. Serial communication

Serial communication is one of the fastest growing technologies in the communications field. With the ability to transmit from a few hundred to millions of data bits per second, serial communication can service nearly all digital communication needs. The basic principle of serial communication is that data is transmitted in a chain-like manner along a single pathway.

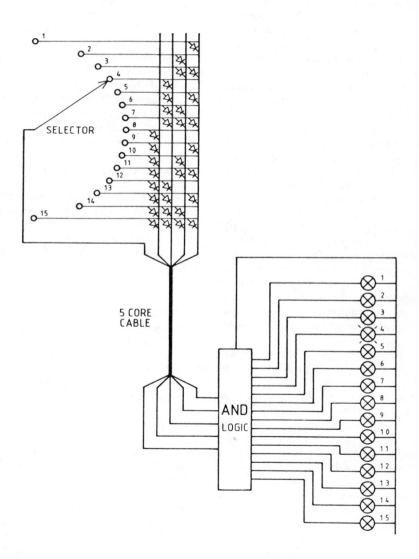

**Figure 16.3.** Simple multiplexer.

This pathway can vary from a standard twisted-pair cable to low-loss coaxial cable, and more recently to fibre-optic cables.

The principles of serial communication are fairly simple; however, the application of some complicated multiplexing and modulating techniques enable higher communication speeds and dense packing of data in these communications.

Before going on to describe the various serial communication systems it is worth reviewing some of the principles. There are three categories of communication link: simplex, half-duplex and full-duplex. Simplex is a two-wire transmission of a signal from Device A to Device B where Device B is not capable of transmitting back to Device A. Half-duplex provides transmission from Device A to Device B and from Device B to Device A, but not simultaneously. This transmission can operate on a two- or four-wire system. The two-wire system has a turnaround facility such that when transmission from Device A to Device B is complete the path is electrically switched to permit transmission from Device B to Device A. Full-duplex is the simultaneous transmission of signal from Device A to Device B and from Device B to Device A over either a two- or four-wire cable. The use of the four-wire is easy to appreciate. However, by the use of modulation techniques it is possible to transmit simultaneously in both directions over a two-wire link.

For any of the above types of communication link when a transmission is made, for example from Device A, Device B will not be aware of this transmission until the first part of the transmission arrives. It is important that the receiver should be ready for the communication and be capable of receiving and identifying each frame or set of data as it arrives. One method for ensuring this is to precede every frame of data with a start bit, and terminate that frame with a stop bit. The receiving device then knows exactly where the data starts and can read in the transmitted data; the receiving device will have been set to expect a frame of a certain number of bits. The incoming data bits will

be counted and when the correct number has been counted the device then checks that a stop bit is present. If a stop bit is not detected the device will ignore the data received, set an error flag to indicate that a framing error has occurred, and wait for the next start bit. This technique is known as asynchronous communication.

Because of the necessity in asynchronous communication to check start and stop bits for every data frame, the transmission rate is restricted to $1200 \text{ bit s}^{-1}$ or below. For rates greater than $1200 \text{ bit s}^{-1}$ synchronous communication — which uses a technique of internal clocking — is used. The transmitted data is preceded by a synchronizing character which acts as a clocking pulse at the receiver, the receiver will then 'clock in' each bit of data. To ensure that synchronization is maintained throughout the transmission an occasional synchronization character is included. The tranmission will end after a predetermined time, and will be confirmed by the receiver checking that a message termination character is present on the line.

The selection of which type of serial communication is used is dependent upon a number of criteria.

The type of communication channel which is fitted to communicating equipment.

Distance over which data are to be transmitted.

Rate at which data are to be transmitted.

Number of data channels to be transmitted over a single transmission line.

Most process-control computer- or microprocessor-based systems have a serial communication interface inbuilt. This could typically be an EIA RS232-type interface which is suitable for the transmission of digital data at rates up to $20 \text{ Kbit s}^{-1}$ over relatively short distances, i.e., less than 15 m. Although a reduction in the transmission rate can mean transmissions over a long distance, the RS232 interface should, where possible, be restricted to short transmission distances.

There are two reasons for this. Firstly, the RS232-type transmission is vulnerable to electrical interference, which can corrupt the transmitted data. By keeping transmission cables short this interference can be minimized. The second reason concerns the properties of the transmission cable. The resistive and capacitive parameters of the cable have an attenuating and smoothing effect on the digitally encoded data. This effect will increase with cable length and higher transmission rates; as a result the sharpness of the individual data bits will deteriorate until some data bits merge into others causing transmission corruption.

The EIA have long recognized the restrictions imposed by the RS232 interface, and as a result of developing technology they have developed as alternatives the RS449/RS422 and RS449/RS423 standards for balanced and unbalanced interfaces respectively which will, in time, replace the RS232 interface. However, they have been specified so that although they can offer higher transmission data rates and longer transmission distances, they are electrically compatible with RS232 and can therefore be interfaced with existing RS232 equipment. The only restriction is that RS232 data rates and transmission distances must be used. Figure 16.4. shows approximate transmission distance against transmission rates. However, to attain the higher transmission rates it is necessary to ensure that the rise time for each bit is within the limits specified in the standards.

Although many process control systems are fitted with one of the EIA interfaces, some companies provide one of the complementing CCITT V Series interfaces. The RS and equipment V Series interfaces are practically identical and can, therefore, be interconnected with little or no interface customizing. A comparision of the RS and V Series interfaces is given in Table 16.1.

The normal method of interconnecting the communicating devices is by means of multipin connectors and twisted-pair cables. The RS232/V28 interface uses a 25-way D-type connector. For asynchronous communication as few as three of the 25 pins are used (data transmitted, data received, common return). Synchronous communication requires connection to five pins, the two additional pins providing the timing pulses. The rest of the connector pins provide various control and test circuits. The RS422, RS423, V10 and V11 interfaces utilize a 37-way connector as standard, each of the pins being either assigned to data, control and testing or for future use. As with the 25-way connector, it is most probable that only a few of the interface connections would be used for any simple transmission. In addition to the 37-way connector a 9-way connector is also specified. This connector provides a secondary communication channel which enables either slow speed communication or interrupt facilities whilst the main channel is being used for data transfer. Table 16.1 details the applicable ISO connector standards.

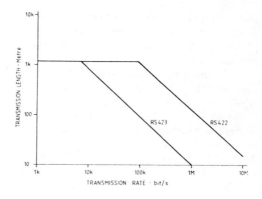

**Figure 16.4.** Transmission rates vs. transmission distances for RS422 and RS423 interfaces.

**Table 16.1: Comparison of interfaces**

| CCITT | | EIA | | ISO |
|---|---|---|---|---|
| Functions | Electrical | Functions | Electrical | Connectors |
| V24 | V28 | RS232 | RS232 | 2110 |
| V24 | V10 | RS449 | RS423 | 4902 |
| V24 | V11 | RS449 | RS422 | 4902 |

One of the reasons why RS422 and RS423 interfaces can achieve longer transmission distances than the RS232 interface is that the transmission and receiving cable cores do not share a common return core but have their own separate return cores. Because of this, twisted-pair cables can be used effectively to reduce electromagnetically induced noise, which would otherwise corrupt the transmitted signal. An alternative to the twisted-pair cable is the coaxial cable; this can be particularly beneficial where the transmission is appoaching the limits of distance or data rate. The coaxial cable, when correctly terminated, has less of an attenuating and smoothing effect and can offer better protection against electrical noise especially if a layered-type screen is used. If coaxial-type cables are used, the associated control and test signals are usually routed via a separate cable.

Where serial communication is to be transmitted through unscreened cables which are either alongside or part of power and plant control cables, as could be the case when the receiving device is located in an overhead crane cabin and the only physical transmission route is via the crane catenary cables, it is unlikely that successful communication even at relatively low speeds will be possible using either RS422 or RS423 interfaces. In this type of environment the only real solution is a current-based signalling system which is relatively immune to the electrical interference. Fortunately, there is a time-served current serial-link transmission system which has been used successfully for teleprinter communication. This transmission method is generally known as current-loop transmission and consists of switching between 0 and 20 mA, a space state being represented by zero current flow and a mark state by 20 mA. Data rates, however, are normally restricted to a maximum of 4800 bit s$^{-1}$. The current-loop transmission is also restricted in that there are not usually any control or test lines. Therefore, the only way to check if the remote device is switched on and ready to receive data is

to actually communicate with it. Another disadvantage of current loop is that it does not provide a standard computer interface. Therefore, if current-loop communication is required extra hardware may be necessary to convert from the standard RS or V series interface. However, the extra cost of these converters must surely be outweighed by the satisfaction of knowing that the transmitted data will reach the receiver, albeit slower than some other techniques, but for this type of application high-speed communication is not always a requirement.

The digital transmission techniques discussed above will normally be used to transmit data from one device to another. However, in some applications it is necessary for a number of devices at one location to communicate with a similar number of devices at another location. For this application it is not always necessary to install dedicated lines as it is possible to transmit a number of data channels along a single transmission line by using a time division multiplexing technique. The details of this technique will be discussed later.

So far the communication techniques discussed have been more suited to the shorter transmission lengths. Even though the RS422/V11 interfaces are specified as being capable of transmitting at high data rates over distances up to 1.2 km in the industrial environment, this is not always practical. Where there is a requirement to send information over longer distances at high data rates, alternative transmission techniques can be used. These alternatives are all centred around a technique known as modulation.

The digital signals from the transmitting device are converted (modulated) on to an analogue waveform. The converting device which is known as a modem (MOdulator/DEModulator) changes the '1's and '0's into either differing wave amplitudes, frequencies or phases. Examples of these three techniques are shown in Figure 16.5.

Amplitude modulation (AM) is where

the amplitude of a carrier wave is varied according to the digital bit pattern, typically between zero and a maximum value.

Frequency modulation (FM), or frequency shift keying (FSK) as it is often called, is the switching from one frequency to another within a particular frequency band to represent the digital bit pattern.

Phase modulation (PM), which is also known as phase-shift keying (PSK), involves changing the phase of the transmitted wave by 180° to represent the digital bit pattern. The example shown changes the phase when a 'space' state is transmitted.

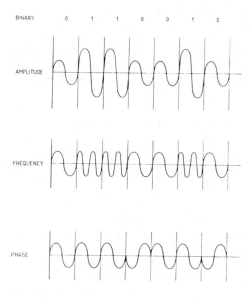

**Figure 16.5.** Amplitude, frequency and phase modulation.

The choice of which technique is used is dependent upon transmission length, transmission rate and cost restraints.

Amplitude modulation is usually restricted to the slower transmission rates and shorter lengths. Frequency and phase modulation are used for the higher rates and long transmission distances. In terms of cost, an amplitude modem can be purchased for a few hundred pounds, whereas frequency and phase-shift modems are priced generally in the range £2000–£6000. The restriction on transmission length for amplitude modulation is primarily due to the effects of electrical noise which could affect the signal amplitude and result in corrupted data. [ . . . ]

Another technique which is used with modems is to multiplex a number of signals into one signal. There are two methods which are normally used, time division multiplexing (TDM) and frequency division multiplexing (FDM). With TDM each channel is allowed to transmit down the communication line for a short time period. With this technique a number of slow-speed channels can be sent over one high-speed channel, e.g., two 4800 bit s$^{-1}$ signals can be transmitted at 9600 bit s$^{-1}$ or two 2400 bits/s$^{-1}$ and a 4800 bit s$^{-1}$ signal can be transmitted at 9600 bit s$^{-1}$.

FDM, as the name implies, is a technique which involves assigning a band of frequencies to each of the signals to be transmitted. The number of signals or channels that a cable can simultaneously carry is dependent upon the bandwidth which the cable is capable of transmitting and upon the data transmission rate. A coaxial cable can support a bandwidth of up to 100 MHz. The individual signal frequency bands will be divided across this bandwidth, the spacing between each channel being related to the data transmission rate. For example, a 10 Kbit s$^{-1}$ transmission rate requires an 80 kHz channel separation whereas a 2Mbit s$^{-1}$ transmission rate requires a 6 MHz channel separation. Therefore the coaxial cable would be capable of carrying 1250 channels transmitting at 10 Kbit s$^{-1}$ or 16 channels at 2 Mbit s$^{-1}$.

Serial communication is clearly a very versatile communication technique, having the capability to transmit data at rates from a few hundred bit s$^{-1}$ to several Mbit s$^{-1}$. There are very few process control requirements that could not be satisfied; however, for the higher data rates there is a cost penalty which will hopefully be short-lived as these devices become more 'chip' orientated.

## 16.5. Fibre optic

Fibre optics are heralded as the answer to all of the process control engineer's instrumentation and communication problems. With the great number of advantages which fibre optics are said to offer we are told that it is only a matter of time before fibre-optic systems will start replacing the conventional electronic field instrumentation and transmission systems. Many companies are now investing in the development of instruments which will measure the more commonly measured variables such as temperature and pressure using optical techniques, and which will produce an optical output signal. Therefore, the problems associated with electrical crosstalk and electrical hazards in flammable areas will no longer exist. However, it will probably be 1990 before an optical sensing loop is commercially viable. In the meantime, the process control engineer will have to satisfy himself with the currently available fibre-optic transmission systems.

Before discussing the application of fibre optics to transmission systems it is worth reviewing some of the basic principles of fibre optics.

There are basically three components in an optical system: the optical transmitter, the optical receiver and the fibre-optic medium to connect these two devices. The transmitter will use either a laser source (coherent light) or a simple light-emitting diode (LED) source (incoherent light). The laser sources are capable of connecting up to a few milliwatts of power into an optical fibre, whereas an LED can only couple up to a few hundred microwatts. The receiver uses photodiodes to detect the received light and converts this into an electrical signal. The fibre-optic medium connecting the transmitting source to the receiver can be constructed from either glass or plastic-based materials.

Optical fibres consist of a core surrounded by a cladding, which has a refractive index lower than that used for the core. This combination ensures that most of the transmitted light is retained within the core. There are two categories of glass fibre, multimode and monomode. The term mode refers to the number of optical field patterns which are able to propagate through the fibre. A monomode fibre allows only one mode to be transmitted; this is achieved by using a core with a very small diameter. The cladding, which has a lower refractive index, produces an index step and therefore this type of fibre is known as a monomode step index fibre. Multimode fibres permit a number of optical field patterns to propagate through the fibre. There are two types of multimode fibres, step index and graded index. The step-index fibre is similar to the monomode step index fibre but has a larger core diameter. The graded-index fibre uses a core which has a varying refractive index. This results in the transmitted light being constantly deflected towards the core centre. Figure 16.6 shows the fibre types and the light propagation paths.

Plastic-based fibres have a step-index construction. They have higher attenuation losses than glass fibres and are therefore restricted to shorter distances. However, plastic fibres have the advantages of being less susceptible to fracture and simpler and more reliable to interconnect.

The glass or plastic fibres are made up into cables with PVC or similar sheaths with strengthening members to ensure that the fibres are not damaged during installation or use. A particular requirement of fibre-optic cables is that they must be kept dry, as moisture can increase fibre losses.

The transmitter and receiver are interfaced to the fibre-optic cable with various lens arrangements to focus the light on to the end of the fibre. The most important part of this interface is the connector, as it is here that optical losses can be at their greatest. Similarly, the jointing of fibre-optic cables, be it by connector or splicing, must be carefully made to minimize losses. A quality connection of a fibre-optic connector to the fibre-optic core is becoming easier to achieve, especially with the larger fibre sizes. The connection involves either carefully stripping back the plastic coatings and polishing the fibre ends to ensure that when the connectors are jointed the fibre ends will butt up precisely to each other, or by using a technique known as cleaving which is similar to the technique for cutting sheet glass. Alternatively, cores can be joined without using connectors by using a splice and fusion technique where the cores are aligned using a jig and then fused together using a small electric arc. Another method is to join the prepared cores using a special epoxy adhesive. It is particularly important to ensure that the connectors or core connections are resistant to the ingress of water as the presence of moisture can result in increased signal attenuation. Other than connectors and fibre losses, the transmitted signal can also be attenuated by small-radius cable bends; if bends are too tight, light will escape through the cladding causing appreciable attenuation losses.

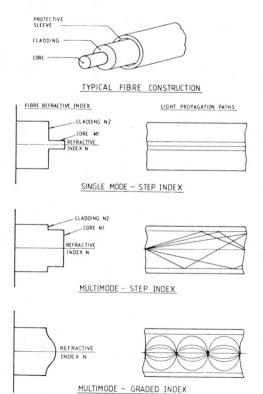

**Figure 16.6.** Optical fibre types and propagation paths.

Fibre-optic cables can be used for either analogue or digital transmission. However, they are more suited to digital transmission as more use can be made of the available bandwidth. A single fibre is capable of carrying a bandwidth which is measured in gigahertz. There is therefore a potential for transmitting a number of high-data-rate channels along a single fibre using a wavelength multiplexing technique. Alternatively, because of the wide bandwidth available, a single wavelength is likely to provide communication at a very high transmission rate, and would be ideally suited to time division multiplexing.

The use of fibre optics as a transmission medium in the process-control environment is certainly an attractive proposition. Although there are many claims about the versatility of fibre optics, the majority of potential users are restricted to using fibre optics in conjunction with standard electrical interfaces. There are now a number of systems available which interface the standard RS232 signal to fibre-optic cables. However, a cost advantage is only realized when a number of channels are multiplexed into one fibre-optic cable. If electrical noise immunity needs to be guaranteed then the additional cost may well be regarded as worthwhile. The real benefit of fibre optics will start to be realized when computer and peripheral manufacturers provide optical interfaces as a standard interface. This is unlikely to occur widely for quite some time as there are currently few agreed standards for connector types or fibre sizes. However, individual companies will probably introduce interfaces to enable connection of their own equipment.

## 16.6. Networks

So far, transmission and communication techniques for sending data from one point to another have been explained. In real life the requirements of communication systems are not always so simple. Some systems may require that the same data is communicated to a number of devices which are geographically separate; other systems require that a number of devices intercommunicate with each other and yet others may require to communicate in a selection of ways with a number of devices.

The obvious solution of connecting all instruments or computer systems to all other systems would, of course, result in prohibitive connection costs and would probably require additional equipment to permit large numbers of transmissions to interface with each device. As a result of the need to communicate in differing ways, 'networks' have been developed which utilize the transmission techniques that have been discussed and provide a number of standard solutions.

Networks are not only intended for use with complicated multicomputer systems, but can also be effectively used for communication between just two devices. The advantage of using a network is that further devices can be added into the system with minimal reconfiguration.

There are four basic network con-

figurations (Figure 16.7): star, tree or hierarchical, highway and ring.

The star network is probably the most common. It is also the most expensive to configure, as each device requires a separate cable. The star configuration is also probably the most likely system to fail completely, as the failure of the central device immobilizes the total system. However, the star configuration does mean that each of the peripheral systems has direct access to the central supervising device, but they do not have access to other peripheral systems other than via the central device. The tree or hierarchical network is really a variation on the star network theme. This network has the advantage of being less costly to cable, and also that a failure of the supervisory unit does not render the system totally inoperable. The tree or hierarchical network is the architecture that traditional distributed control systems have been modelled around.

The highway network is one of the more recent configurations. The com-

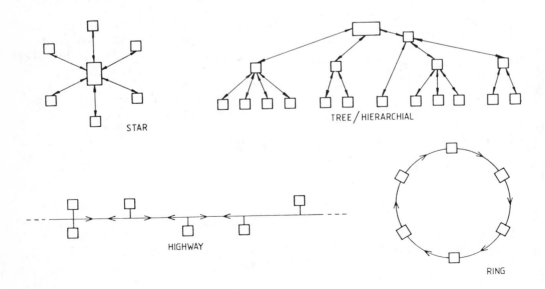

**Figure 16.7.** Network configurations.

municating devices all share a common highway, which means that only one interconnecting cable is required; therefore, a cost advantage can be achieved. Another advantage of this configuration is that any device can communicate with any other device connected to the highway, thus producing a very flexible system. Because there is the possibility that more than one device will want to transmit at any time, a controller is required to either allocate time slots to each device or referee the contention on a priority basis. There are two methods of controlling access to the highway, active and passive. An active highway is one where a separate dedicated device is used to control highway access. It has the distinct disadvantage that if the highway controller should fail, the highway becomes inoperable. A passive bus does not require a separate bus controller as each device connected to the highway takes it in turn to control the highway accessing. The device which has mastership of the highway can control access to the highway either on a deterministic (time-sharing) basis or by a statistical technique.

With a deterministic technique the highway mastership is passed from one device to another. Each device will have a set time to transmit to any other device. This technique ensures that all devices are serviced within a certain time period, but it can be wasteful if one device has a large quantity of information to transmit, as the transmission will have to be interrupted while other devices on the highway are serviced. The main advantage of this type of system is that there is a guaranteed maximum time for all devices to be serviced. In process control this is usually a requirement as the occurrence of an alarm in the field needs to be detected and acted upon within a certain time.

Where a guaranteed response time is not required, a statistical type of highway control provides a much more efficient means of controlling highway access. The statistical highway does not use a single highway controller. Instead, each device connected to the highway monitors the state of the highway and when a device requires to transmit, providing the highway is not in use, that device will immediately transmit. If two or more devices should transmit simultaneously a collision will occur and transmission will be aborted. Each device will then operate a contention algorithm which will enable one of the devices to transmit before the others. With this type of highway, devices will only access the highway when they need to; it also means that a device with considerable data to transmit can do so without interruption. This statistical method of access is known as carrier sense multiple access with collision detection (CSMA/CD).

A statistical highway network has the further advantage that the physical failure of the interconnecting cable is the only reason that will cause a failure of the network. The reliability of the network can be further increased if a redundant cable is added to the highway.

The ring-type network is a technique in which data packets are passed around a loop from one device to another in the same way that a relay baton is passed from one runner to another. A fixed number of packets which continuously circulate around the ring are examined in turn by each of the devices connected to it. When a device wishes to transmit data, it fills up the next empty packet and returns it to the ring. Each of the other devices will receive, interrogate and retransmit the data packets. When a packet arrives at its destination the device will read in the data and then retransmit the packet but add a marker to indicate that the data has been received intact. A different marker is added if the data is not received error free. The packet then continues around the ring until it reaches the original transmitting device where it will be either removed, if the data was correctly received, or retransmitted if the error marker has been set. If one of the devices in the ring should fail, a watchdog will automatically switch in a bypass facility.

The four basic network configurations can be considered as model systems. Manufacturers will base their designs around these models, but will also form hybrids and extend the models to match their needs.

The equipment required to build the networks discussed will vary depending upon the network adopted. Similarly, the communication firmware/software will vary from simple transmission acknowledgement routines to the more complex error-checking and addressing protocols. The more complex protocols now incorporate error-checking codes based on polynomial cyclic redundancy checks (CRC). Common CRCs include the Bose–Chauduri–Hocquenghen (BCH) code and the high-level data-link control (HDLC) code.

From the above it is evident that there are a great many combinations of highway configurations and associated protocols. As a first step towards standardization the ISO have developed a seven-tier reference model (Table 16.2). The aim of this model is to provide a framework around which network designers will build their systems. It would seem that this aim has a chance of being realized as already companies are describing how their systems comply.

With all of the networks described the most important factor is the ability to be able to interconnect devices in a network with the most suitable transmission medium. There is no reason why similar networks should not use differing transmission techniques. The choice must reflect the requirements of a network. Within a single network it may be that a combination of twisted-pair, coaxial and fibre-optic cables is the most effective solution. For some networks or inter-network communication these transmission media might not provide a total solution. Alternative techniques like some of the more 'pure' communication techniques of using line of sight laser or microwave links may also be required.

Of the networks discussed in this section, the ring and highways networks are now becoming very attractive, both in terms of cost and flexibility. It is perhaps not surprising to find that these two networks have been adopted by the commercial users for office communication. In this

**Table 16.2: ISO reference model**

| Layer | Class | Functions | |
|-------|-------|-----------|---|
| 7 | Application | End user<br>Data entry | System user |
| 6 | Presentation | Data formatting | |
| 5 | Session | Start, stop and maintain network connections between applications packetisation | |
| 4 | Transport | End to end data assurance<br>Preserve correct sequences<br>'Circuit' maintenance | |
| 3 | Network | Node to node routing | Network |
| 2 | Data link | Framing<br>Frame flow control<br>Error control | |
| 1 | Physical | Mechanical + electrical | |

field these networks originated from the Cambridge ring, which was developed at Cambridge University, and the Ethernet Highway, developed from a system known as Aloha Net used for computer communication in the Hawaiian Islands. These are now becoming office network standards such that many peripheral and systems manufacturers are supporting these networks which should allow the user to configure a system with equipment of his choice rather than being confined to one manufacturer's equipment. The process control industry has recognized for a number of years that there is a need to standardize on a communication technique, as a result the International Electrotechnical Commission (IEC) have been coordinating an attempt to develop a process control data highway known as Proway.

Proway will enable devices of differing manufacturers to communicate with each other using a standard interface and protocol which will require no other complicated interfacing systems. By developing this standard data highway, the doors will then be open for the smaller specialist suppliers to provide distributed control subsystems which can be integrated into a total system. Therefore, at long last, it may be possible easily to interconnect the instrumentation and systems that the customer wants without being confined to the components offered by one of the major suppliers. There is also an added cost benefit in that a single cable could satisfy all of the communication needs for a multistation distributed control system.

Proway being a process control data highway is quite different from any of the telecommunication type transmissions in that it is critical that data are received uncorrupted. A single transmission error which passes undetected could result in costly downtime and could affect operational safety. Therefore Proway incorporates sophisticated error and integrity checking protocols to ensure reliable communication.

The highway transmission line can be of twisted-pair, coaxial or even fibre-optic construction. The distributed devices, including multi-loop controllers, computer systems and any other instrument system or single instrument will communicate with each other via the highway provided each is fitted with the Proway interface. Although the highway will provide a means of intercommunication between any instrument or system connected to the highway, it is not intended that the highway will be used for mass memory transfer but will be confined to the passing of control information between devices. The devices that are connected to the highway will need to be intelligent in that they must be able to respond to requests for data from other devices and be capable of demanding data from the other devices. The Proway highway will not have a dedicated bus access controller to control which devices communicate with each other as the passive deterministic principle is used. The various devices connected to the bus take turns at controlling and using the highway for a limited period; this ensures that alarms trip and emergency conditions are detected and serviced within a certain time.

A number of the instrument companies have already developed distribution control system highways which are not dissimilar to the proposed Proway highway. These include the German PDV bus as used by Siemans and three other German suppliers, the Brown Boveri Partner bus, Honeywell's TDC Data Highway and Foxboro's Foxnet. The real benefits of the Proway highway will only be realized after its publication and the process-control equipment industry as a whole develop systems which will directly connect to the highway.

## 16.7. Conclusion

A wide variety of transmission and communication techniques is available to the process-control engineer. The standard analogue pneumatic and electrical techniques will serve adequately many process-control systems. However, changes are in sight. The next transmission revolution is likely to be as a result of either the development of optical measurement techniques for transmitters and the subsequent use of optical fibres for interconnection, or the application of the Proway system enabling instrumentation to communicate along a single highway.

Parallel transmission is likely to be restricted to the laboratory or for short intercomputer links where a very high data rate is required.

The various serial transmission techniques provide communication at a selection of data rates over various distances. In the industrial environment the greatest problem to all electrical digital transmission techniques is the effect of electrical interference. Until transmission techniques which are immune to this interference are developed, e.g. fibre optics, the engineer must painstakingly consider each transmission technique, transmission medium, the routing of cables and the application of error checking and correcting protocols, to ensure that the transmitted signal will be received uncorrupted and at the required data rate.

For the majority of process plants the

cable-linking systems currently available will satisfy most of the communication requirements but there is no doubt that in the future communications of all types will find their way into process control systems. As a result it will be necessary for instrument and process control engineers to understand the main principles of communications to ensure integration into process control systems and into the industrial environment.

## References

Bass C. (1981) Local area networks — a merger of computer and communication technologies. *Microprocessors Microsyst.* **5**(5), 187–192.

Cook B. E. (1972) Power supplies for process transmitters. *Meas. Control* **5**(3), 373–376.

Cook B. E. (1979) Electronic noise and instrumentation. *Control* **12**(8) 326–335.

French R. (1981) Telecommunications for computers. *Internat. Syst.*, September, 31–37.

Goldberger A. (1981) A designer's review of data communications. *Comput. Design*, May, 103–112.

Hanlon P. and Weston R. (1981) A distributed industry. *Internat. Syst.*, September, 22–24.

Midwinter J. E. (1978) Optical communication today and tomorrow *Electron. Power* **24**(6), 442–447.

Wilson H. S. (1978) Future trends in instrumentation — where are we going? *Meas. Control* **11**(6), 202–206.

## Appendix   Glossary of terms

*Amplitude modulation (AM)*
Data is impressed on a carrier by varying its amplitude.

*Asynchronous*
A mode of transmission in which the completion of one operation initiates another.

*Band*
A narrow spectrum of frequencies.

*Bandwidth*
The range of frequencies that can be transmitted, carried or received.

*Baton passing*
A technique of passing data around a network. [ . . . ]

*Binary*
Two state.

*Bit*
The element of digital data represented by a logic '0' or logic '1'.

*Bus*
Communication path which is shared by a number of devices.

*Byte*
A sequence of eight bits.

*Carrier*
A transmission signal which is modulated by the data signal.

*Channel*
A path along which information flows.

*Digital transmission*
A transmission technique where the data signal is transmitted using a number of discrete levels.

*Error*
A condition which is at difference to the expected condition.

*Frame*
A sequence of bits grouped together which represent a logical unit, sometimes referred to as a block.

*Highway*
Communication path which is shared by a number of devices.

*Mark*
The '1' state in a binary signal.

*Modem*
A device for modulating and demodulating a carrier wave with the data signal.

*Peripherals*
Devices which service a central system, eg printer.

*Port*
Device input and/or output connection point.

*Rise time*
The time for an electrical pulse to rise from one-tenth to nine-tenths of its final value.

*Space*
The '0' state in a binary signal.

*Synchronous*
A mode of transmission in which clock pulses control the timing of data transmission.

# Paper 17

# Measurement errors and instrument inaccuracies

## M. J. Cunningham

Any instrumentation system is concerned with the measurement of some physical parameter. No transducer, amplifier, display device or recording device is perfect; all introduce errors into the process of measurement. This paper discusses systematic and random errors, and introduces some of the mathematical concepts by means of which the likely effects of errors can be predicted.

The paper uses some simple statistical concepts including mean, standard deviation and normal distribution. Readers who are not familiar with these concepts may find the latter part of section 2 of the paper hard going. Subsequent sections of the paper can usefully be studied without a full understanding of the error statistics. (Eds.)

After a brief introduction to the activity of measurement, the meaning of a measurement error is discussed. The concept of systematic errors is described, examples given and comments made on the difficulties of ensuring that all significant systematic errors have been considered. After an introduction to random errors, the reasons are given why it is possible to state the range of values within which the mean value of a quantity subject to random variations will be from a small sample of measurements and to give a probability that the statement is true. The method of randomizing systematic errors is given. Various ways are described for stating the result of a measurement where both systematic and random errors are significant.

The statement of the performance of a measuring instrument is reviewed and several proposals to unify the method of statement by manufacturers described. The advantages and disadvantages of these proposals to manufacturers and users are discussed. Examples of the specification of instrument performances are given. The terms used are discussed and a method of obtaining the likely operating error stated.

## 17.1. Measurement

The physical world is a world of change and apparent complexity. Heraclitus, the 5th-century BC Greek philosopher, said (Wheelwright 1959) 'You cannot step twice into the same river, for other waters are continually flowing on'. It is therefore always a surprise that the behaviour of much of this physical world can be described by a very few simple formulations of law. Heraclitus also said (Wheelwright

Originally published in *J. Phys. E Scientific Instruments* Vol. 14, 1981.

1959) 'The hidden harmony is better than the obvious'.

Many of these formulations of law are written in the language of mathematics.

Galileo said (Drake 1957) that science, 'is written in this grand book, the universe, which stands continually open to our gaze. But the book cannot be understood unless one first learns to comprehend the language and read the letters in which it is composed. It is written in the language of mathematics and its characters are triangles, circles and other geometric figures without which it is humanly impossible to understand a single word of it; without these, one wanders about in a dark labyrinth'.

The simplicity of the mathematical-type formulations of law is astonishing. For example, according to Coulomb's law, the force of attraction between two electric charges is inversely proportional to the square of their separation. It might be that the exponent is not exactly 2 but just nearly 2. It can be called $(2 + q)$ where $q$ is a small positive or negative number. Recent measurements (Williams et al. 1971) show that $q$ is less than $1 \times 10^{-15}$. The 'inverse square law' does appear to be an inverse square law.

Once such simple formulations are available, two activities are possible. The first is further investigation of the nature of the physical world, an activity called science. The second is that of effective application and design, an activity called engineering. For example, large parts of the electrical and electronics industries are based on the observations and formulations of Faraday and their subsequent extension and rendering in mathematical form by Maxwell.

Both of these activities, science and engineering, rely on measurement. Measurement can be defined as the assignment of a number to some quality of an object in terms of an arbitrarily agreed unit value in such a way that the number faithfully represents the properties of that quality of the object.

Measurement thus is a means of gaining access to the world of the formulation of law and the process of reasoning which accompanies it. Feynman (1965) says, in relation to the mathematical statement of law, 'mathematics is a language plus reasoning'. Measurements are therefore made in terms of the framework of the laws and made possible by the realization and maintenance of a satisfactory system of units. The satisfactory establishment of the unit is in itself an interesting, if largely unnoticed, activity (Vigoureux 1971). None of the subsequent activities of measurement can take place without the existence of the one.

The meaning of the definition of measurement is clear enough for a quality directly observable by the senses. For example a metre rule can be used to ascribe a number to the quality length of an object in terms of the agreed unit, the metre.

What about qualities not observed by the senses directly, or only vaguely sensed, such as electric current? A device has to be found to make the quality observable. Such a device is called a measuring instrument. Although the human body is equipped with five powerful senses, sight is the sense used almost exclusively for measurement.

One of the main reasons for this is that the visual length is one of the very few stimuli (Schiffman 1976) that has a linear relation between perceived stimulus and the magnitude of the stimulus itself. This means that half and double the length of a line can be judged by eye in a very satisfactory manner. The same is not true of most other stimuli.

Examples of instruments which transform the quantity to be measured into length are very numerous and would include mercury-in-glass thermometers, barometers, deflection voltmeters, ammeters, cathode ray oscilloscopes, analogue clocks, pressure gauges and flow meters. An enormous number of measurements can be illustrated by Figure 17.1.

Another vast range of measurements is made by transforming the quantity to be measured into discrete things and counting these, usually over a specified period of time. Examples of this are digital voltmeters, digital multimeters and quartz thermometers. These can be represented diagrammatically by Figure 17.2.

Other implementations of measurement can be represented in a similar way.

**Figure 17.1.** Measurement by conversion to length.

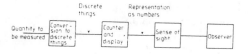

**Figure 17.2.** Measurement by conversion to counting.

## 17.2. Measurement errors

### 17.2.1. Introduction

Since the quantity to be measured, the instrument by which the quantity is brought to the awareness of the observer via the senses and the state of the observer are in continual change, what is the significance of the result of a measurement? From one point of view every measurement is right since it gives rise to the number ascribed to that quantity by that instrument by that observer at that instant. This approach is not fruitful, however, because the aim of the measurement is to relate it to other measurements by other instruments and other observers at other times.

An International Organisation of Legal Metrology definition of the error of a measurement (OIML 1968) is the discrepancy between the result of the measurement and the true value of the quantity measured.

Since, to modify the first quotation of Heraclitus, 'You cannot make the same measurement twice', even with a perfect instrument and observer, the result of a measurement would never be exactly the same twice, except for the trivial example of counting objects. The instrument and observer also introduce errors.

The result of the measurement is what is available, but the 'true value' of the quantity, free from any error, is what should be communicated. The 'true value' is unknowable but it is necessary to state from the result of measurement within what range of values the 'true value' is likely to occur and the confidence that can be placed on the validity of this statement. The procedure is cumbersome but inevitable. The statement of the result of a measurement without any indication of the range of values within which, with a certain confidence, the 'true value' can be assumed to be is almost worthless.

Initially the way to state the results of a measurement will be discussed assuming errors owing to perturbations of the quantity to be measured, the instrument and the observer. In section 17.4 the special case of specifying the performance of an instrument will be considered.

### 17.2.2. Systematic errors

It is convenient initially to divide errors into the two classes of systematic and random, although, as will be shown later, the distinction is not always as distinct as might at first appear.

OIML define systematic error as (OIML 1968) 'an error which, in the course of a number of measurements, made under the same conditions, of the same value of a given quantity, and which either remains constant in absolute value and sign, or varies according to a definite law when the conditions change'. The examples given for a systematic error are as follows:

The error which results from a weighing by means of a weight whose mass is taken to be equal to its nominal mass of 1 kg whereas its conventional true mass is 1.010 kg.

The error which results from using, at an ambient temperature of 20 °C, a rule gauged at $0\,^{\circ}C$ without introducing a suitable correction.

The error which results from the use of a thermoelectric thermometer whose circuit suffers from parasitic thermoelectric effects.

There will be in general a large number of effects which can change the result of the measurement in this way. A great deal of intelligence and experience is required to decide which are the significant effects. Some components of the systematic error are revealed by performing experiments in which the conditions of the measurement are changed in an orderly way and noting the changes in the result of the measurement. It is, however, always possible that a significant systematic error will remain unsuspected. This seems to be common and frequently the results of a measurement are stated with extreme optimism.

Since a systematic error is simply an error much care must be taken in designing the measurement to reduce the size of systematic errors as much as possible and then to try to discover the size of the remaining significant systematic errors. It becomes increasingly difficult, as the tolerable size of the error becomes smaller, to say with complete confidence that there are no systematic errors of signficant size that have not been considered.

Sometimes it is possible to measure the quantity using quite a different principle, which will have mostly different systematic errors, to reveal an unsuspected systematic error in the original measurement or to increase the confidence in the statement of the maximum systematic error.

As an example of this at the highest level, the various methods of realizing the ampere are at present under the suspicion of suffering from unknown systematic errors (Taylor 1976). In practice, the ampere is maintained at standard laboratories such as the National Physical Laboratory in the United Kingdom by means of stable voltage sources and standard resistors. The ratio, $K$, between this maintained unit of

current and the SI ampere can then be established in one of two basic ways: either directly, by using some form of current–force balance, or indirectly by the appropriate combination of the values of various fundamental constants.

There are three main indirect methods:

the gyromagnetic ratio of the proton by both low and high field methods
the faraday
the Avogadro constant

The values of $K$ found by these three methods not only differ among themselves, but also disagree with those given by the direct current balance method to such a degree that unsuspected systematic errors may be present (Petley 1979).

The General Conference on Weights and Measures (CGPM 1979) recently recommended because of these discrepancies 'the continuation and intensifying of research on both the direct realization of electrical units and the indirect realization through the determination of physical constants'.

Systematic errors can be significant at any level of measurement and the user of measuring instruments, particularly digital ones, needs to develop the frame of mind whereby he does not easily believe the numbers presented to him to be acceptably close to the true value of the quantity he would like to measure.

### 17.2.3. Random errors

If a quantity is measured with an instrument with good discrimination, sometimes called resolution, it is often found that repeated readings are not identical, as would be the case with only constant systematic errors. In such a case it is interesting to consider how best to convey the result of the measurement.

In order to indicate the result of a series of measurements of the 'same' quantity to others, all the results can be presented in tabular form. Rather more easy for the mind to work with is a graphical presentation. Vigoureux (1966) has suggested a simple method in which the results are entered on a scale as they are taken. Whenever a value is repeated, the point is placed on the line above the previous line of that value. This is illustrated in Figure 17.3, and this method was used by Vigoureux to present the results of his determination of the ampere (Vigoureux 1965). All the results and their scatter are clearly presented. This display is really a particular form of a graphical presentation of scattered values called a histogram.

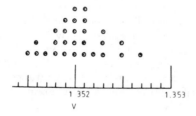

**Figure 17.3.** Vigoureux's method for displaying results.

### 17.2.4. The histogram

When large numbers of measurements are made of the 'same' quantity, it is convenient to represent the results as a frequency distribution, a particularly convenient form of which is the histogram shown in Figure 17.4. The area of each rectangle is proportional to the number of results within the limits given by the edges of that rectangle. If the widths of the rectangles are made equal, that is difference between the limits of the classes are equal, then the heights of the rectangles are proportional to the frequency of the results of measurements between the limits shown.

### 17.2.5. The limiting mean of the results of measurements

Although the histogram displays much of the information of the results of the measurements, it is often useful to describe the series of results by numbers. One such number is the mean of the series of results.

Dorsey (1944) said 'The mean of a

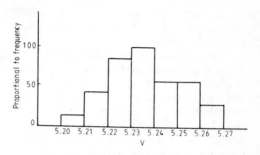

**Figure 17.4.** The histogram.

family of measurements — of a number of measurements for a given quantity carried out by the same apparatus, procedure and observer — approaches a definite value as the number of measurements is increased. Otherwise, they could not properly be called measurements of a given quantity. In the theory of errors, this limiting mean is frequently called the true value, although it bears no necessary relation to the true quaesitum, to the actual value of the quantity that the observer desires to measure. This has often confused the unwary. Let us call it the limiting mean'. Eisenhart (1963) has called this the 'postulate of measurement'.

### 17.2.6. The standard deviation

Following from this postulate it is possible to obtain a number which gives an indication of the scatter of the results of the measurements. There are several ways of doing this. The most commonly used is called the standard deviation, which is defined as the value approached by the square root of the average of the sum of the squares of the deviations of individual measurements from the limiting mean as the number of measurements is indefinitely increased.

In symbols

$$\left( \frac{1}{n} \sum_{i=1}^{n} (x_i - m)^2 \right)^{1/2} \to \sigma$$

as

$$n \to \infty$$

where $\sigma$ is the standard deviation, $m$ the limiting mean and $x_i$ the $i$th measurement. Variance is defined to be the square of the standard deviation $\sigma^2$.

This definition of standard deviation as the root mean square deviation from the limiting mean is an acceptable way of specifying the scatter of the results of measurement, although the word 'standard' does seem to give this definition perhaps an unwarranted appearance of fundamentality. The term probable error, defined to be the deviation from the limiting mean which is just as likely to be exceeded or not, although acceptable in principle, has an unfortunate name. Moroney (1951) gives the comment that 'it is neither an error nor probable'. This term will not be used in the remainder of this article.

The limiting mean and the standard deviation can be given for any very large number of measurements of the 'same' quantity.

### 17.2.7. Samples of results of measurements
Of course, usually only a few repeated measurements of the 'same' quantity are made and so it might appear that the concepts of limiting mean and standard deviation are of little practical use. Perhaps surprisingly this is not the case.

A theorem exists called the central limit theorem. Mood (1950) describes this theorem as the 'most important theorem in statistics from both the theoretical and applied points of view'.

The theorem states:

'If a population has a finite variance $\sigma^2$ and mean $m$, then the distribution of the sample mean approaches the normal distribution with variance $\sigma^2/n$ and mean $m$ as the sample size $n$ increases.'

Nothing is said in the theorem about the form of the population distribution function, which is just as well since it cannot be known from a few measurements. The theorem does state that the distribution of sample means of independent

measurements will be approximately normal about the population limiting mean with a standard deviation for this distribution of $\sigma/n^{1/2}$. $\sigma/n^{1/2}$ is called the standard error of the mean.

For a series of measurements with a more or less symmetrical histogram and small deviations from the mean compared with the magnitude of the mean, the distribution of sample means is very close to normal even for $n$ as small as 3 or 4 (Ku 1969). These conditions apply for a large number of measurements.

### 17.2.8. The normal distribution
The normal distribution referred to in the central limit theorem is well known in statistics. It was first given by De Moivre (1733), while he was living in England as a consultant on gaming and insurance. The distribution is illustrated in Figure 17.5.

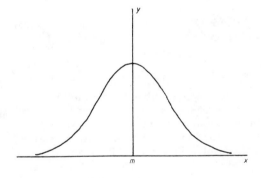

**Figure 17.5.** The normal distribution.

Mathematically the normal distribution is given by

$$y = \frac{1}{\sigma(2\pi)^{1/2}} \exp[-(x-m)^2/2\sigma^2].$$

The normal distribution is one for which the relative frequency of the data having a value between $x$ and $x + \delta x$ is $y\,\delta x$, where $y$ is given by this equation. $y$ is therefore called the probability density of the distribution.

The area under the curve between two $x$ values is therefore the probability that the quantity will occur between these two values. The total area under the normal distribution is unity.

### 17.2.9. Statement of the population mean from a sample of measurements

The central limit theorem says that to a good approximation, the means of fairly small samples will lie on a normal distribution with mean equal to the population mean and standard deviation equal to $\sigma/n^{1/2}$ where $\sigma$ is the population standard deviation.

The central limit theorem is powerful because it enables statements to be made on the probability that the population mean is less than a specified deviation from the sample mean.

Table 17.1 gives the probability that a normally distributed quantity will be within $r$ standard deviations of the mean.

**Table 17.1.**

| $r$ | % probability |
|--------|--------------|
| 0.6745 | 50 |
| 1 | 68.27 |
| 2 | 95.43 |
| 3 | 99.73 |
| $\infty$ | 100 |

So if the sample mean is $m_s$ and the population mean, the quantity we are seeking, is $m$, then in more than 95% of samples, the population mean will be within two standard deviations of the sample distribution.
So

$$- 2\sigma/n^{1/2} < m_s - m < + 2\sigma/n^{1/2}$$

or

$$m_s - 2\sigma/n^{1/2} < m < m_s + 2\sigma/n^{1/2}$$

will be correct for over 95% of a large number of sets of samples.

This statement would allow statements to be made, from a single small sample of measurements, on the probability of the population mean being more than say $2\sigma/n^{1/2}$ away from the sample mean. Unfortunately, it is necessary to know $\sigma$, which is the standard deviation of the whole population and is found by a very large number of measurements!

All is not lost, however, because the population standard deviation $\sigma$ can be estimated from the sample standard deviation, $\sigma_s$. Of course, $\sigma_s$ underestimates the population standard deviation and so the probability statement has to be rewritten that

$$m_s - t\sigma_s/n^{1/2} < m < m_s + t\sigma_s/n^{1/2}$$

is true for a certain percentage of samples. The multiplier $t$ depends on $n$ and the probability level chosen and is called Student's $t$. (The distribution of $t$ was obtained mathematically by Gosset (1908) and published under his pen name of 'Student'.)

Values of $t$ have been evaluated for the normal distribution and are given in Table 17.2.

To find $t$, it is necessary to know the number of degrees of freedom of the calculated sample standard deviation $\sigma_s$. Since $m_s$ is found from the same $n$ results, the last value is fixed by $m_s$ and the other $(n-1)$ values. The degree of freedom is therefore $n-1$. Strictly the estimated standard deviation is given by

$$\left(\frac{1}{n-1}\sum_{i=1}^{n}(x_i - m)^2\right)^{1/2}.$$

However, except for very small $n$, the change is small (Müller 1979).

For example, if a series of five measurements of current give a mean of 1.123 A and a standard deviation of 0.005 A, then for a 99% probability level and degree of freedom 4, the table shows that $t = 4.6$.

The statement can therefore be made that there is 99% probability that the population mean will be in the range

$$1.123 - \frac{4.6 \times 0.005}{5^{1/2}} < m < 1.123$$

$$+ \frac{4.6 \times 0.005}{5^{1/2}}$$

or

$$1.113 < m < 1.133.$$

The ability to assign a probability that the population mean will be within a specified range of the mean of a fairly small sample is the basis of the application of statistics to measurement.

This above coverage of this very complicated subject is of necessity a simplified one. The standard text books on the subject, of which there are a vast number, give a more full treatment, for example Brownlee (1960), Mood (1950).

*17.2.10. Randomizing systematic uncertainties*
It is not always clear whether an influencing quantity produces systematic or random errors. For example, for a sample of

measurements taken close together in time, humidity might well produce a systematic error, whereas if readings are taken for the sample over a long time-scale humidity might well produce random errors.

This suggests one method that can be used to reduce the effect of or reveal systematic errors. By intelligent experimental technique it is possible to convert some systematic errors into random ones. At least their presence is then obvious!

## 17.3. The statement of errors

It will be clear that to be of maximum use, the statement of the result of a measurement or series of measurements should include a statement about the errors involved. For accurate measurements this should certainly be done very fully, however for routine low-accuracy measurements this would be unnecessary. Nevertheless every statement should include enough information, often in the form of a sentence, which makes clear the intention of the author. This is necessary owing to the lack of agreed procedure in the statement of errors.

**Table 17.2. Values of Student's *t***

| Degrees of freedom | $P = 68.3\%$ ($1\sigma$) | $P = 95\%$ | $P = 99\%$ | $P = 99.73\%$ ($3\sigma$) |
|---|---|---|---|---|
| 1 | 1.8 | 12.7 | 64 | 235 |
| 2 | 1.32 | 4.30 | 9.9 | 19.2 |
| 3 | 1.20 | 3.18 | 5.8 | 9.2 |
| 4 | 1.15 | 2.78 | 4.6 | 6.6 |
| 5 | 1.11 | 2.57 | 4.0 | 5.5 |
| 6 | 1.09 | 2.45 | 3.7 | 4.9 |
| 7 | 1.08 | 2.37 | 3.5 | 4.5 |
| 8 | 1.07 | 2.31 | 3.4 | 4.3 |
| 9 | 1.06 | 2.26 | 3.2 | 4.1 |
| 10 | 1.05 | 2.23 | 3.2 | 4.0 |
| 15 | 1.03 | 2.13 | 3.0 | 3.6 |
| 20 | 1.03 | 2.09 | 2.8 | 3.4 |
| 30 | 1.02 | 2.04 | 2.8 | 3.3 |
| 50 | 1.01 | 2.01 | 2.7 | 3.2 |
| 100 | 1.00 | 1.98 | 2.6 | 3.1 |
| ∞ | 1.00 | 1.96 | 2.58 | 3.0 |

It is always possible to state the systematic and random errors separately. Campion et al. (1973) have advocated this approach.

They recommend that the components of the random uncertainty should be listed in sufficient detail to make it clear whether they would remain constant if the experiment were repeated. The components of the systematic uncertainty should be listed, expressed as the estimated maximum value of that uncertainty. The method used to combine systematic uncertainties should be made clear. The authors do not recommend the combination of random and systematic uncertainties.

On the other hand Müller (1979) recommends random and systematic errors should have applied to them the general propagation law of errors. Each error is represented by the measured standard deviation, or the best estimate available, and these are combined using this law. Müller suggests this leads to a 'natural and unambiguous evaluation of the overall uncertainty to be associated with an experimentally determined quantity'.

This method, the author states, should be applicable to any 'level of metrology', thus unifying the different procedures hitherto used.

In another recent publication on the subject by Hayward (1977) there are recommendations that the total random uncertainty and the total systematic uncertainty should always be stated separately. If a value for the overall uncertainty is also required, then the total random uncertainty should be added to the total systematic uncertainty in quadrature.

Since there cannot be a theoretically justifiable way of combining random and systematic errors, it is not surprising that several methods have been proposed (Wagner 1980).

The International Bureau of Weights and Measures (BIPM) is actively seeking to find international consensus on the statement of the results of measurement (Giacomo 1979, BIPM 1980).

Even if a somewhat arbitrary method is used to combine random and systematic errors, most would agree that the components should also be given in full and the method of their combination stated.

The minimum acceptable statement of the result of a measurement would be:

'The current was 1.06 A. All errors are negligible to this degree of rounding.'

### 17.3.1. *Rounding the statement of the result of a measurement*

The number of decimal places to which the result of a measurement is stated in the absence of any other statement implies the level of errors expected. With decimalization reaching consumer products in the United Kingdom the examples of poor rounding are numerous. For example, the information on a packet of seeds states that the plants will grow to 91.5 cm high. This implies the height will be between 91.45 cm and 91.55 cm high. Before decimalization, the plants were stated to grow to 3 ft high, implying the height will be between 2 ft 6 in and 3 ft 6 in. It has to be assumed that a new super-constant-height plant has not been developed!

## 17.4. The specification of the errors of an instrument

When a potential user would like to select an instrument from those available from the manufacturers it is necessary to assess the errors introduced by the instrument in the particular measurements concerned. Since the conditions under which the instrument is used will vary from user to user, the manufacturer faces an almost impossible task in making general statements about the error introduced by the instrument. There are clearly commercial pressures to state the error that might be expected in favourable circumstances. It is therefore a fairly lengthy task for the users to deduce from this the error to be expected in their own circumstances. Since manufacturers do not all state the errors and circumstances in the same way, the comparison of the expected performance of instruments from several manufacturers can be no light task.

The International Electrotechnical Commission (IEC) has over the last 10 years attempted several times to suggest ways in which the errors of an instrument could be stated. These will now be briefly described to illustrate the range of approaches that are possible in this difficult area.

IEC Publication 359 (1971) suggested a method for describing the performance of electronic measuring equipment. It was hoped that the adoption of this method by the industry would lead to easier and better comparison of equipment specifications by the user.

The publication specifies a large number of influence quantities. An influence quantity is defined to be any quantity, generally external to the apparatus, which may affect the performance of the apparatus. Influence quantities include climatic conditions such as ambient temperature, relative humidity of the air, velocity of the ambient air and so on, mechanical conditions such as operating position, ventilation, vibration, mechanical shock and so on, supply conditions such as mains supply voltage, frequency and so on and fields and radiations.

The publication also states ranges of many of these influence quantities. The manufacturer is then asked to state the error given by the most adverse combination of all the influence quantities.

This 'worst case' approach has much to commend it from the user's point of view. However, since many instruments are significantly affected in performance by several influence quantities, the testing needed to verify the stated error is considerable. Also, a manufacturer quoting the inevitably large error would be at a commercial disadvantage.

By constrast IEC Publication 51 (1973), for example, suggests another approach. In this, errors, called intrinsic errors, are determined under a narrow range of influence conditions called reference conditions. These errors can be easily checked by most users. Some indication of the actual error to be expected in use is gained from stated 'variations'. Variation is the change in error produced by each influence quantity separately when changed from its reference value to an extreme of a stated operating range, while all the other influence quantities are kept at their reference values. There is therefore no indication of what might happen in actual use when several of the influence quantities would not be at their reference values.

The IEC has been considering the revision of IEC Publication 359 with the aim of combining the advantages and avoiding some of the disadvantages of the above two approaches. As part of these considerations the IEC published a Draft Standard (1980). In this there is an attempt to give an indication of the actual error likely to occur while under operating conditions while on the other hand allowing the relatively easy checking of the stated errors.

In order to simplify, several rules were proposed which include the following. For most applications only three influence quantities cause significant error. The effects of these three influence quantities are independent. The effect on performance by an influence quantity is linear or a second-order effect. A probable operating error is stated from the square root of the sum of the squares of the intrinsic error and the variations produced by the influence quantities.

It must be stressed that these last suggestions are only at a draft stage. A meeting of the IEC technical committee TC66 decided in September 1980 in the light of comments on the whole draft to rewrite it instead as a general document. This would primarily define regulations for the uniform specification of errors of all apparatus covered by the standard.

Nevertheless, the general thinking behind these latest proposals does offer some basis for a method of stating the performance of instruments acceptable to user and manufacturer alike. Any move towards the adoption of a common method of stating the performance of instruments would be of assistance to the user.

Hayward (1977) has given an interesting proposed standard procedure for measuring the repeatability and estimating the accuracy of industrial measuring instruments.

## 17.5. Some examples of instrument inaccuracy

There are a number of terms used to specify the performance of an instrument which are confusing to the mind. An example of such a term is the 'instrument accuracy'. It seems reasonable to understand by this term the accuracy of the instrument. This turns out not to be the case. The quantity specified under this heading is part of the inaccuracy of the instrument and usually relates closely to the intrinsic error discussed in the previous section.

Examples of instrument specifications will now be given and the method for arriving at the likely operating error discussed.

### 17.5.1. A deflection multimeter
The following specification, although fictitious, is typical of this sort of instrument:

accuracy (DC voltage and current)
    ±2% full scale deflection
    (AC voltage and current)
    (50 Hz) ±3% full scale
    deflection

sensitivity DC voltage 20 000 Ω/V
    AC voltage 2 000 Ω/V

frequency response <3% variation
    from reading at
    50 Hz
    (10 Hz–10 kHz)

temperature effect 0.2% °C.

The loading effect of the voltmeter, often referred to by the ambiguous name sensitivity, can give rise to a very large systematic error. Usually the desired result of the measurement is the value of the measured quantity without the measuring instrument present. For a voltmeter, the presence of the instrument can significantly change the potential difference to be measured. For example, consider the circuit shown in Figure 17.6. Without the voltmeter connected the potential difference between points $A$ and $B$ would be 5.00 V. When the specified voltmeter in its 5 V range is connected to points $A$ and $B$ a 100 kΩ resistance will be introduced between $A$ and $B$. The effective resistance between $A$ and $B$ would be 33 kΩ. The potential difference between $A$ and $B$ would therefore be 3.75 V. Even if the voltmeter had no other source of error, the percentage error of this measurement would be 25%. It is clear that this systematic error must be considered carefully whenever a measuring instrument is used. Since the voltmeter is the means of finding the value of the potential difference, obtaining the value in the absence of the voltmeter does raise some difficulties. Of course a higher-resistance voltmeter can always be used. If this is not available, a similar voltmeter can be connected to the same points as the original voltmeter. If the reading of the original voltmeter changes, then the original voltmeter must have significantly affected the potential difference to be measured.

The frequency of an AC voltage can be regarded as an influence quantity when the amplitude of the AC voltage is re-

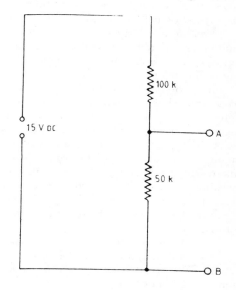

**Figure 17.6.** The loading effect.

quired. When being used in this way for measurements other than at 50 Hz, a systematic error of 3% has to be assumed. More information from the manufacturer would be useful since it could be argued that the measurement would never be exactly at 50 Hz and so the 3% systematic error must always be assumed.

Another systematic error occurs owing to temperature changes. The calibration temperature is not stated, but presumably could be found by contacting the manufacturer. Then a systematic error of 0.2% for every °C deviation from that temperature must be assumed.

The only remaining figure supplied is called the 'accuracy' of the instrument. In the terminology of the last section this is the intrinsic error. Although little of the random component of the result of a measurement is due to the instrument, such as there is, for example noise, will be included in this term. This term should also include the maximum error owing to all other influence quantities. That it does is often hard to believe. Such things as linearity, hysteresis, instrument attitude, electric and magnetic fields and temperature gradients in their worst combination are likely to contribute errors in excess of 2% of the full scale deflection. In the absence of further information, the user is forced to vary any unstated influence quantity suspected of introducing significant systematic errors in order to reveal the size of the error.

It is possible to combine the intrinsic error and all the variations produced by significant influence quantities. The method suggested in the last section to find an estimate for the operating error was by taking the square root of the sum of the squares of the intrinsic error and all the variations produced by influence quantities. Of course the error in a particular measurement can be larger than this estimate.

Clearly the instrument cannot be said to have an inaccuracy of 2% on DC ranges. The error in an operating condition can be much larger than this. The error of the instrument is not a constant amount. Users should ask themselves 'What is the error of the instrument now?'

### 17.5.2. A digital frequency counter

The following are the specifications of a fictitious, but typical, frequency counter:

| | |
|---|---|
| input impedance | 1 MΩ in parallel with 25 pF |
| temperature stability | ±4 parts in $10^6$ over 20 °C to 40 °C |
| ageing rate for crystal | ±2 parts in $10^6$ per month, 2 months after delivery. |

The loading effect, although not as serious as for the deflection multimeter, can still give rise to problems by changing the frequency to be measured on connecting the instrument, for example, by changing the characteristics of an oscillator. This effect is most likely to produce an error at high frequency since the input impedance of the instrument is much lower in this case than for low-frequency measurements.

Over the specified temperature range, the contribution to the error will be less than the stated amount. The statement is rather pessimistic for normal environments and a smaller error contribution could presumably be obtained for a small temperature range.

The ageing error gives some indication of the error from this source. With regular calibration, this error can be made very small.

The only other error is the normal ±1 error of a digital instrument. This can be made small by including a sufficiently large number of counts.

It has to be assumed that all other errors, such as random errors, are negligible. The comments on obtaining the operating error also apply to this instrument.

### 17.6. Conclusions

Except for the trivial case of counting objects, it is not possible to give exactly the value of a quantity from the result of a measurement of it. When the result of a measurement is stated the aim must be to communicate as much as possible, and this means a statement of the probability that the true value lies within a stated range. This statement would be very full for high-accuracy measurement but could be as simple as 'the errors are negligible to this degree of rounding'.

The statement of the performance of an instrument is a subject with many technical and commercial considerations. There are recent proposed procedures for specifying the performance of an instrument which appear to overcome some of the objections to previous proposals. Any move towards a common method of statement of instrument performance would be welcomed by the users.

## References

BIPM (1980) *Rapport BIPM 80/3.*

Brownlee K. A. (1960) *Statistical Theory and Methodology in Science and Engineering* (New York: Wiley).

Campion P. J., Burns J. E. and Williams A. (1973) *A Code of Practice for the Detailed Statement of Accuracy* (London: HMSO).

CGPM (1979) The Sixteenth Conference was reported by Giacomo P. (1980) News from the BIPM *Metrologia* **16**, 55–61.

De Moivre A. (1733) Approximatio ad Summam Terminorum Binomii $(a+b)^n$ in Serien Expansi *Miscellenea Analytica de Seriebus et Quadraturis* (Second Supplement).

Dorsey N. E. (1944) The velocity of light. *Trans. Am. Phil. Soc.* **34**, 1–110.

Drake S. (1957) *Discoveries and Opinions of Galileo* (New York: Doubleday).

Eisenhart C. (1963) Realistic evaluation of the precision and accuracy of instrument calibration systems. *J. Res. Nat. Bureau of Standards* **67C**, 161–87.

Feynman R. (1965) *The Character of Physical Law* (London: BBC).

Giacomo P. (1979) News from the BIPM *Metrologia* **15**, 51–4.

Gosset W. S. (1908) The probable error of a mean. *Biometrika* **6**, 1–25.

Hayward A. T. J. (1977) *Repeatability and Accuracy* (London: Mechanical Engineering Publications).

IEC (1971) *Publication 359 Expression of the Functional Performance of Electronic Measuring Equipment.* This document was the basis for the British Standards Institution publication BS 4889 (1973) British Standard method for specifying the performance of electronic measuring equipment.

IEC (1973) *Publication 51 Specification for Direct Acting Indicating Electrical Measuring Instruments and their Accessories.* This document was produced in identical form as the British Standard BS 89 (1977).

IEC Draft Standard (1980) *Draft Standard for the Expression of the Performance of Electrical/Electronic Equipment for Measurement, Control or Analysis.* This document was issued by the British Standards Institution as Draft for Public Comment 80/250 77 DC.

Ku H. H. (1969) Statistical concepts in metrology. *National Bureau of Standards Special Publication 300* **1**, 296–330.

Mood A. M. (1950) *Introduction to the Theory of Statistics* (New York: McGraw-Hill).

Moroney M. J. (1951) *Facts from Figures* (London: Penguin).

Müller J. W. (1979) Some second thoughts on error statements. *Nucl. Instrum. Meth.* **163**, 241–51.

OIML (1968) *Vocabulary of Legal Metrology.* The British Standards Institution issued an unofficial English translation in PD 6461.

Petley B. W. (1979) The ampere, the kilogram and the fundamental constants — is there a weighing problem? *NPL Report QU52.*

Schiffman H. R. (1976) *Sensation and Perception* (New York: Wiley).

Taylor B. N. (1976) Is the present realization of the ampere in error? *Metrologia* **12**, 81–3.

Vigoureux P. (1965) A determination of the ampere. *Metrologia* **1**, 3–7.

Vigoureux P. (1966) Errors of observation and systematic errors. *Contemporary Phys.* **7**, 350–7.

Vigoureux P. (1971) *Units and Standards for Electromagnetism* (London: Wykeham Publications).

Wagner S. R. (1980) Combination of systematic and random uncertainties. *Conf. Precision Electromagnetic Measurement* (New York: IEEE).

Wheelwright P. E. (1959) *Heraclitus* (Princeton: Princeton University Press).

Williams E. R., Faller J. E. and Hill H. A. (1971) New experimental test of Coulomb's law; a laboratory upper limit on the photon rest mass. *Phys. Rev. Lett.* **26**, 721–4.

# Paper 18

# How to calibrate flowmeters and velocity meters

*A. T. J. Hayward*

All instrumentation systems need calibrating to allow removal or correction of systematic errors introduced into the measurement by transducer and other systems component inaccuracies. In some cases, where system components are subject to drift, or where extreme accuracy is required, a calibration procedure may be used many times a day. In other cases, where long-term stability of components is good, calibration may only occur once a week or once a month. Without such calibration, credibility in the results of the measurement cannot be established.

This paper is included as an example of the calibration process, in this case the calibration of flow meters, and illustrates clearly that calibration is not a trivial task, and requires measurements to be made under closely controlled laboratory conditions. A full calibration facility allows comparisons to be made with more and more accurate standards, eventually with standards maintained by a government agency such as the National Physical Laboratory. (Eds.)

---

*CALIBRATE, verb trans: Find calibre of*
*CALIBRE, noun:*        *Internal diameter of gun*

> Concise Oxford Dictionary
> (1946 edition)

Where anything more than a moderate standard of accuracy is required, flowmeters need calibration. Sometimes, as in the case of large turbine meters used for the fiscal metering of petroleum, the meters may have to be calibrated every day against a built-in calibration device. At the other extreme, some meters may only need to be taken out of the line for calibration once every couple of years or so, in which case the user will probably prefer to send them to an independent laboratory rather than to maintain his own calibration facilities.

Either way it is useful for every flowmeter user to know something of the calibration techniques in common use, of which all the more important ones are described below.

---

An extract from: Flowmeters — A Basic Guide and Sourcebook for Users. A. T. J. Hayward (Macmillan Press Ltd, London & Basingstoke, 1979)

## 18.1. Methods of calibrating flowmeters with liquids

### 18.1.1. Volumetric tank used in the standing-start and finish mode

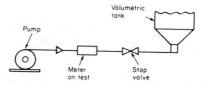

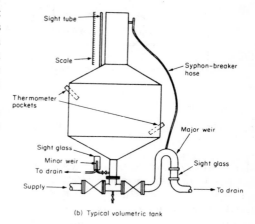

(a) Principle of the method

(b) Typical volumetric tank

This method (Figure 18.1) is particularly suited to the calibration of total quantity meters, such as the positive displacement meters commonly used for metering oils and fuels. It is not suitable for use with flowrate meters, or with instruments such as turbine meters, which permit large quantities of unregistered fluid to slip past at low flowrates.

At the start of a calibration the pipework from the pump to the base of the volumetric tank is completely filled with liquid, the stop valve is in the closed position, and the dial of the meter is on zero. Then the pump is started, after which the valve is opened as fast as possible and allowed to remain open until the volumetric tank is full, when it is rapidly closed. For accurate results, steady flow should exist during at least 99% of the duration of the test. The indication of 'total volume passed' on the dial of the meter at the end of the test is compared with the volume known to have flowed into the volumetric tank, after making any necessary corrections for thermal expansion in both the liquid and the tank.

Volumetric tanks of many different designs are in use, of which one of the most common is illustrated in Figure 18.1b. This design ensures that the level of liquid in the bottom neck of the tank at the start of a test is always the same, and the calibrated scale against the sight tube at the top of the tank enables the operator to read directly the quantity passed into the tank during the test. Tanks of this type are supplied as a standard commercial item. Other designs are also in use in which the tank is filled from the top.

An accuracy of ±0.05% of total volume measurement is easily obtainable with this method, and if great care is taken higher accuracies should be possible.

**Figure 18.1.** Calibration by the standing-start-and-finish method, using a volumetric tank.

### 18.1.2. Gravimetric tank used in the standing-start-and-finish mode

A cheaper alternative to a volumetric tank for standing-start-and-finish-calibrations is a gravimetric system. This can easily be assembled by the user, who merely needs to provide a tank sitting on a weighing machine as shown in Figure 18.2, and to arrange for it to be filled with a 'swan-neck' constant-level device as shown.

The main difficulty in using such a system is that the density of the test fluid must be known for each calibration, so that the weight reading (after applying a correction for air buoyancy) can be converted to a volume. This is a very simple procedure when the test fluid is water, whose density does not change rapidly with changes in temperature, and can be determined from tables after making a temperature measurement. Hydrocarbon fuels, however, have a high coefficient of thermal expan-

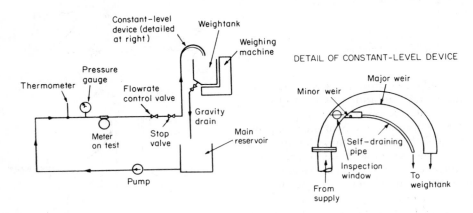

**Figure 18.2.** Calibration by the standing-start-and-finish method, using a gravimetric tank.

sion, and having to determine their density accurately on each occasion is a nuisance when this method is used. On the other hand, viscous hydrocarbon oils cannot readily be used with a volumetric tank because they cling to the walls of the tank, and this problem is eliminated when a gravimetric tank is used instead.

On balance, if it is intended to use a standing-start-and-finish method of calibration, it is probably better to use the gravimetric method if the test fluid is water or a viscous oil (above, say, 5 cSt), and to use a volumetric tank with low viscosity hydrocarbons (below, say, 5 cSt).

An accuracy of ±0.05% of total volume determination can be obtained, provided that sufficient care is taken to measure densities accurately.

### 18.1.3. Gravimetric flying-start-and-finish method, with static weighing

This method is widely used when flowrate meters have to be calibrated with water. It is less appropriate for the calibration of total volume meters, or for meters which are being calibrated with oils or fuels.

The test liquid after passing through the flowmeter passes through a control valve and into a fishtail which produces a fan-shaped jet; as shown in Figure 18.3 a diverter is arranged so that this jet can be

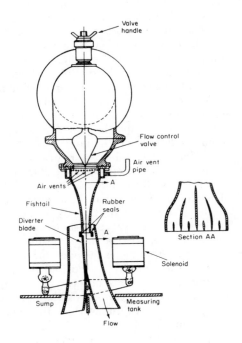

**Figure 18.3.** A typical assembly of flow control valve, fishtail and diverter.

made to pass at will into either a reservoir or a weightank. A switch fitted to the diverter plate is connected to an electronic timer, which indicates the length of time during which the flowing jet is diverted into the weightank. The gravimetric flowrate is equal to the mass collected in the weightank divided by the diversion time, and this can be converted to a volumetric flowrate by dividing it by the density of the test liquid at the appropriate temperature.

Because it takes a finite time for the diverter to pass through the flowing jet, the times at which diversion is said to begin and end must be chosen arbitrarily. That is to say, the mechanical switch attached to the diverter blade which starts and stops the timer must be set at some precise points in the travel of the diverter, which are shown as E and H in Figure 18.4. Bearing in mind that the area under a flowrate/time graph is equal to total volume, it is evident that when the first diversion (into the weightank) takes place, some liquid (volume a) passes into the collecting tank before the timer starts, and some liquid (volume b) passes into the sump instead of the weightank after the timer has started. If the timing switch is correctly positioned, these two volumes, (a) and (b), will be identical and therefore self-compensating.

Similarly, in the final diversion (out of the weightank at the end of the test) there will again be two volumes (c) and (d) which

should be self-compensating if the timer switch is correctly positioned. Because the motion of the diverter and the shape of the flowing jet are not likely to be entirely symmetrical, it is fallacious to imagine that the correct position for the timer switch is at the geometrical centre of the diversion. The ideal position for the switch can only be obtained by experiment.

In such an experiment the flow must be set to a constant value, and a number of diversions of alternately short and long duration must be made. Then the measured flowrate can be plotted against the reciprocal of the diversion time as shown in Figure 18.5a. If the line joining the experimental points is sloping, as in this figure, this shows that the timer switch is incorrectly set and must be adjusted. When the setting of the switch is right the line joining the experimental points will be horizontal as in Figure 18.5b.

Failure to set diverter switches correctly in this way causes inaccuracy in many flow calibration systems of this type. In a well designed system the correct setting for the switch will be independent of flowrate, but in some systems when a timer switch is set correctly for high flowrates it is in the wrong position for low flowrates. The experiment described above to check the correct setting of the timer switch should therefore always be carried out at both high and low flowrates.

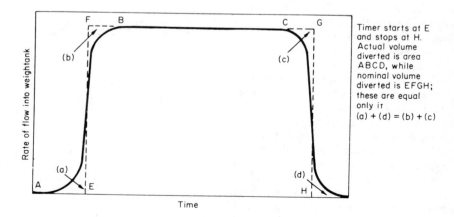

Timer starts at E and stops at H. Actual volume diverted is area ABCD, while nominal volume diverted is EFGH; these are equal only if

(a) + (d) = (b) + (c)

**Figure 18.4.** Variation of flowrate entering weightank in a flying-start-and-finish calibration.

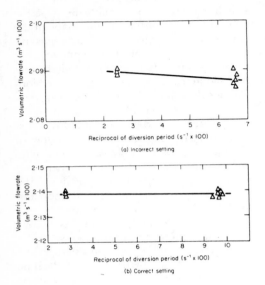

**Figure 18.5.** Results of diverter adjustment tests carried out at NEL: (a) diverter trigger incorrectly set, (b) diverter trigger correctly set.

If care is taken to set the diverter trigger correctly, to measure density accurately and to service and recalibrate the weighing apparatus at regular intervals, an accuracy of ±0.2% in flowrate measurement can be attained.

### 18.1.4. Gravimetric flying-start-and-finish method, with dynamic weighing

This method (Figure 18.6) is employed chiefly for the calibration of flowrate meters at low and moderate flowrates. At very high flowrates inertia effects are likely to create unacceptable errors.

After passing through the flowmeter on test the liquid falls into a weightank, and to begin with it passes out from the weightank through the opened dump valve at its base. When a test is required, the dump valve is quickly closed, whereupon the level in the weightank begins to rise. As soon as it reaches a certain preset value, the electrical output from the weighing machine automatically starts a timer and/or flowmeter pulse counter. The level in the weightank continues to rise, and

when another value of weight is reached the timer and/or pulse counter are automatically stopped by the weighing machine output. To avoid the possibility of overflowing the tank it is convenient if the dump valve is then automatically opened by the same electrical impulse, operating after a short preset delay.

For accurate results the liquid entering the weightank must do so as a vertical free-falling jet, the weightank must have vertical walls, and its overall height should not be much greater than its width or breadth. Also, the weighing machine must not be affected by the inertia of the weightank being much greater when it is full than when it is empty. This last requirement can best be met by using a load-cell type of weighing machine.

The dynamic weighing method lends itself to the 'substitution' or 'comparator' method, which is illustrated in Figure 18.7. In this method an accurately known weight sits on the weighing machine beside the weightank, and this weight is removed after the rising level in the weightank has automatically started the timer. Another automatic circuit stops the timer when enough liquid has entered the tank to compensate for the loss of the weight which has been removed. In this way the total 'before' and 'after' inertias of the weightank system are identical, and the weighing machine is, in effect, recalibrated

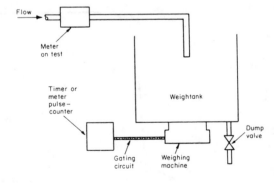

**Figure 18.6.** Principle of the dynamic weighing method.

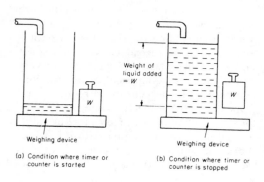

Weighing device

(a) Condition where timer or counter is started

Weight of liquid added = W

Weighing device

(b) Condition where timer or counter is stopped

**Figure 18.7.** Principle of the 'substitution' or 'comparator' method of dynamic weighing.

every time a test is carried out. Thus weighing errors are kept to a minimum and the maximum calibration accuracy is obtained. It is not difficult to measure flowrate with an accuracy of $\pm 0.2\%$ by this means.

### 18.1.5. Pipe provers

Pipe provers are used very widely in the petroleum industry and to a limited extent elsewhere, for the accurate calibration of liquid quantity meters and especially of large turbine meters. They are extremely expensive to install, but very convenient

and rapid in operation, and consequently have low operating costs. They also have the advantage of possessing a higher repeatability than any other device for calibrating liquid flowmeters. Although they can be used with flowrate meters, in practice their use is largely confined to the calibration of quantity meters. Large petroleum metering stations frequently have a pipe prover permanently built in; it is then known as a 'dedicated prover'.

There are several types of pipe prover, of which currently the most popular is the bi-directional sphere-type (Figure 18.8). A hollow sphere of synthetic rubber is inflated with water under pressure until its diameter is bout 2% larger than that of the epoxy-lined pipe from which the prover is constructed. When the sphere is forced into the pipe it seals it, and acts as a kind of piston which is capable of going round bends. A four-way valve is used to control the flow of liquid after it has passed through the meter which is to be calibrated, by causing the liquid to travel through the prover either from left to right or from right to left. At either end of the prover there is a chamber of enlarged diameter to receive the sphere at the end of each trip.

Prior to a calibration run the flow is allowed to bypass the prover. At the start

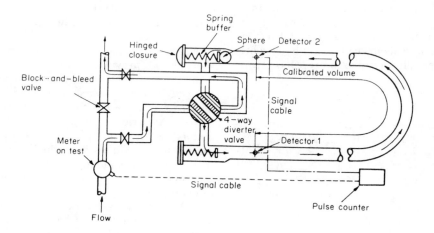

**Figure 18.8.** Principle of the bi-directional pipe prover.

of a test the flow is directed through the prover in such a way that the sphere travels the whole length of the prover. Soon after the start of its run it passes a sphere detector, which operates an electrical gating circuit and causes the electrical pulses from the meter on test to be counted.

Near the end of the run a second sphere detector is operated which stops the count of pulses. The pulse count from the meter is then compared with the known volume of the prover between sphere detectors, which has been determined from a previous static calibration of the prover. Additional accuracy is gained by totalising the pulse counts and the prover volumes during two successive runs, one in each direction. Directional effects in the sphere detectors are thus largely cancelled out.

The accuracy of the bi-directional prover when it is maintained in first-class condition may be as high as ±0.1% on flowrate, and between ±0.05 and ±0.02% on total volume.

In the uni-directional sphere-type prover (Figure 18.9) the sphere travels round the prover in one direction only, and always returns to its starting point through a sphere-handling valve. Because it is not possible in this type to eliminate directional effects in the sphere detectors, the uni-directional prover is slightly less accurate than the bi-directional type.

A third alternative is the bi-directional piston-type prover, which employs a long length of straight pipe and a metal piston with elastic seals instead of a sphere. This type generally occupies more space than the sphere-type provers, because of the need for a long length of straight pipe with subsidiary connecting pipework, and is rather more expensive. For this reason it is mostly used in special circumstances, for instance, where refrigerated liquids or very hot liquids are being used, or in situations where very high accuracy is required and expense is no object. Some early type provers employing ordinary pipeline 'pigs' travelling in ordinary pipe are still in use, but this simple type of piston prover is generally regarded as obsolescent nowadays.

A proprietary variant on the piston-type is known as a ballistic flow calibrator. It employs compressed air from a separate source to propel the piston and thus displace liquid from the calibrator through the meter on test. Its use is largely confined to small fuel meters, in applications where moderate flowrate variations during the test can be tolerated.

Another proprietary type of piston prover utilises a glass tube containing a loosely fitting piston, at the centre of which an annular ring of mercury, held in place by surface tension, acts as a practically frictionless seal. It is used specifically for calibrating very small gas flowrate meters.

### 18.1.6. Master meters

The simplest and cheapest way of calibrating a flowmeter is to put it in series with another flowmeter of higher accuracy and to compare their readings. This can give reasonably accurate results over a short period, provided that care is taken to install the two meters sufficiently far apart to ensure that the downstream meter is not affected by the wake from the upstream meter. A serious disadvantage of the method is that the performance of the master meter will itself gradually change with time; consequently, recalibration of the master meter will be needed at intervals.

Also, if the master meter should suffer any kind of sudden mechanical wear its performance could change markedly in a very short period without the operator being aware that anything was amiss. As a

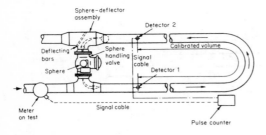

**Figure 18.9.** Principle of the uni-directional pipe prover.

safeguard against this happening it is possible to use two master meters in series (again, taking care to ensure that the downstream meter is not affected by the wake from the upstream meter) as shown in Figure 18.10. As long as the two meters continue giving the same reading the operator can be fairly confident that all is well, but as soon as one master meter gives a significantly different reading from the other he knows that one meter is malfunctioning. The rig must then be shut down while both meters are removed for recalibration.

To avoid having to shut down the rig in circumstances like this, three master meters can be installed in series. In this case, the malfunctioning meter will im-

mediately betray itself by showing that it is out of step with the other two. The defective master meter can then be taken out of the line and sent away for recalibration, while the operator continues to use his rig with the other two master meters in operation.

A third alternative is to use two master meters in parallel, with high-fidelity shut-off valves, as shown in Figure 18.11. This system is suitable only when meters of high repeatability, such as large turbine meters, are to be calibrated, because the meter on test provides the only way to compare the performances of the two master meters. One master meter is used as the main calibration device, with the other master meter, denoted in Figure 18.11 as a 'super-master meter', being used alternately with the first master meter on infrequent occasions as a check that the master meter remains unchanged. As soon as a change in the performance of the master meter is observed it is removed for recalibration, after which it is regarded as the supermaster meter, while the other meter then becomes regarded as the master meter until such time as another recalibration is called for, when the roles of the two meters are again reversed.

A well maintained master meter calibration system should, if based on the best available meters, provide accuracies of ±0.2% (or, in special circumstances, ±0.1%) on flowrate, and ±0.1 or ±0.05% on volume.

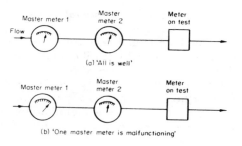

**Figure 18.10.** The use of two master meters in series.

### 18.1.7. Combination of master meter and calibrator

A very convenient method of calibrating flowmeters is obtained by installing a master meter in combination with one of the calibration devices described above. The meter can then be calibrated against the calibrator at frequent intervals to ensure that its characteristics have not changed, and meanwhile the meter can be used to calibrate flowmeters directly. In this way the accuracy of the basic calibration device is combined with the speed of operation and convenience of the master meter. Moreover, if the master meter is of the positive displacement type and has a high-

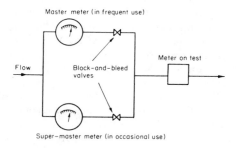

**Figure 18.11.** The use of two master meters in alternate parallel.

frequency pulse-generator fitted to it, and if the calibrator is of the standing-start-and-finish variety (sections 18.1.1 and 18.1.2), then the master meter can be calibrated by this highly accurate method and subsequently used in flying-start-and-finish tests to calibrate flowrate meters. In this way, a reasonably inexpensive calibration system for flowrate meters can be obtained, with an accuracy of $\pm 0.1\%$ on flowrate, which is probably unattainable with any other system of comparable cost.

This method is particularly useful when a portable calibration system is required, since both calibrator and master meter can be mounted on one vehicle or trailer, as shown in Figure 18.12.

**Figure 18.12.** Combination of volumetric standard and master meter (photo: Moore, Barrett and Redwood Ltd.).

### 18.1.8. Indirect methods for calibration at high flowrates

The calibration devices described above become prohibitively expensive when very high flowrates — say, above about 1 m³ s⁻¹ — have to be dealt with. In this range indirect methods of calibrating flowrate meters are much less costly.

Both the tracer-velocity and tracer-dilution technique [Hayward 1979] can be used for the calibration *in situ* of very large flowrate meters. Various types of tracer, each with an appropriate detection system, can be used: dyes, detected by colorimeters or fluorimeters; salt solutions, detected by conductivity meters or by chemical analysis; and radioactive materials, detected by radiation meters. Skilled operators can measure flowrates by these means with an accuracy of $\pm 1\%$, or perhaps $\pm 0.5\%$ under the most favourable conditions.

Velocity-area integration techniques [Hayward 1979] are also used for calibrating large flowrate meters to an accuracy of $\pm 1\%$.

### 18.1.9. Points to watch when calibrating liquid flowrates

(a) Good upstream flow conditions are even more essential in a calibration circuit than elsewhere. If necessary, carry out a velocity traverse at the entry to the test section to check that there is a good velocity profile there, and use a multi-hole pitot probe or a laser velocimeter to verify that there is no swirl.

(b) Steadiness of flow is also important. If possible, the test section should be fed by gravity from a constant-head tank; if the supply is directly from a pump be sure that the pump used has a steady output. Keep the number of pipe bends — and especially sharp bends — to a minimum. Pulsations of high frequency can sometimes be reduced to an acceptable level by inserting a coarse wire-mesh screen into the pipework at a flanged joint well upstream of the test section.

(c) Avoidance of bubbles in the test section is essential. Follow the rules for avoiding cavitation and air entrainment. A ring-shaped sight-piece of transparent plastic inserted between two flanges, as shown in Figure 18.13, provides a convenient way of confirming that no air is present in the flowing liquid.

(d) When a gravimetric method is used, a

**Figure 18.13.** A sight-piece of transparent plastics for the detection of bubbles and pockets of air in a liquid calibration system (photo: NEL).

correction for air buoyancy must be made in order to convert weight into mass of liquid. Failure to do this will introduce a systematic error of about 0.1%.

## 18.2. Methods of calibrating flowmeters with gases

### 18.2.1. Soap film burettes

The soap-film burette (Figure 18.14) is the gasman's counterpart of the pipe prover described in section 18.1.5. It is suitable only for low flowrates, since it is difficult to form a stable soap film across a burette of more than about 50 mm diameter. The film is made to act as a frictionless 'piston' which travels freely with the flowing stream of gas, so that the velocity of the film is a good indication of the velocity of the flowing gas.

The soap-film burette can be used to calibrate both flowrate and quantity meters. It will measure flowrate with an accuracy of ±0.5% and volume with an accuracy between ±0.5 and ±0.2%.

### 18.2.2. Bell provers

The bell prover, as shown in Figure 18.15,

is analogous to the volumetric standing-start-and-finish method of section 18.1.1. In this method a previously calibrated cylinder is lowered into a bath of water at a controlled rate, thus displacing a known volume of gas through the meter on test. Like all standing-start-and-finish methods it is best suited to the calibration of total quantity meters, although if the bell used is large enough the acceleration and deceleration errors can be ignored in a calibration of a small flowrate meter. Physical limitations on size restrict the use of the bell prover to fairly small meters, such as domestic gas meters. It will deliver its stipulated

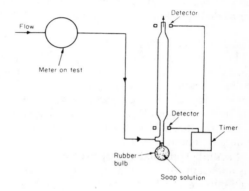

**Figure 18.14.** Principle of the soap-film burette.

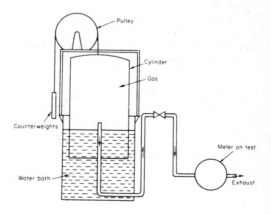

**Figure 18.15.** Simplified schematic arrangement of bell prover system for calibrating small gas quantity meters.

volume with an accuracy approaching $\pm 0.2\%$, under the most favourable conditions of operation.

### 18.2.3. The Hyde prover

The Hyde prover is probably the most accurate device available for calibrating small gas quantity meters, since an accuracy of $\pm 0.1\%$ is claimed for it. It consists essentially of a metal pipette-shaped vessel with a volume of approximately 30 litres which is enclosed in a constant temperature bath. Water is used to displace a known volume of air from this vessel through the meter on test.

This is essentially a tool for standards laboratories, because its operation calls for considerable skill and experience.

### 18.2.4. Sonic venturi-nozzles

The sonic venturi-nozzle (Figure 18.16), or critical-flow venturi-nozzle as it is sometimes called, is a very convenient device for calibrating a gas flowrate meter at one volumetric flowrate. It depends upon the fact that in the throat of a nozzle the gas cannot travel faster than the speed of sound. Provided the upstream pressure is sufficient to ensure that sonic velocity is actually reached, the flowrate through the nozzle will therefore always have a fixed value for a given gas at a specified temperature and pressure. The tapered venturi section downstream of the nozzle plays no part in controlling the flowrate; its function is merely to assist in recovering some 90–95% of the initial pressure, thus conserving energy.

If a number of sonic venturi-nozzles of various sizes are used in succession a gas flowmeter can be calibrated over a range of flowrates with them. When the highest possible accuracies (about $\pm 0.5\%$ on flowrate) are called for it is usual for the venturi-nozzles to be calibrated against a primary gas flow standard before they are used as calibration devices. If slightly lower accuracies are acceptable their performance can be predicted fairly reliably from a knowledge of their dimensions.

### 18.2.5. Gravimetric flying-start-and-finish gas-meter calibrator

The gravimetric flying-start-and-finish calibrator is a highly sophisticated and expensive device, but is regarded as the best available primary system for calibrating secondary high-pressure gas flow standards. It can also be used directly for the calibration of high-pressure gas meters. It is broadly similar in principle to the gravimetric flying-start-and-finish system with static weighing for liquids described in section 18.1.3, but with one important difference: in the liquid system the meter being calibrated is upstream of the flow diverter, but in the gas system it is downstream and in a line venting to atmosphere. This enables the meter to be read under steady-state conditions, thus overcoming the problem of diminishing flowrate which occurs while the weighing vessel is being filled.

As shown in Figure 18.17 a critical venturi-nozzle is used to maintain a constant flowrate through the test system.

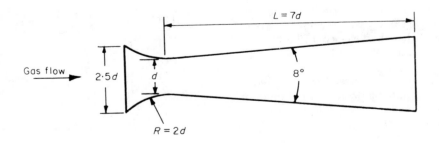

**Figure 18.16.** Standard sonic venturi-nozzle.

In the first part of the test the flow downstream of this nozzle is diverted through the meter being calibrated, while its reading is noted. Then the flow is diverted into a lightweight spherical pressure vessel for a measured time, and the measurements of the weight of this sphere before and after diversion are used to calculate the mass flowrate during the diversion period. By varying the pressure upstream of the critical venturi-nozzle a fairly wide range of mass flowrates can be covered with this system, and by using several alternative nozzles of different sizes

an almost unlimited range can be obtained.

The accuracy of flowrate measurement in a system of this kind is between $\pm 0.5$ and $\pm 0.2\%$.

### 18.2.6. PVT system

The PVT (pressure-volume-temperature) method is used mainly as a primary standard, to calibrate reference meters and sonic venturi-nozzles which can thereafter be used as secondary calibration devices. In this sytem, illustrated in Figure 18.18, a storage vessel of known volume is charged

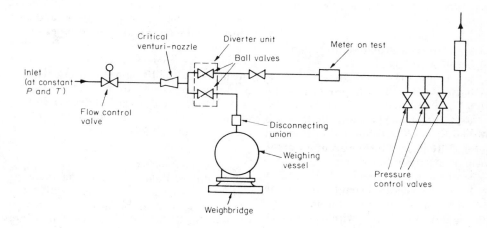

**Figure 18.17.** Simplified arrangement of a gravimetric system for compressed gases.

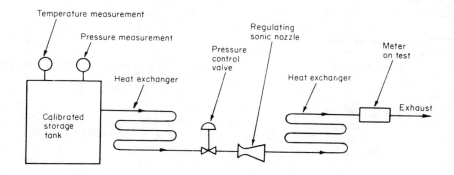

**Figure 18.18.** Principle of the PVT system for calibrating gas meters.

with gas at high pressure. The pressure and temperature of the gas in the vessel are first measured, then the gas is allowed to flow out through a regulating sonic nozzle in series with the meter on test, and finally the pressure and temperature of the gas in the storage vessel are measured again at the end of a measured period of time.

From these measurements the mass flowrate through the system during the test period can be calculated. Heat exchangers and an upstream pressure controller are used to control the conditions during flow and thus the performance of the meter on test can be determined over a wide range. Accuracies of flowrate measurement between $\pm 0.5$ and $\pm 0.2\%$ can be achieved.

### 18.2.7. Master meters

Master meters are frequently used for calibrating other gas meters. A useful combination is obtained by linking a calibration device such as a bell prover (section 18.2.2) or a Hyde prover (section 18.2.3) to a wet gas meter [Hayward 1979] which can then be easily recalibrated at frequent intervals. By this means the accuracy of the prover can be combined with the convenience of the master meter.

It is possible to extend greatly the range that can be covered with one prover by using it in conjunction with a very large gas bag and a fairly small wet gas master meter. In this way a large gas meter can be calibrated by passing the flow from it into the gas bag. When this bag has been filled it can subsequently be collapsed by the application of external pressure, so as to expel the gas through the wet gas master meter which was previously calibrated against the prover. In this way the small master meter operating at a relatively low flowrate is used indirectly to calibrate the meter on test, which shortly before had passed the same total mass of gas at a much higher flowrate.

The wet gas meter is a total volume meter and its use as an accurate master meter is therefore restricted to total volume meters. Used directly its accuracy can be in the region of $\pm 0.25\%$; used along with a

gas bag as described in the previous paragraph, the overall accuracy is probably limited to about $\pm 0.5\%$.

Flowrate meters such as orifice plates are sometimes used as master meters to calibrate other gas flowrate meters, the accuracy being between $\pm 1$ and $\pm 0.5\%$. However, they look like being displaced from this role by critical nozzles, which will do much the same job rather more effectively.

### 18.2.8. Tracer methods

Tracer-velocity and tracer-dilution techniques can also be used for calibrating gas flowrate meters. Radioactive tracers have considerable advantages over other types of tracer when used with gases, and they are now the most widely used types. Their main disadvantage is that handling radioactive materials safely is a job for experts, and calibrations by this method are usually carried out by nuclear physicists.

Under favourable conditions flowrate can be measured in this way with an accuracy of $\pm 0.5\%$.

### 18.2.9. Pitot traversing

A useful method of calibrating medium-sized or large flowrate meters with air at pressures near ambient is that of pitot traversing, followed by velocity-area integration [Hayward 1979]. The advantage of this method is that it requires only the relatively simple equipment shown in Figure 18.19, and this makes it particularly suitable for general laboratory use.

The accuracy obtainable by a skilled operator is about $\pm 1\%$.

### 18.2.10. Points to watch when calibrating gas flowmeters

(a) Because they have such a big effect on gas density, the pressure and temperature in a gas-meter calibration system must be measured very accurately. Variations with time and with position in the system must be studied and allowed for, if necessary.
(b) Pulsations can be a serious problem. Fans and compressors with good delivery

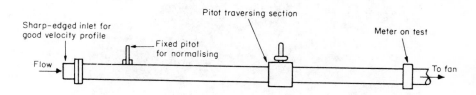

**Figure 18.19.** Typical pitot traversing system for calibration with air at ambient pressure.

characteristics should always be selected for flowmeter calibration systems, and stilling chambers and/or screens should be added if necessary to attenuate any severe pulsations which cannot be eliminated at the source.

(c) Avoid condensation.

(d) Avoid bad upstream pipework conditions. In installations like that of Figure 18.19, where the flow conditions should be well-nigh perfect, do not make the all-too-common mistake and spoil things by putting the sharp-edged inlet too close to the floor, the ceiling, a wall or an obstruction — or by letting a careless experimenter stand nearby!

## 18.3. Calibrating velocity meters

Calibrating a fluid velocity meter is either a very easy job or a very difficult one.

It is difficult if one wishes to start from scratch and make an absolute calibration. For current meters this normally involves fixing them to the travelling carriage of a ship-testing tank, and observing their performance when moving through still water at a known speed. Anemometers are usually fixed to the end of a long rotating arm and moved around a toroidal duct. In both cases corrections have to be made to allow for the fact that the test conditions during calibration are not the same as those when the meters are subsequently used, and the whole procedure becomes rather complicated. So it is just as well that

such determinations are normally carried out only in a few national laboratories.

The ordinary user will almost certainly calibrate his velocity meters the easy way. That is, he will purchase an anemometer probe or a current meter which has been calibrated at a national laboratory and use this as a master velocity meter against which to calibrate his own working meters.

Calibration then consists of placing the two velocity measuring devices, master and working, one after the other in the same stream of air or water and comparing their readings. Fluctuations in the stream velocity are allowed for by 'normalising' [Hayward 1979]. A pitot tube is shown being calibrated in a free jet of air in Figure 18.20.

## 18.4. Where to learn more

A survey paper on methods of calibrating flowmeters with liquids was published by the author[1], and a parallel paper on calibrating gas meters by Brain[2]; both of these contain useful bibliographies. [ . . . ]

The whirling-arm method of calibrating anemometers has been described by Cowdrey[4], and the towing-tank method of calibrating current meters is covered by an ISO standard[5]. Pipe provers are the subject of an API standard[6] and a simple code of practice issued by the IP[7]. There are useful papers on soap-film burettes[8], sonic venturi-nozzles[9] and a gravimetric gas

**Figure 18.20.** A pitot tube being calibrated in a free jet of air (photo: NEL).

calibration system[10]. The American Society for Testing and Materials[11] and the Instrument Society of America[12] have both published codes of practice for calibrating rotameters. Hyde has described his prover in the book by himself and Jones[13].

The procedure for calculating the air buoyancy correction required for converting weight to mass is given in Kaye and Laby[14] along with a table of values.

## References

*A field code of practice for proving turbine displacement meters with prover pipes* (1979) (London: Institute of Petroleum).

*API Standard 2531* (1963) Mechanical displacement meter provers (now being superseded by Chapter 4.2 in the new API Manual).

*ASTM Standard D 3195-73* Recommended practice for rotameter calibration.

Brain, T. J. S. (1978) Reference standards for gas flow measurement. *Measurement and Control* **11** No.8, 283–288.

Collins, W. T. and Selby, T. W. (1965) *Report No. K-1632, A gravimetric gas flow standard — Parts 1 and 2* (Oak Ridge: Union Carbide Corp., Nuclear Division).

Cowdrey, C. F. (1950) A note on the calibration and use of a shielded hot-wire anemometer for very low speeds. *J. Sci. Instrum.* **27** No. 12, 327–29.

Harrison, P. and Darrach, I. F. (1967) *NEL Report No. 302, Air flow measurement by the soap-film method* (East Kilbride, Glasgow: National Engineering Laboratory).

Hayward, A. T. J. (1977) Method of calibrating flowmeters with liquids — a comparative survey. *Measurement and Control* **10** No. 3, 106–116.

Hayward, A. T. J. (1979) *Flowmeters — a Basic Guide and Sourcebook* (London: MacMillan).

Hillbrath, H. S. (1971) The critical flow venturi: a useful device for flow measurement and control. *Symposium on Flow*, Paper no. 1-3-205 (New York: American Society of Mechanical Engineers).

Hyde, C. G. and Jones, M. W. (1960) *Gas calorimetry*, 2nd edn. (London: Ernest Benn).

*ISA Standard RP 16.6* (1961) Methods and equipment for calibration of variable area meters (rotameters).

*ISO Standard 3455* (1976) Liquid flow measurement in open channels — calibration of rotating-element current meters in straight open tanks.

*ISO Standard 2975* Water flow (in seven separate parts); and *ISO Standard 4053* Gas flow (in four separate parts).

Kaye, G. W. C. and Laby, T. H. (1975) *Tables of physical and chemical constants*, 14th (Revised) edn. (London: Longmans).

## Paper 19

# Noise suppression and prevention in piezoelectric transducer systems

## *Jon Wilson*

This paper is concerned with electrical noise which introduces errors into piezoelectric vibration transducers. While some of the included calculation is specific to a piezoelectric accelerometer/charge amplifier combination, the description of sources of noise and how to reduce noise pick-up is applicable to a wide range of measurement situations using electrical transmission of the transducer output signal. (Eds.)

A common problem in shock and vibration measuring systems is noise generated by ground loops and by pick-up from electrostatic and electromagnetic fields. This article reviews the mechanisms of noise generation and suggests how such noise can be prevented or suppressed.

### 19.1. Noise generation

#### 19.1.1. Capacitively coupled noise
Presence of a varying electrostatic field (difference in potential) between two conductors coupled electrically by some stray capacitance is a prominent source of noise in piezoelectric transducer systems. For example, two pins in a connector act as plates of a capacitor and the mounting insulator is the dielectric. The pins are susceptible to noise pick-up if they are not properly shielded.

#### 19.1.2. Magnetically coupled noise
A varying magnetic field in the vicinity of a signal path is a source of noise in almost all measurement systems. If one of the pins

above is conducting an AC current, there exists a magnetic field around it which can induce noise into the signal being carried on the other pin.

#### 19.1.3. Current-coupled noise
Installations where the signal current and other currents use a common path are also a prominent noise source. Any impedance in this path causes non-signal currents to induce noise into the signal currents. For example, if a conductor were used as both the 'low' side of the signal path from a transducer and the 'low' side of the AC exitation for another device, the common resistance would develop a component of the AC that would appear with the signal.

Originally published in *Sound and Vibration* (April 1979).

## 19.2. Noise reduction

### 19.2.1. Shielding for capacitive coupling

Charge amplifiers are widely used in vibration measurement systems because they offer so many advantages. But they are susceptible to electrostatic coupling because they respond directly to charge input. Consider the circuit in Figure 19.1 showing the schematic of a typical piezoelectric vibration measuring system.

Assume that the transducer has a sensitivity of $10 \ \mathrm{pC \ g^{-1}}$ and that the vibration level to be measured is approximately 1 g. Assume also an AC electrostatic voltage ($e_1$) present of 100 V (this can be typical). The charge amplifier measures the total charge appearing at its input, which in this example would be

$$Q = Q_a \times g + e_1 \times C_1$$
$$= 10 \ \mathrm{pC \ g^{-1}} \times 1 \ \mathrm{g} + 100 \ \mathrm{V} \times C_1$$
$$= 10 \ \mathrm{pC} + 100 \times C_1 \ \mathrm{pC}$$

where $C_1$ is the capacitance, in pF, between the interference source and the charge amplifier input.

This equation indicates that a coupling capacitance of only 0.01 pF will produce an error of 10% in the measurement. To ensure rejection of the electrostatic voltage, the signal lead must be shielded at all times with the shield connected to the input common of the amplifier.

### 19.2.2. Shielding for magnetic coupling

It is generally more difficult to shield magnetically than to shield electrostatically. The instrumentation engineer must take precautions to ensure that instruments and cables are not located in the vicinity of power transformers, solenoids or motors; and that signal cables are not located in cable troughs with current-carrying conductors. To understand the problem, refer to Figure 19.2.

Circuit (a) illustrates a conductor carrying a current $i$, lying beside the input signal lines. The flux produces voltages $e_1$ and $e_2$ which have the polarities shown (at

one particular instant). The magnitudes are

$$e_1 = \mathrm{d}/\mathrm{d}t(N_1\phi_1) \qquad e_2 = \mathrm{d}/\mathrm{d}t(N_2\phi_2)$$

where $\mathrm{d}/\mathrm{d}t$ is the rate of change with respect to time, $N$ is the number of turns being coupled (essentially 1 unless there is an excess of signal lead that has been neatly coiled!), and $\phi$ is the number of lines of flux coupling the conductors (a function of the magnitude of $i$, how close the conductors are located over how great a length, and the effect of magnetic shielding). If $e_1$ and $e_2$ were exactly equal, there would be no effective error signal. But this is seldom the case. A difference voltage usually exists, and it is induced into the measurement signal. Note that induced voltage will be larger at higher frequencies, due to the faster rate of change.

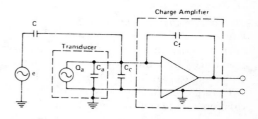

**Figure 19.1.** Electrostatic noise is generated by stray capacitance $C_1$ which couples voltage $e_1$ into the measurement system.

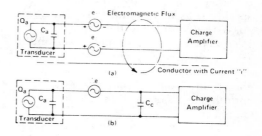

**Figure 19.2.** Electromagnetic noise is caused by current-carrying cables (a) or electrical machinery which induces noise into signal cables (b).

Pick-up can be suppressed by more direct cable runs through problem environments and by eliminating loops and turns. A large improvement is achieved when the current-carrying conductor is tightly twisted with its return conductor. This results from the cancellation effect of having essentially equal and opposite electromagnetic fields in close proximity, with the resultant reduction of $\phi$. Similarly, the magnitude of the induced voltage can be reduced by locating a remote charge converter near the transducer and using twisted-pair cables between it and the amplifier. The amplified output also reduces the net effect of induced voltage. Remote converters are also beneficial because their lower output impedance reduces electrostatic coupling problems, but they may reduce signal-to-noise ratio because of their limited maximum signal input.

### 19.2.3. Isolation for current coupling

Current-coupled noise occurs when a current other than the vibration signal is introduced into the measurement system. Referring to Figure 19.3 there would be no current flow in $R_c$ if there were no potential difference between grounds 1 and 2. But grounding is never perfect, and it is usually less than desirable. In instrumentation jargon, the path from ground 1 through $R_c$ to ground 2 and back to ground 1 through $e_g$ and $R_g$ is called a ground loop. Such a ground loop can occur when a transducer is grounded directly through a test specimen, as well as indirectly though the coaxial cable shield to the amplifier which is grounded to the instrument rack and to the power system ground. When this is allowed to happen, the current flowing in the ground loop will couple a noise signal into the high side of the signal line to the amplifier.

The ground loop path can be suppressed by breaking the circuit at $A$, $B$ or $C$ so that the system is grounded at only one point. A break at $A$ is achieved by using an insulating stud or washer. A break at $B$ is obtained by using an accelerometer in which the low side of the

transducer element and the cable shield are insulated from the transducer case. A break at $C$ is accomplished by amplifier techniques rather than by simple insulation. A study of system grounding techniques and the relative achievements of each is provided in later sections.

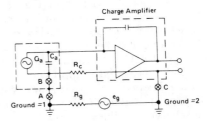

**Figure 19.3.** Ground loops occur when measuring systems are grounded at more than one point, causing noise to be introduced into signal cables.

## 19.3. Amplifier configurations

Amplifiers may be identified by the manner in which they handle the grounding of transducer signals. Single-ended amplifiers are most widely used. They have a common path for one side of the signal source, the amplifier input and amplifier output. This may be referred to as the low, common or ground side. Referring to Figure 19.4($a$), the low side of the amplifier is 'grounded' and in 19.4($b$) 'floating'.

A single-ended grounded input should be used when the transducer low signal is connected to its case and the transducer is insulated from the specimen. Insulation can be accomplished by the use of an insulating mounting stud. This system provides excellent protection against noise pick-up.

When the transducer low signal is grounded to its case, which is in turn grounded through the test specimen, the system should not be grounded again at the amplifier. A single-ended floating amplifier eliminates the amplifier grounding, but a large capacitance to ground may still exist.

Charge Amplifier

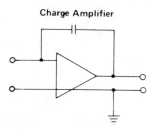

Single-ended, Grounded

(a)

Charge Amplifier

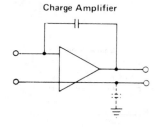

Single-ended, Floating

(b)

Charge Amplifier

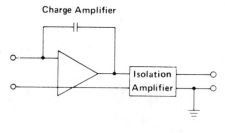

Single-ended, Floating, Isolated

(c)

**Figure 19.4.** Single-ended amplifiers are available with the input common path either grounded or floating.

Therefore, there may be poor protection against ground loops. The amplifier in Figure 19.4(c) is a better solution because the isolation amplifier opens the path from the transducer/amplifier common. This not only breaks the DC path, but it also minimizes the AC (capacitive) path.

Floating amplifiers do not provide as good isolation as single-ended grounded amplifiers, yet they are sometimes the only practical way to provide isolation.

If the transducer cabling and charge amplifier are properly designed and correctly grounded, the system usually has a good signal-to-noise ratio. But these precautions may be inadequate for low-level measurements in the presence of excessively high electromagnetic noise or when using very long cables. In these situations, remote charge converters are available for location near the sensor so that the majority of the line is driven from a low-impedance source. Remote charge converters, like the circuits in integral electronic accelerometers, have limited dynamic range and limited slew rate as well as temperature limitations. In applications where these factors are critical to performance of the system, a so-called differential charge amplifier may be required.

Differential amplifiers reject common mode signals: i.e. non data signals present on both conductors of a signal line. Unlike single-ended amplifiers which amplify the total signal between their input terminals, differential amplifiers amplify only the difference signal. Traditional differential amplifiers shown in Figure 19.5(a) and (b) use symmetrical capacitive feedback circuits, but the symmetry is easily disrupted by a change in gain or change of cable or transducer. And they require use of specially-balanced differential accelerometers.

Figure 19.5(c) illustrates the differential charge amplifier. It overcomes symmetry problems and has proven to be highly satisfactory. Unlike common laboratory differential amplifiers, differential charge amplifiers do not reject the common mode signal. They equally amplify the two individual signals with respect to ground and differentially sum the resultants. The first amplifier (charge converter) is floating and is shielded with respect to the output ground. It provides a stable high-level source for the second amplifier which is a 'differential in, single-ended out' voltage amplifier. Because the two amplifiers are integrally assembled, symmetrical configuration is maintained, and the higher operating level allows lower impedance circuitry so that lack of symmetry due to reactive effects is greatly reduced. Therefore, a high degree of common mode suppression is achieved with much less difficulty than with conventional differential amplifier systems. This is the design used in Endevco charge amplifiers, such as the 2735 and 6633. These amplifiers can also be operated in the single-ended grounded input mode.

## 19.4. A study of grounding configurations

### 19.4.1. Case-grounded transducer/insulated mounting/single-ended grounded amplifier

Generally, a case-gounded accelerometer with insulated mounting gives the best protection against pick-up induced by ground-loop currents. The system should terminate with a single-ended grounded amplifier. The electrical equivalent of this system is shown in Figure 19.6.

The accelerometer has an inherent capacitance $C_h$ between the case and the signal side of the crystal (actually the crystal capacitance). It would normally be susceptible to electrostatic pick-up, but the insulated mounting restricts ground-loop currents to small values and to a path through signal ground as illustrated. The error introduced is negligible, typically less than 0.001 g (equivalent) $V^{-1}$ for 60 and 400 Hz power line frequencies.

Manufacturers offer a variety of insulated mounting studs and washers as well as cementing studs which may be attached to the specimen with an insulating adhesive. Some case-grounded accelerometers have an integral insulating stud or washer.

Because insulating studs lower the mounted resonance frequency and therefore the upper frequency limit by 15 to 30% compared to that available with grounded studs, it is sometimes necessary

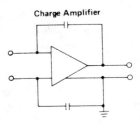

Charge Amplifier

Differential in Single-ended Out (Semi-Differential)
(a)

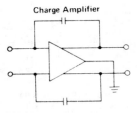

Charge Amplifier

Differential in, Differential Out (True Differential)
(b)

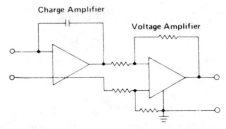

Charge Amplifier

Voltage Amplifier

Single-ended Floating Charge Converter to Differential In, Single-ended Out Amplifier
(c)

**Figure 19.5.** Differential amplifiers provide common mode rejection. The design shown in (c) is best suited to industrial applications.

to use a grounded stud. Or, high temperatures may prevent the use of insulating adhesives.

### 19.4.2. Case-isolated transducer/grounded mounting/single-ended grounded amplifier

Accelerometers with their electrical connectors insulated from the case may be more susceptible to noise pickup than case-grounded accelerometers. The electrical schematic of such a system is shown in Figure 19.7. There is no significant effect from coupling $E_g$ through $C_{in}$. But the possibility of capacitance $C_h$ from the case to the signal side of the crystal makes the circuit highly susceptible to pick-up induced by $E_g$. This requires careful design of the accelerometer to minimize $C_h$.

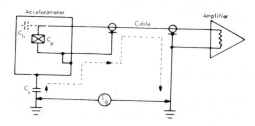

**Figure 19.6.** Case-grounded accelerometer with insulated mounting stud and single-ended grounded amplifier provides excellent protection against pick-up induced by ground loops and capacitive coupling.

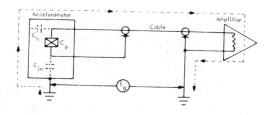

**Figure 19.7.** Case-isolated accelerometer with grounded mounting and single-ended grounded amplifier is susceptible to noise pick-up through internal stray capacitance.

### 19.4.3. Case-grounded transducer/grounded mounting/single-ended grounded amplifier

This system shown in Figure 19.8 has two grounds and is, therefore, highly susceptible to ground loops. It is the most noise-prone method of installing and connecting accelerometers. If a potential $E_g$ exists between the two grounds, and it probably does, there is no alternative except to break the ground-loop circuit.

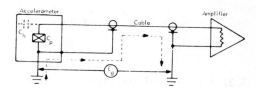

**Figure 19.8.** Measuring systems grounded at the transducer and at the amplifier are highly susceptible to noise induced by ground loops.

### 19.4.4. Case-grounded transducer/grounded mounting/floating amplifier

If a case-grounded transducer must be grounded through the test specimen, a floating amplifier should be used to ensure a single system ground, see Figure 19.9. The system may still be susceptible to ground loops because of imperfect grounding at the amplifier output unless an isolation amplifier is used, as in Figure 19.4($c$). This feature is incorporated in the Endevco 2735 Charge Amplifier, whose differential amplifier configuration shown in Figure 19.5($c$) also protects against common mode noise. This is the best alternative to the system consisting of case-grounded transducer, insulated mounting and single-ended grounded amplifier.

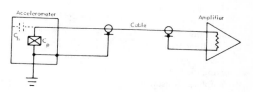

**Figure 19.9.** Case grounded transducers with grounded mounting require floating or differential charge amplifiers.

### 19.4.5. Case-isolated transducer/insulated mounting/floating amplifier

This system has no ground and is the logical recipient of spurious noise and ground loops from instruments, motors and other electrical equipment on the same power line.

### 19.4.6. Case-grounded transducer/insulated mounting/floating amplifier

This system also lacks a ground connection and is susceptible to pick-up from other equipment on the same power line.

## 19.5. A quantitative evaluation of noise suppression

Various instrument systems and ground configurations have been discussed qualitatively. Here, let us perform a quantitative comparison.

Consider a system consisting of a transducer, a coaxial connecting cable and a charge amplifier which exhibit the following parameters:

| | |
|---|---|
| transducer capacitance | 1000 pF |
| cable capacitance | 15 000 pf (500 ft) |
| cable resistance | 8 Ω |
| resistance between grounding points | 10 Ω |
| insulating stud leakage resistance | 100 MΩ |
| capacitance | 70 pF |
| floating transducer coupling capacitance to common | 5 pF |
| coupling capacitance to signal lead | 5 pF |
| coupling capacitance unbalance | 0.5 pF |
| differential charge amplifier feedback capacitance | 500 pF |
| floating charge amplifier resistance between grounds | 50 MΩ |
| capacitance between grounds | 200 pF |

Assume a 50 Hz interface voltage and a voltage difference between grounding points of 1 V.

A comparison of the current coupled noise magnitude for various system configurations is given in Table 19.1. This is a typical set of values. The magnitudes will differ significantly for other test conditions, but the relative magnitudes serve to rank order the effectiveness of the various systems in protecting against noise pickup.

Table 19.1 indicates that a case-grounded transducer with an insulating stud and a single-ended grounded amplifier provides excellent protection. It is the simplest system available for laboratory and field use. If insulated stud mounting is not practical, use a case-grounded transducer with grounded mounting and a differential charge amplifier. With either system, a remote charge converter will significantly improve signal-to-noise ratio.

## 19.6. Summary

A few simple rules will help minimize noise pick-up in most vibration measurement systems.

1. Use high-quality coaxial cable and coaxial connectors at each electrical connection.
2. Ground the system at a single point. Preferably, choose a point where the ground can be controlled and positive.
3. Avoid placement of cables in cable troughs or harnesses that contain power or high-current conductors.
4. Avoid running cables close to large electromagnetic sources, and avoid long cable runs in difficult environments.
5. Use a remote charge converter as the susceptibility to noise pick-up is high.
6. Last, but not least, choose a high sensitivity accelerometer to reduce the effect of noise pick-up on signal to noise ratio.

**Table 19.1.**

| System | Case-grounded transducer | Case-isolated transducer |
|---|---|---|
| Grounded mounting, single-ended, grounded charge amplifier | 3600 pC | 5 pC |
| Grounded mounting, single-ended, charge amplifier with remote charge converter (Endevco 2735 and 2731B) | 134 pc | 5 pc |
| Insulating stud, single-ended, grounded charge amplifier (Endevco 2735) | $1.8 \times 10^{-3}$ pC | 5 pC |
| Insulating stud, single-ended charge amplifier with remote charge converter (Endevco 2735 and 2731B) | $6.8 \times 10^{-5}$ pC | 5 pC |
| Grounded mounting, true differential charge amplifier | 500 pC | 0.5 pC |
| Grounded mounting, semi-differential charge amplifier | $2.4 \times 10^{-2}$ pc | 0.5 pC |
| Grounding mounting, differential charge amplifier (Endevco 2735) | $1.0 \times 10^{-2}$ pC | 0.5 pC |
| Grounded mounting, differential charge amplifier with remote charge converter (Endevco 2735 and 2731B) | $1.8 \times 10^{-4}$ pC | 5 pC |

Paper 20

# Process measurement in the food industry

## D.J. Steele and I. McFarlane

This is the first of three papers which examine instrumentation aspects of particular industries. This one is concerned with the food industry, in particular the baking processes in biscuit manufacture.

Of interest is the range of different measurands, and the range of different techniques used to make the measurements in the particular conditions pertaining. (Eds.)

This article reviews a selection of familiar techniques for in-line measurement of temperature, moisture, flow, and level, with descriptions of some transducers adapted for particular applications in the food industry. There is also a number of illustrations of method for measuring other variables, and descriptions of some special devices for foreign-body detection, dimension gauging, and viscosity measurement. The review will conclude with some discussion of future trends in food process plant instrumentation.

### 20.1. Introduction

In this article we review the more common on-line devices used by food manufacturers, with particular reference to the baking process and to the impact of microelectronics on food process instrumentation. Applications are to be found in the food industry for almost all the standard techniques for measuring physical quantities on-line; this article looks particularly at the way these techniques have been adapted to suit particular conditions.

There are very large numbers of applications in the industry for the measurement of temperature, moisture, flow, weight, and level (or contents), and there are sections on each of these topics. There then follows a composite section on the slightly more specialized needs for measurement of pressure, force, pH, humidity, density, and colour. Finally, there is a group of applications for specially developed transducers, which tax the skill and efforts of the food industry instrument engineer. Some examples are given of dimension gauging, foreign-body detection, and viscosity measurement.

All these measurements are preliminary to any form of closed-loop automatic control. The very large scale of many modern food processes provides

Originally published in *Measurement and Control* **14** (1981).

plenty of incentive for control schemes which improve yield and productivity. The rate of introduction of automatic integrated control schemes is limited partly by the generally low number of measuring devices in use, and partly by the very small number of instrument people employed in the industry to install and maintain such equipment. The concluding section of the paper looks forward to new developments which will make it easier for the food manufacturers to introduce automatic control on continuous process plant.

One of the aims of this paper is to show that instrumentation does exist which can provide the reliable sensing equipment needed for integrated control schemes. Manufacturers have been rather slow so far to invest in instrumentation and control, but such schemes are likely to become more common throughout the food industry, due to the increasing cost of raw materials, and pressures for efficient use of plant and labour.

## 20.2.  Temperature

The importance of temperature measurement for the preparation and processing of food products is self-evident. Apart from traditional devices such as thermocouples, platinum resistance thermometers and thermistors, specialised instruments have been developed for specific applications. Some of the special devices are listed in Table 20.1, and described below.

One of the problems in the industry is non-contact measurement, especially for the temperature range $-10\,°C$ to

$+120\,°C$; for example, chocolate mould temperature during coating. Probably the method most tried is the measurement of the thermal radiation from a body. The ability of an object to emit radiation is termed its emissivity; different materials at the same temperature will give a different output of radiation. Therefore in order to obtain an actual temperature measurement, compensation must be made for the emissivity of the material. Proprietory radiation pyrometers measure radiation over a band of wavelengths. An optical system is used to focus the infrared radiation onto a detector.

Another non-contact instrument is the thermal camera, which can be used to identify poor lagging, excessive heat losses and other important features. These are revealed in a thermal image displayed on a television screen. The temperature can be displayed as a six-colour isothermal pattern.

In some processes, for example, in the canning industry, where it is important that a product is known to have been fully processed, use is made of colour change. Special paints are obtainable which change colour at specific temperatures. The colour change can be permanent or revert to its original form when cooled. Waterproof stick-on tabs or tapes are also available, which turn black at stated temperatures.

Radio telemetry is sometimes used to extract temperature measurements from closed or continuously moving equipment. (Steele 1972, 1975a,b, 1978). A schematic diagram showing the basic component parts of a radio telemetry system designed by the British Food Manufacturing Indus-

## Table 20.1:  Special temperature-measuring devices

| Device | Range | Accuracy |
|---|---|---|
| Radiation pyrometer | 20 to 4000 °C (expensive below 80 °C) | ±2% FSD |
| Thermal Image system | 50 to 80 °C | ±1.5 °C |
| Colour patches | Paint: 30 to 700°C | ±1% of nominal value |
| Radio telemetry systems | −20 to +250°C | ±0.5 °C |
| Data store systems | −20 to +420 °C | ±0.5 °C |

tries Research Association (BFMIRA) is shown in Figure 20.1. It uses frequency modulation superimposed on an RF carrier. Both single- and multichannel transmitters have been designed to operate at temperatures from $-20\,^{\circ}$C to $+130\,^{\circ}$C without the need for thermal protection while still maintaining calibration. Figure 20.2 shows the design of a unit for in-can temperature measurement in a retort. Units have also been designed for use in bread ovens, biscuit ovens (having suitable thermal protection) and for measuring Easter egg and other types of chocolate moulds. In one instance, a transmitter was used in a blast freezer at a temperature of $-40\,^{\circ}$C.

Another way of monitoring inaccessible temperatures is to store the data in a miniature recorder. BFMIRA has developed systems of this type as well. These are battery-powered self-contained units designed to pass through the process plant. Temperatures as measured by suitable sensors are 'read' at regular periodic intervals and the temperature readings are stored in an electronic solid-state memory. The data storage unit is retrieved at the end of the process and connected to a playback unit. A number of commercial units are available; the data store units need efficient thermal insulation if they are to be used at ambient temperatures in excess of 45 to 50°C. BFMIRA has designed a unit capable of operating at temperatures from $-20\,^{\circ}$C to $+125\,^{\circ}$C without the need for thermal insulation while still maintaining accuracy. A typical unit is shown in Figure 20.3, while Figure 20.4 is a block diagram

showing the basic operation of the system, and Figure 20.5 shows the playback function.

**Figure 20.2.** Radio telemetry transmitter.

### 20.3. Moisture

The moisture content of food products is important for the quality of the product and for shelf life. The moisture at different stages of processing also has a bearing on processing costs.

Food materials are chemically and physically complex, and this imposes

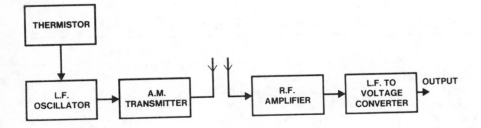

**Figure 20.1.** Radio telemetry system (schematic).

limitations on moisture measuring techniques. Moisture standards are set in the laboratory using Karl–Fischer titration or, more commonly, loss in weight on drying under prescribed conditions. A comparison of these two methods has been published by Christen and Richardson (1976) with reference to moisture measurement in cheese. On-line instruments are always calibrated with reference to laboratory standards.

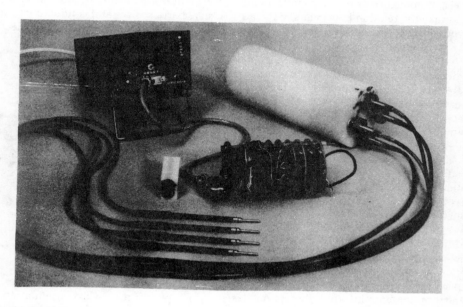

**Figure 20.3.** Data store.

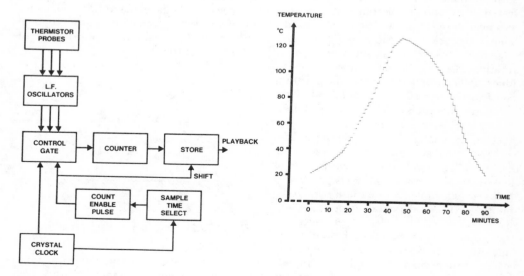

**Figure 20.4.** Data store (schematic).

**Figure 20.5.** Data store recorder trace.

The difficulties in making measurements on-line arise firstly from moisture gradients within the product, and secondly from the chemical bonds which form between some of the water molecules and molecules of the food substance. At low levels of moisture, these chemical bonds can make the relation between, say RF absorption and oven-drying measurements very non-linear.

The dielectric constant is a good indicator of moisture content for many materials. The dry matter has a dielectric constant in the range 2 to 5, while for free water it is about 80. Small changes in water content therefore have a strong effect on the observed dielectric constant. The effect is slightly temperature sensitive, and readings are also a function of density. A capacitance moisture meter usually comprises an RF bridge circuit working at 0.5 to 20 MHz. Material passes between or over plates which form one arm of the bridge. The method is suitable for moisture below 25%.

If moisture at or near the surface is sufficiently representative of overall moisture content, infrared absorption has the advantage of offering non-contact measurement. Free water has a strong absorption band at 1.94 $\mu$m, in the near infrared. The ratio of reflectance at 1.94 and at a nearby reference wavelength is found, usually by looking at the product through two narrow-band filters mounted on a rotating filter wheel. The ratio of reflectance is a simple function of moisture content. Best results are obtained with powdered materials, such as soup powder, icing sugar, or starch. The reading can be affected by changes in particle size. The method is usually used for low moisture contents, but good results have been obtained at high moisture levels for some foods, for example, potato mash with 70% moisture. It is not suitable for materials with dark or highly reflecting surfaces.

Another group of techniques use microwaves in various ways, measuring either attenuation on transmission through bulk material, or changes in radiation loss from strip lines, or drying by microwave heating. Figure 20.6 is a photograph of a microwave transmission cell. Material is kept out of the transmitting and receiving horns by windows which are transparent to microwaves. The method is suitable for powdered and granular material with up to 40% moisture content, but packing density, particle size, and conductivity (due to salt content for example) all affect the measurement.

**Figure 20.6.** Microwave transmission cell.

Strip lines are another configuration which permit microwave attenuation to be used to measure moisture (Steele 1975c, 1976). A strip transmission line gives attenuation due to interaction in a fringing electric field on the exposed side of the line; the other side is shielded by a ground plane (Figure 20.7). Sensitivity is of the order of 10 dB% $H_2O$/m of line.

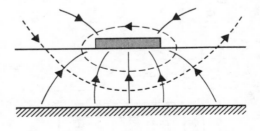

**Figure 20.7.** Strip line — field pattern.

Microwave heating offers a rapid alternative to conventional laboratory oven drying for loss-in-weight off-line testing.

Commercial microwave ovens can be adapted for this, but they require extra loading to prevent damage to the generator when the product is completely dry. One supplier offers an oven with integral gravimetric weighing for this application. BFMIRA have developed a resonant cavity for microwave heating using variable power. The cavity is shown in Figure 20.8. It can measure the moisture in, for example, comminuted meat (moisure content 60–70%) to within 0.5%. Resonant cavity techniques are also used where the increase in the dielectric constant due to the moisture present changes the resonant frequency of the cavity.

A method suitable for measuring the proportion of water in another liquid uses the conveniently linear relation between concentration and acoustic velocity (Blitz 1963). The ultrasonic devices developed for non-destructive testing of metals provide a method which can readily be adapted to measure acoustic velocity in two-liquid mixtures.

## 20.4. Liquid flow

The more common types of flowmeter which measure a pressure difference across an orifice or other restriction in a pipe are not widely used in the food industry, probably because so many liquid ingredients contain suspended solids, which settle out in dead spots causing a hygiene hazard.

Variable-area flowmeters (gapmeters) are satisfactory if suspended solids are finely divided and the viscosity is low enough, but they do not normally produce a transmitted signal. Turbine meters are usable with broadly the same range of fluids, with viscosities up to about 100 cP (400 s Redwood). This type of meter requires laminar flow conditions for best accuracy, which is about ±0.2% FSD. Some viscosities of liquids at various temperatures are shown in Table 20.2.

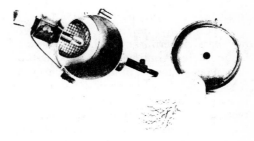

**Figure 20.8.** Resonant cavity.

**Table 20.2: Liquid viscosities**

| Liquid | Temperature, °C | Viscosity, cP |
|---|---|---|
| Water | 15 | 1.14 |
| Salt solution (20%) | 15 | 1.73 |
| Sucrose solution (30%) | 15 | 3.75 |
| Sucrose solution (50%) | 15 | 19 |
| Sucrose solution (75%) | 15 | 4000 |
| Corn syrup | 15 | 2000 to 3000 |
| Corn syrup | 60 | 12 to 18 |
| Olive oil | 15 | 90 |
| Olive oil | 60 | 20 |
| Olive oil | 100 | 7 |
| Rape oil | 15 | 100 |
| Rape oil | 60 | 20 |
| Rape oil | 100 | 9 |
| Castor oil | 15 | 1400 |
| Castor oil | 60 | 100 |
| Castor oil | 100 | 17 |

More viscous liquids (but not slurries) can be metered using the 'positive displacement' of a rotating or reciprocating piston. This is not a pump, but a chamber containing a piston or rotating disc arrangement which is driven by the fluid. Inlet and outlet ports are arranged so that a fixed volume is transmitted in each cycle.

Metering of slurries is usually possible with a magnetic flow meter. This consists of a straight pipe with coils fitted outside, to produce a magnetic field within the pipe. As liquid flows, an EMF is generated proportional to volumetric rate of flow, and this is detected by a pair of measuring electrodes in the wall of the pipe. The liquid must have some conductivity; not less than about $1 \mu S\,mm^{-1}$. This prevents the meter being used for pure chocolate, which is totally non-conducting.

A recently developed flowmeter of novel design permits even chocolate flow to be measured continuously, in a U-tube with no obstructions in the bore. This tube is vibrated, and when liquid flows the tube is slightly twisted by a Coriolis force; the twist is measured optically. The output is linear with mass flow over a wide working range. A full description has been published by Plache (1977).

The meters described allow liquid flow to be measured with sufficient accuracy in almost any food application. In addition, vortex flow meters and ultrasonic flowmeters have recently been introduced for some purposes; these share the advantage of being without moving parts.

## 20.5. Weighing of dry materials

Continuous weigh-feeding of dry ingredients is an increasingly common requirement, usually associated with the need for controlled blending of two or more streams. A closely related technique is the in-line weighing of containers of finished product while in motion on conveyors. The latter measurement is often the most significant in the whole process, because it is the means by which the manufacturer ensures that legal regulations on packet weights are satisfied for pre-packed goods, and it is also the point at which the manufacturer monitors the overall yield of the manufacturing process.

Packet checkweighing is sufficiently important in biscuit manufacture to have justified the development of fast microprocessor-controlled checkweighers (Figure 20.9). In this type of device, items to be weighed are fed onto a carefully designed carrier for transport across the weigh-head

**Figure 20.9.** Packet checkweigher.

(Hope 1979). Speeds of up to 400 items per minute have been achieved. Although this type of weighing is only indirectly related to process control in the continuous part of the process, the development has highlighted two attributes which assist any in-line weighing of dry material:

zero-displacement weigh-heads greatly assist speed and accuracy of weighing; for best accuracy, the material to be weighed needs to be transferred to a short length of special conveyor for transport over the weigh-head.

The first requirement, zero-displacement, has always been possible to achieve at some cost, using force-balance techniques. Recent improvements in load-cell technology have given performance which allows load-cells to be used for all but the most demanding accuracies. The second requirement, a separate conveyor, is always bound to be more costly than the installation of transducers in the supports for an existing belt. Papers at a recent conference on weighing (IMC 1979) indicated that a separate conveyor can give better than ±0.5% absolute precision over several weeks between recalibration, compared with ±2% for transducers fitted to existing belts.

Continuous weigh-feeders have been used successfully for some time in the baking industry. A unit weighing flour at up to 1500 kg h$^{-1}$ is shown in Figure 20.10.

Sometimes it is not possible to use any form of belt weighing. In these circumstances there are two other techniques worth considering. The first is a simple and relatively inexpensive impact plate flow-meter, in which the material is allowed to fall under gravity onto an inclined plate. The horizontal force on the plate is proportional to the mass flow, but it is also sensitive to the impact and fall angles, and to frictional effects. The best accuracy obtainable is about 1%.

Another alternative to belt weighing is to differentiate a continuous weight measurement from a feed hopper, to obtain a measure of feed rate; this is known as the 'loss-in-weight' technique. The accuracy of this method is set by accuracy of the hopper weighing system, which may be limited by noise from the mechanism used for discharging the hopper. The method is best suited to installations where the feed rate is normally constant for long periods.

Finally, there is the possibility of nucleonic belt weighing. It has been used with some success in very demanding environments, for example, the coal in-

**Figure 20.10.** Flour weigh-feeder.

dustry, using an arrangement as shown in Figure 20.11. It appears to be capable of about the same accuracy as the impact plate flowmeters. It is perfectly possible to use the equipment in a food factory (the material measured does not become radioactive), but there is a natural reluctance to introduce equipment containing a radioactive source if there is some alternative.

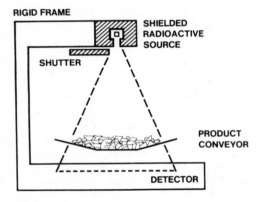

**Figure 20.11.** Nucleonic absorption gauge.

## 20.6. Level (or vessel contents)

The knowledge of the level of a material in a vessel, hopper or tank may be required for a number of reasons. It may be necessary for control purposes, so that the level is maintained at a constant value, or to prevent under- or overfill. It may be required for stock control or to meter specific quantities into a mix.

There is a variety of methods and devices to measure the levels of both liquid and solid materials, and the choice of a suitable detector depends upon the application for which it is to be used, the environmental conditions, the type of material and the accuracy required. In basic terms, level can be determined from measuring either the height of a surface above a datum line or by the weight of the material. In many

cases, food manufacturers prefer to determine the quantity of material in a tank or hopper by placing the container on load cells and weighing. In determining the level of solids, the variable surface configuration of the material make this a more difficult measurement than for liquids. The siting of level sensors must be carefully considered. The effects introduced during filling and emptying hoppers, for example, bridging, coning, angle of repose and funnelling must be borne in mind.

For some food ingredients such as flour, an electromechanical plumb line is sometimes used. This system comprises a metal or plastic weight suspended by a cable from a motor-driven winding drum (Figure 20.12). The weight is automatically lowered from a datum line at periodic intervals. During the descent of the weight, a counting device is actuated which continuously subtracts depth increments from the datum line measurement. When the weight comes into contact with the surface of the material, the loss in cable tension is sensed by a balancing device which locks the depth reading.

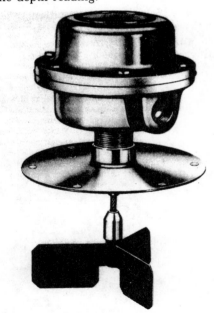

**Figure 20.12.** Mechanical flour level sensor (Photo by courtesy of Auxitrol).

Capacitance probes are to be found in many processes. A metal rod can be used as the electrode when used with non-conducting liquids (e.g. oils) but for conductive liquids, an insulated rod must be used. The insulation can be a thin coating of PTFE, which becomes the dielectric of the capacitor. The capacitance changes with level. Vertical mounting of the electrode is recommended, although alternative means must be used where stirrers make this necessary. Horizontal electrodes are normally used as high/low sensors. Where the vessel wall cannot be used as an electrode, or the dielectric constant of the material to be measured is low, a concentric tube can be fitted around the normal electrode to act as the other conductor. This type of system can be made rugged and easily cleaned; it can be used for liquids and particulate solids. It must be borne in mind that compensation is needed for changes in the dielectric constant of the material with temperature.

Ultrasonic gauges offer convenient non-intrusive level sensing for liquids. Ultrasonic pulses are transmitted towards the liquid surface to be reflected back to the transducer acting as a receiver. The time taken for a pulse to travel the return distance is measured electrically and converts the time to a measure of distance. An example on a chocolate tank is shown in Figure 20.13. Alternatively, the ultrasonic pulse is transmitted upwards through a liquid from the base of a tank, which is preferable if there may be froth on the surface.

Systems using microwaves to measure level in large vessels are similar in operation to the ultrasonic method but using much higher frequencies. A nominal frequency of 10 GHz gives reasonably high accuracy for deep containers.

## 20.7. Other standard physical measurements

### 20.7.1. Pressure

The most widely used pressure transducer in the food industry is the Bourdon-type

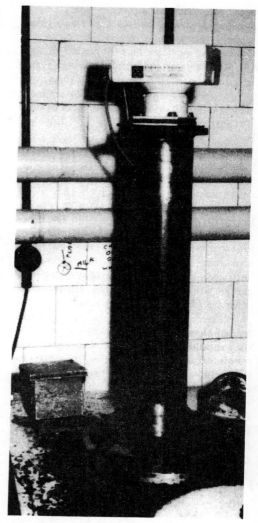

**Figure 20.13.** Ultrasonic level sensor.

gauge giving a dial reading. Probably the next most common are pressure transducers which are used to give an electrical output from strain gauges. The pressure transducers can be compensated for changes in ambient temperature up to approximately 120 °C. However, the compensation is only effective when the transducer is temperature equilibrated. A temperature differential across the transducer can cause erroneous readings.

It is becoming increasingly important to know the internal pressures developed in cans and packages (both flexible and semi-rigid) during processing. For a given process temperature these pressures are higher than saturated steam pressures, and are the result of the combined liquid and air contained in the package. In cans, one factor which influences the internal pressure is the amount of non-condensible gases present in the head space of the can. With a knowledge of internal pressures the suitability of package material and effectiveness of seaming can be assessed.

Adams and Owen (1972) have described a method by which they have measured can pressures using a potentiometric pressure transducer in conjunction with a radio telemetry system; Hawkins and Cooper (1970) hve described a pressure transducer used in conjunction with a Hartley oscillator, so that changes in carrier frequency are used to transmit pressure variations within a can.

The main requirements of pressure transducers for this application are that they should be capable of operating at high temperatures in steam/water, and have a negligible effect on the volume of the head space. It should also be possible to connect the transducer into a can after seaming, or else the transducer must be put into the can or package before seaming, and be capable of passing through the seamer. These devices can operate up to temperatures of 130 °C.

### 20.7.2. Force

Some of the force transducer applications have already been mentioned, such as requirements for weighing vessels and material on conveyors. There are also applications for strain-gauge torque transducers on mixer drive shafts, although the problem of retrieving the signal from the moving part makes this rather complicated and expensive. It is more common to use a power measurement based on electrical measurements in the main drive, with some approximate allowance for power losses in the drive mechanism.

### 20.7.3. Chemical composition

One of the manufacturers of ion-selective electrodes has published a review of applications for the company's products in various parts of the food industry (FPI 1977a). The measurement of pH is particularly important in the gelation of jam. It also has wide implications in fruit-juice processing, in sugar refining, and in dairying. The effects of pH on proteins are observed in the processing of most foods, and particularly in the behaviour of enzymes. Mathason (1977) has described some of the detailed implications of pH control for yeast-leavened products.

Measurement of sulphur-dioxide concentration is another example of important on-line compositional monitoring; in the future, the growing science of food chemistry is bound to set new requirements for monitoring and control of chemical constituents. There will be many applications not just for pH transducers but for a whole family of in-line analytical devices.

### 20.7.4. Humidity

Ambient relative humidity affects the storage of all food kept in unsealed containers. Hand-held wet-and-dry bulb hygrometers are widely used for monitoring storage conditions. It is also quite straightforward to measure humidity automatically up to 90% RH and at temperatures up to 60 °C, using one of a variety of types of sensor.

The most common automatic relative humidity sensors are of the metal oxide type. A porous oxide layer covers a metal base, which forms one electrode of a capacitor. The other electrode is a permeable layer of gold deposited on top of the oxide layer. Water vapour penetrates the oxide structure, causing changes in capacitance which are a function of water vapour pressure. Other popular sensors use hygroscopic layers of absorbing chemicals such as lithium chloride, and a further group of devices measure changes in conductivity in polymer films. All these types of sensor have been used successfully to monitor ambient conditions in food stores. Some care is needed to relate the measurements to absolute levels of humidity.

A different sort of transducer is required to monitor humidity at the higher temperatures used in ovens and driers. In a recent review of humidity measurement in baking ovens (McFarlane 1979) it was shown that a sensor which uses thermoelectric cooling to maintain the temperature of a small pad exactly at the dewpoint has proved the most suitable transducer for dewpoints in the range 40 to 80 °C. Figure 20.14 shows how such a sensor can be mounted in the exhaust flue of an oven.

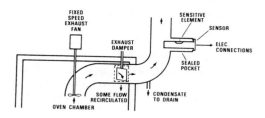

**Figure 20.14.** Humidity sensor in flue.

### 20.7.5. Density

The well-known vibrating tube is the most accurate method for measuring fluid density, and there is some need for such accuracy in the food industry. A short account of the development of in-line density measurement has been published by Harrison (1979), where a compact vibrating spool variation of the vibrating tube density meter is described. The density of air–fluid mixtures can be measured using a radioactive absorption gauge clamped to a pipe; but temperature compensation is necessary, and some mixtures such as batters do not retain a homogeneous dispersion when pumped. Any unevenness in bubble distribution affects the absorption measurement.

### 20.7.6. Colour

The colour of food products is obviously important, and various on-line colour sorting devices are available. For example, nuts and fruits are commonly inspected automatically for colour. These instruments normally work on the basis of comparing reflectance or absorption at two wavelengths, or via two colour filters. This provides compensation for varying intensity in the light source, or varying sensitivity in the detectors, and variations in surface conditions.

There are also instruments for measuring depth of colour. These are used in baking. The characteristic colour of biscuits is made up of a reflectance spectrum as shown in Figure 20.15. Over- or under-baking shifts the reflectance curve throughout the spectrum. The depth of baked colour can therefore be measured by comparing either the overall reflectance or the reflectance at some part of the spectrum with respect to a reference standard.

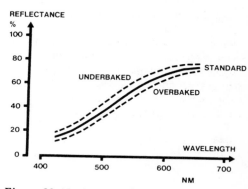

**Figure 20.15.** Biscuit reflectance spectrum.

## 20.8. Special transducers

### 20.8.1. Foreign-body detection

Food manufacturers take every precaution to prevent contamination of their products with foreign bodies, but there are a number of problems in detecting foreign bodies and indigestible components in process streams. X-ray techniques are used in some instances, in particular to detect bones in comminuted meat product. In general, X-ray techniques are expensive both in capital and maintenance, and detection capabilities are limited. Optical techniques involving visible light and photo-electric devices are used to determine foreign bodies or flaws in glass containers, and metal detectors are used in many factories to indicate and warn of the contamination of food products by both ferrous and non-ferrous materials.

A metal detector comprises two basic units — a search head through which the product passes and a control unit. A metal object passing through the sensing head alters the electrical coupling factor between two coils forming part of an oscillating circuit. Most metal detectors employ phase detection to give optimum discrimination; the sensitivity is a function of the volume inspected, and of the shape and orientation of the metal particles within the product. Lowering the frequency increases the sensitivity to ferrous metals but decreases the effect due to non-ferrous materials.

An instrument for detecting a ferrous material in a foil-wrapped product has been developed. In this instrument any metal particle is first magnetized by a strong permanent magnetic field. The particle then moving through a coil induces a small current in the coil.

### 20.8.2. Dimension gauging

Many food manufacturers face the problem of product which is sold by weight having to fit into packages of fixed volume. This is particularly the case for biscuits sold in pile-packs and roll-packs. Automatic measurements of product dimensions at intermediate stages of the process assist operators in keeping the product within specification.

Figure 20.16 shows the operating principle of a device which has been developed to measure biscuit thickness on line, and the device is shown in use in Figure 20.17. It works on the principle of intersecting light paths, and is capable of resolving to within 0.01 mm in 10 mm field of view. A full description has been published (McFarlane 1978).

Various semi-automatic inspection systems are available, based on closed-circuit television images. These are used, for example, in sorting potatoes, where process operators watch the potatoes on a screen, 'marking' rejects with a light pen. These are then separated from good product automatically.

Corn-cobs have also been sorted by on-line optical gauging. In this case the object is to get the pieces in the right orientation for subsequent treatment.

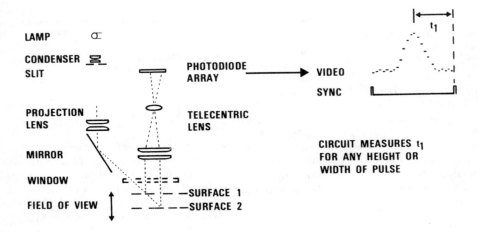

**Figure 20.16.** Heightscan — principle of operation

**Figure 20.17.** Heightscan head

Linear photodiode arrays are available with diodes at 25 $\mu$m spacing which can be scanned at rates as fast as 0.1 $\mu$s per diode and still give an analogue output proportional to light level. The large amount of data collected in this way is potentially valuable for on-line image analysis, which may one day relieve the tedium of visual inspection of food conveyors for defective product.

### 20.8.3. Viscosity

Several well-known viscometers are available for use on food products which measure the force required to rotate a spindle or disc in a cup of fluid, usually over a range of shear rates. The viscosity of food products covers a wide range and many products exhibit non-Newtonian behaviour. A fundamental analysis of shear rates has been published by Bagley (1957) and some special viscometers are described in a review article (FPI 1977b).

The viscosity of food materials is a sufficiently useful guide to quality to have made it worthwhile to develop some very specialized test equipment. This includes equipment to measure the extensibility of flour doughs, which are pseudo-plastic, with viscosity decreasing with shear rate. A model of this behaviour has been described by Remsen and Clark (1978).

In chocolate technology, various special viscometry tests have been devised to monitor the complicated process of chocolate tempering. These have been reviewed by Robbins (1979), who shows that Macmichael viscosity tests and Casson flow values for chocolate can be related to standard Brookfield viscometer measurements.

### 20.9. Future trends

The range of in-line measuring instruments available to food manufacturers is already extensive, and new devices are being developed every year. In this review there has been space only to give a bare outline of the equipment already in use.

Two major trends are apparent. The first is associated with the use of microelectronics in field-mounted measuring systems. The cost of the special items of equipment of weigh feeding, flow measurement, dimension gauging, and other purposes mentioned above is commonly in the range of £1000 to £10 000, and each one incorporates some custom application of standard electronic components. Future versions of these systems will be more compact and reliable, and easier to install and maintain, with the increasing introduction of micro-electronics. Some of them may become cheaper, but even at present costs they can pay for themselves very quickly by improving both productivity and yield, thereby reducing waste.

The second trend is towards the introduction of integrated plant control systems, arising from continued developments in micro-electronics. Microprocessors offer the possibility of control on small process plants, and the future trend will be towards the installation of microprocessor control in smaller factories. Control does not necessarily have to become centralized, in the sense that every motor drive and valve positioner has to receiver its command from control equipment at one central location. This has scarcely ever been a real economic possibility, and the risk of total shut-down of a whole plant is too great. In future, small sections of plant such as blenders, ovens, and packing machines will each be self-regulating. They will have local sequence controllers, temperature control loops, and similar functions to control their routine systems. They will transmit information such as ingredient usage, production rate, and plant status to a central monitoring system; they will in turn receive commands needed to match the throughput at different stages of the process and to trim the plant for optimum performance.

These latter commands may in some cases be computer-generated in a central system, but in many more cases they will be inserted by process operators who have been retrained to do this, using keyboards and displays specifically sited for easy access.

The local plant controls will be very simple and highly reliable. They will operate autonomously, with their own inbuilt control programme, under supervision from a computer-based monitoring system. Any malfunction in the computer system will not affect the plant directly, and the plant can remain in operation while the computer is isolated and the fault rectified. The loss of facilities during computer downtime will reduce the overall efficiency of the plant to some extent, but this can usually be tolerated for short periods. Conversely, if a malfunction occurs in any of the local controls, that part can easily be isolated from the rest of the control system, and again the plant can continue to operate. This is why there is a preference for distributed controls within an integrated scheme for the entire plant.

Moreover, the operators to carry out the new function do not need to be highly trained. Experience with some modest integrated control schemes which already exist has shown that operators of existing plant readily adapt to the new techniques, and this should not pose serious problems in plant management. There will be a need to retrain maintenance staff, and this should begin at once, in preparation for the arrival of new technology.

## References

Steele, D. J. BFMIRA Research Reports Nos 186 (1972), 216 (1975a), 237 (1975), and 295 (1978b).

Christen, G. and Richardson, G. (1976) *J. Assoc Off Anal Chem*, **62** (4), 828–831.

Steele, D. J. BFMIRA Research Reports Nos 223 (1975c) and 255 (1976).

Blitz, J. (1963) *Fundamentals of ultrasonics*, Butterworth, pp. 17–126.

Plache, K. O. (1977) ASME publication WA/FM – 4.

Hope, V. E. (1979) *Int Flavours & Food Additives*, 10 (4).

Proceedings of Inst MC Conference (1979) *Weightech*.

Adams, H. W. and Owen, W. (1972) *Fd Tech*, July, 28–30.

Hawkins, R. E. and Cooper, J. E. (1970) *Chem and Ind*, Nov, 1426–1427.

*Food Processing Industry* (1977) Jan. 30–34.

Mathason, I. J. (1977) *Bakers Digest*, Feb. 52–56.

McFarlane, I. (1979) *Advances in Instrumentation*.

Harrison, R. (1979) *Control and Instrumentation*, July, 45.

McFarlane, I. (1978) *Proc SPIE*, **145**, 50–54.

Bagley, E. B. (1957) *J. App. Phys*, **28**, 624.

*Food Processing Industry*, (1977) Jan. 34.

Remsen, C. H. and Clark, J. P. (1978) *J. Fd Proc Eng*, **2** (1), 39–64.

Robbins, J. W. (1979) *Manf Confectioner*, May, 38–44.

# Paper 21

# Some instrumental techniques for hostile environments

## E. Duncombe

The hostile environment referred to in the title of this paper is the environment within nuclear reactors. Both this and the final paper are therefore about instrumentation in the nuclear power industry.

Accurate measuring instruments are often regarded as delicate devices and are sometimes housed in environmentally controlled areas to preserve their accuracy. In many industrial applications the measurement must be made in conditions of high temperature, high humidity, in the presence of aggressive chemicals, or in the presence of intense radiation. The nuclear power industry presents an extreme case of the need for reliable measurements in a hostile environment, and this paper contains some impressive examples of rugged transducers. (Eds.)

Metrology and process measurement sensors for use in the aggressive conditions within nuclear reactors in general, and within liquid-metal-cooled fast reactors in particular, are outlined. Materials and constructional techniques suitable for high temperatures are discussed. References are made to sensors for strain, displacement, pressure, temperature and flow measurement in the 300–600 °C range of ambients, more especially to less-well-known designs.

## 21.1. Introduction

Whereas in industrial processes and in major experiments it is not often necessary to station sensors (other than those for temperature) within the equipment, it is a relatively frequent requirement in nuclear plant, and in sodium-cooled fast reactors in particular. Most of such sensors located within nuclear reactors must thus be designed to withstand, both internally and externally, the more usual conditions of high temperature, high pressure or chemical attack, as well as to resist the effects of high level radiation.

Although a typical work of reference dealing with instrumental measurements (e.g. Sydenham 1980) will describe many types of sensor, few in general use can operate immersed in aggressive high-temperature atmospheres. For this reason, the nuclear industry has had to evolve sensor designs to cater for its own special high-stress environments, as well as making

Originally published in *J. Phys. E. Scientific Instruments* Vol. 17, 1984.

use of the few suitable types commercially available. This article outlines some of the high-temperature sensors for process measurements, commercially available or otherwise, used in nuclear reactors, especially in sodium-cooled fast reactors with ambient temperatures in the 300–600 °C range. Suitable materials and constructional methods are mentioned, in the belief that these might be of more general interest.

## 21.2. Principles of measurement at high temperatures

Although by no means unusual elsewhere, high ambient temperatures (regarded here as those above 300 °C) are virtually universal application conditions within the shielding of nuclear reactors. This temperature condition alone imposes severe limitations in the materials that can be employed, and so greatly reduces the number of principles of measurement that can be exploited for process variables in high-stress environments. If it is assumed that none of the ever-expanding number of optical methods can be employed, then the basic techniques reduce virtually to conventional pneumatic, electrical and ultrasonic principles. It is not surprising that most of the electrical and ultrasonic devices that have been engineered for use at elevated temperatures are essentially similar in principle to those at lower temperatures but constructed from refractory and corrosion-resistant materials.

Electrical sensors (which form the bulk of those described here) can be of two main types, electromagnetic and capacitative. The former are nominally more suited to liquid environments, the latter to gaseous environments. Unusual materials and constructional techniques are required for both types and these are outlined below.

## 21.3. Sensor construction techniques

### 21.3.1. Connecting cables
Any sensor required to operate immersed in an aggressive environment needs its coupling wiring protected from this environment. One solution for short runs is to have refractory wires, insulated with ceramic beads or quartz-fibre sleeving, contained in small-bore metal tubes. A much more satisfactory solution is to use metal-sheathed mineral-insulated cable (MI cable) which can be obtained in long continuous lengths and in a variety of diameters, sheath and conductor metals and insulating materials. The most readily available types are the well-known thermocouple cables with magnesia or alumina insulant enclosed in a stainless steel sheath, but other conductors, including Nichrome or copper, can easily be obtained. For sensors, single- or twin-conductor MI cable in diameters from 1 to 4 mm are adequate for most purposes, but much larger diameters are feasible. Cables with a single copper conductor up to 10 mm diameter can be obtained for heavy-current purposes. Copper conductor MI cables with a nearly square cross-section have also been made.

Provided the insulant is kept dry, most MI cables can be operated satisfactorily up to 600 °C and up to 1200 °C using the more refractory metals, albeit with reduced insulation values. Cables with single Nichrome conductors (commonly used for trace heating) will operate reliably over long periods at least up to 800 °C with substantial voltage differences to the sheath.

The sheaths of MI cables can be fed through bulkheads and containment walls using compression glands or via ferrules brazed or welded to the sheath. Alternatively, if pipe stubs cooled at their remote ends are available, silicon rubber sealing may be used, especially for multiple cable installations.

### 21.3.2. Materials
Electromagnetic sensors imply the use of ferromagnetic materials. Since laminated magnetic circuits can introduce structural complications and sealing problems, magnetic alloys are most conveniently used in solid form. This demands the use of alloys with low electrical conductivities.

Perhaps the best of these is V Permendur which has a conductivity of $2.5 \times 10^6$ S at 20 °C, and a magnetic Curie point temperature of 980 °C, but the ferritic stainless steels (e.g. EN56A or AISI Type 410) can be used if a lower Curie temperature (about 730 °C) is acceptable. Alloys for sensor cladding or for MI cable sheathing must be chosen to suit the environment but a particular aspect needing to be catered for is the substantial difference in thermal expansion between ferritic and austenitic alloys.

Permanent magnets undergo irreversible changes above 500 °C but Alnico VIII can be used up to 600 °C for short periods without undue loss of energy (Muller and Thun 1980). Insulating coatings can be of flame-sprayed or deposited (McCann 1975) alumina. Ceramic cements such as Brimor or Ceramic-cast, can be used either for coating or for bonding purposes but require a curing process. Soapstone, electrical porcelain, alumina, sapphire or machinable glass may be used for coil formers or bulk insulators or as electrode supports in capacitance devices. As with differing alloys, thermal expansion coefficients have to be allowed for.

*21.3.3. High-temperature inductive windings*
A variety of high-temperature winding constructions have been used from time to time. These include refractory wires (e.g. Nichrome, Chromel or Kanthal) wound on ceramic formers, nickel-clad or stainless-steel-clad copper wires insulated with ceramic textile, and anodized aluminium with ceramic paper interleaving wound on ceramic or ceramic-coated metal formers. In the writer's experience, however, only two winding methods have been found to be reasonably successful (Davidson and Duncombe 1966) and with known long-term reliabilities.

One winding method uses Secon wire, comprising of a noble metal conductor coated with a 'green' semi-flexible ceramic coating. The conductor may be of silver alloy, platinum or gold. The winding may be applied directly onto a ceramic or ceramic-coated former and then 'cured' in a sequence of controlled heat-treatment stages. Alternatively, it can be formed and cured as a self-supporting coil which can be mounted in the sensor body. Flame-sprayed alumina or pasted ceramic cement can be used to anchor the coil and its end connections.

This type of winding can be regarded as an extension of conventional winding methods, but considerable care in design and in handling is required to obtain satisfactory performance. The 400 °C variable-reluctance transducer using gold Secon wire described by Dacey et al. (1971) serves to illustrate the construction used. Another inductive device designed by McCann (1975) also using gold Secon wire has operated for short periods at up to 850 °C.

The second type of winding uses MI cable with twin or single conductors. The windings are formed on austenitic or ferritic steel formers. Long 'tails' are left to act as coupling cables, thus avoiding the complication of junction boxes within the process environment. The cable conductors may be a thermoelectric pair, or single or multiple wires of copper or Nichrome. Provided the sheath is not fractured and conductor-to-sheath voltages are limited to a few tens of volts, Nichrome conductor MI cable seems exceptionally reliable at temperatures up to 800 °C, even in the presence of extremely high radiation levels. It is also the least efficient from an electrical point of view. Copper conductors are limited to the 600 °C region but give the best electrical efficiency, which is still poor by conventional standards and inferior to Secon windings for a given winding volume.

With ordinary care, MI windings can be constructed in any workshop and can be expected to give highly reliable service. Over 100 examples of cable-wound sensors, embracing several designs, have operated in the 300–550 °C region at the Prototype Fast Reactor at Dounreay for over ten years. The very few failures have been due to mechanical damage to the external cable tails.

## 21.4. Strain and displacement

Strain measurements at high temperature in air or in only mildly corrosive gases are relatively common, especially in the aircraft industry, and a number of manufacturers market strain gauges for this purpose. Several types have been discussed by Sharp (1975) and the subject was more recently reviewed by Strong and Procter (1981).

### 21.4.1. Dynamic strain
Most available high-temperature strain gauges are of the resistance type, attached to the measured surface by flame-sprayed alumina or by ceramic cement. This construction obviously is not suitable for corrosive or electrically conducting fluids. A virtually universal (but expensive) solution is to use Ailtech gauges. This encapsulated form of strain sensor comprises a wire gauge embedded in compacted ceramic powder and sealed in a cylindrical metal casing (Figure 21.1). Connections are made with an integral MI cable welded into the casing. The gauge carries a thin metal flange by which it is attached to the metal surface by light spot-welding, an operation requiring only modest skill. Gauges can be obtained in $\frac{1}{4}$-bridge and $\frac{1}{2}$-bridge configurations and with a variety of metal sheathing alloys and MI cables. Operation up to the 800 °C regions is possible with the highest-temperature type. Satisfactory service in 550 °C sodium has

been achieved with the appropriate type of gauge.

### 21.4.2. Static strain
Pneumatic principles can and have been used for high-temperature in-pile creep (Catling et al. 1966) but is not a particularly versatile approach and the need for gas lines penetrating the containment represents an unwelcome safety hazard. Although an accuracy of $1 \times 10^{-3}$ mm over an 0.3 mm range was achieved, other methods are to be preferred.

The Ailtech type of wire resistance gauge can be used for slowly cycling or unidirectional strains, but its usefulness is limited by drifts in the gauge characteristics, particularly at the higher temperatures. Sharp (1975) quotes drift rates of 4 microstrains an hour for the Ailtech design. Gauges suitable for long-term strain and creep measurement include the Hughes, Boeing and CERL/Planer gauges, all depending on capacitance measurement. This latter type, illustrated in Figure 21.2, is reported (Downe et al. 1981) to have an exceedingly small drift rate, of the order of 0.01 microstrains per hour at 600 °C, ignoring drifts for the first ten days or so. Unhappily, none of the three capacitance gauges listed above is available in encapsulated form, so that direct immersion in hostile fluids is not possible.

An alternative but much lower-sensitivity form of capacitance gauge (Pugh 1980) is being investigated, with the objective of evolving an encapsulated construction suitable for operation in the environments in mind. The proposed solution is to use an envelope in the form of a stainless-steel bellows arched into a semicircle (Figure 21.3). Each end of the bellows is welded to a mounting boss, each of which in turn is attached to the measured surface. The semitoroidal volume so enclosed houses a semicircular beam fixed in one boss. The free end of this beam terminates in a vane stationed between an electrode array carried on the other boss, thus transmitting the relative

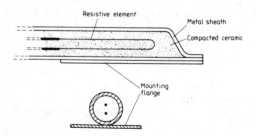

**Figure 21.1.** Active length of Ailtech strain gauge.

displacement (of ± 1 mm) of the two bosses to the capacitance measuring system. This construction is laterally flexible yet able to withstand external pressure without undue forces being applied to the bosses.

A somewhat different means of measuring small strains and creep is to use a radio-frequency resonant cavity device. A convenient form of resonator for this purpose is the relatively low-frequency

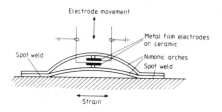

**Figure 21.2.** CERL Planer strain gauge.

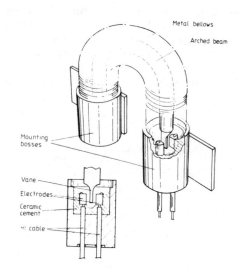

**Figure 21.3.** Encapsulated capacitance strain gauge.

coaxial cavity which has been used for creep of pressure vessels (Irvine 1962). The essential form is shown in Figure 21.4. The resonant frequency is determined by the dimensions of the cavity including the variable capacitance at its flexible end 'window', variation of which alters the resonance frequency. The resonance is excited and detected by inductive loops in the cavity. The construction is easily adapted to high-temperature working, using single-conductor MI cables as low-impedance coaxial lines. Such cables will operate at up to 600 °C, giving acceptable attenuation at between 100 and 600 MHz, depending on the immersed length. High $Q$ values are not attainable with an austenitic stainless-steel construction but are sufficient to resolve with certainty a few tens of micrometres over a 'gauge length' of 100 mm or so.

Although stable and sensitive, the resonator is somewhat bulky for the restricted resonance frequencies in mind and not easy to set up to a standard condition. Another drawback is that the flexible end window can impart a pressure sensitivity to the measurement, avoidable with non-corrosive gases by equalizing the pressure in the cavity via a breather hole. It is possible that a smaller, adjustable form would be feasible, based on the work of Fossheim and Holt (1978).

*21.4.3. Displacement sensors: inductive*
For extensions and displacements greater than those expected with strain and creep, a wider choice of sensor types presents itself. Inductive sensors can be designed for ranges of a fraction of a millimetre up to several metres. Sensor constructions using single Secon windings on unlaminated ferritic armatures and encased in austenitic stainless-steel envelopes result in inductances having a large effective resistive loss component, a component likely to change with radiation-induced (Yancey and Kelsey 1980) or thermally induced metallurgical changes. With suitable precautions, however, satisfactory operation can be obtained. With MI cable wind-

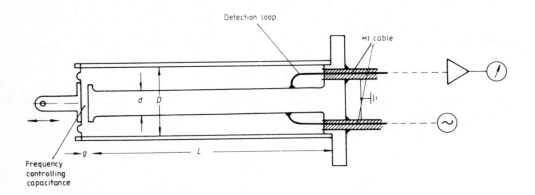

**Figure 21.4.** Irvine RF resonant cavity extensiometer, $f_0 \simeq 600$ MHz; $L = 100$ mm; $D = 30$ mm; $d = 10$ mm; $g = \frac{3}{4}$ mm.

ings, the resistive component tends to dominate the self-inductance of the winding, a $Q$ value of less than $\frac{1}{2}$ being quite usual. For this reason, the mutual inductance change between primary and secondary windings (closely related to self-inductance changes) is a much more satisfactory parameter to monitor, using a constant-current primary excitation and a high input impedance detector to measure the EMF generated in the secondary winding.

A variable-reluctance transducer using Secon wire has been described by Dacey et al. (1971). Other high-temperature displacement transducers of the variable-reluctance type using Secon or similar wires are manufactured commercially. An alternative construction, using a twin-conductor MI cable winding on a solid pot core is shown in Figure 21.5. This sensor operates in conjunction with a ferritic steel 'target' attached to the displaced member, if this itself is not ferromagnetic. Relative movement of the target to the active face of the sensor varies the external magnetic reluctance, causing a similar variation in the self-inductance of the energized conductor of the MI cable. The accompanying change in mutual inductance is indicated by change in the

EMF generated in the second conductor of the cable.

A considerable improvement in linearity and effective temperature coefficient can be obtained if pairs of transducers can be operated in 'push–pull'. This type of transducer has been used to monitor the running clearance of pump shafts rotating in liquid sodium. Predicting the high-temperature calibration by extrapolation from the cold state is far from easy with this simple form of variable-reluctance sensor. This fact, together with its near-inverse-law characteristic, limits its usefulness.

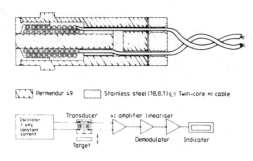

**Figure 21.5.** Variable reluctance transducers wound with M1 cable.

Displacements of the order of a few millimetres can be catered for by a somewhat different and simpler form of sensor, currently under development. This form, due to Roach (1981) is illustrated in Figure 21.6. It comprises a pair of parallel ferromagnetic plates arranged to move transversely at a fixed separation by flexure springs or spacing studs. The opposing faces are grooved to receive a single-turn winding of MI cable. One winding is energized from a constant-current AC supply; the other develops an EMF related to the mutual inductance between these two windings. At a partially overlapping position, this mutual inductance becomes zero, a state governed primarily by the geometry of the system, so giving a stable reference zero. The order of sensitivity of an experimental device using 1 mm diameter MI cable is shown in Figure 21.6.

The minimum range of this design of sensor is determined by the diameter of the cable used. Larger ranges can be obtained by increasing the spacing of the conductors. The operative range can also be extended (at the risk of ambiguous readings) by increasing the number of grooves and windings, giving a series of zero values which allow constant accuracy and high incremental sensitivity over long ranges.

An interesting feature of this design, intended for multiple measurements in reactor structures, is that it lends itself to operation in a matrix array. It also offers the possibility of assembling sensors onto continuous, previously laid cables, loops of these cables being crimped into their grooves as a part of the *in situ* installation process. Provided the cable sheathing and the ferromagnetic plates are not attacked chemically, this device can operate immersed in virtually any process fluid at up to 700 °C.

Larger displacements can be measured by the basic MI-wound sensor arrangement outlined in Figure 21.7, now widely used to monitor the fluid level of liquid sodium metal, as illustrated in the figure. In the application shown in the figure, any variation in sodium level causes

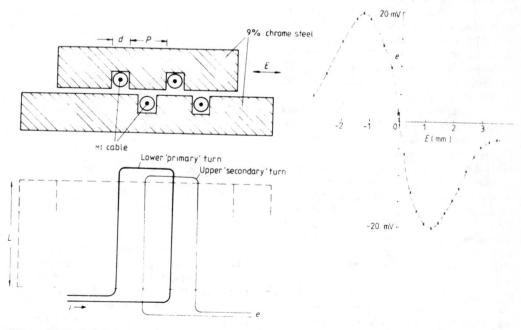

**Figure 21.6.** Parallel plate extensiometer, $P = d = 1.1$ mm; $L = 100$ mm; $f = 20$ kHz; $i = 1$ A.

related changes in the self-inductance of the energized primary winding and hence of the mutual inductance with the secondary winding due to the 'short-circuited turn' effect of the sodium. With a constant-current excitation of the primary, typically in the 2–10 kHz region, the EMF induced in the secondary winding is directly related to level. Even with an interposed stainless-steel pocket, the sensitivity is adequate. Satisfactory operation as a long-range displacement sensor (Davidson and Duncombe 1966) in non-conducting media can be obtained using a metal sleeve or an axial metal plunger attached to the displaced member.

The sensor (Pettinger et al. 1965) consists of two single-layer interposed MI cable windings, wound side by side as a pair and brazed to the central stainless-steel or Inconel former. Lengths up to 2 m are relatively easily constructed, using a lathe as a winding machine. The winding tails are left long to form the coupling cables. This design can operate at up to 700 °C continuously and withstand short excursions up to 900 °C without damage. This basic design of sensor is virtually unaffected by very high pressures.

Although clumsy and of poor electrical efficiency, this design based on MI cable has been shown to be remarkably stable and reliable in service. Substantial temperature compensation is required, so that calibration under actual or simulated application conditions is usually a requirement. The use of transformers, as shown in Figure 21.7, increases the effective reliability in that operation in the presence of earth leakages or with a single earth fault is possible, as well as providing impedance matching and electrical common-mode noise rejection. A major problem with long-range devices when used in aggressive atmospheres is that of providing suitable sliding bearings for the operating collar or plunger.

The MI cable form of construction can, of course, be adapted to the 'differential transformer' form of winding. Plungers consisting of insulated ferromagnetic laminae or wires shrouded in a thin protec-

tive housing, can give greater electrical efficiencies especially for short ranges but not as great as might be supposed from experience with more conventional constructions. Many variants of this construction are possible, e.g. devices for detecting liquid metal levels in ducts (Roach and Dacey 1981).

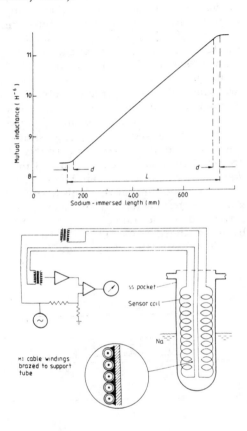

**Figure 21.7.** Sodium liquid level sensor. Coil diameter, $d = 25$ mm; coil length, $L = 600$ mm.

### 21.4.4. Displacement sensors: capacitance
The use of differential charge amplifiers or three-terminal transformer bridges allows very small capacitance changes to be measured at the remote ends of long cables, and these cables can be of the MI variety (Massey 1977) provided the insulant is kept dry. The electrode arrays used can

be simple parallel plates, or of 'differential' form or of the 'fringe field' type (Walton 1977a, b). Some basic configurations of electrodes and charge amplifiers are illustrated in Figure 21.8 but other arrangements, somewhat less easily engineered in the present context, can be made (Hugill 1982). The fringe-field configuration provides a particularly convenient construction with near-linear characteristics. It is, however, the least sensitive. The very high displacement sensitivities achievable (of the order of fractions of micrometres) are exemplified by the Planer/CERL, Boeing and Hughes creep gauges referred to previously. Other forms of creep and displacement gauges are manufactured commercially, mainly for use in air or inert gas.

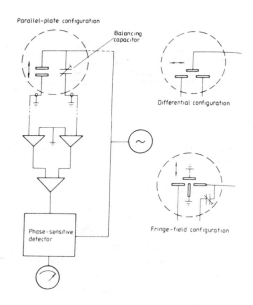

**Figure 21.8.** Basic capacitance electrodes/charge amplifier configurations.

One design problem is that of arriving at a dimensionally stable electrode array capable of withstanding cycling up to high temperatures. Sintered alumina or machinable glass ceramic are satisfactory bases for

such designs. Another, less tractable, problem is that of keeping the process fluid away from the electrode system unless it be a non-corrosive insulant. For this reason, displacement measurements in conductive or corrosive fluids are limited to the range achievable with metal diaphragm or bellows sealing, i.e. a few millimetres. Nevertheless, sensitive short-range displacement detectors can form convenient bases for high-temperature pressure gauges and accelerometers (Walton 1977).

A particular example of an enclosed high-temperature capacitance-measuring system is the high-temperature inclinometer, designed (Thomson et al. 1982) to monitor for changes in verticality of the above-core structure of the Dounreay Fast Reactor, representing a non-specific but sensitive method of measuring dimensional changes. The device is essentially a plumb-bob suspended by a stranded stainless-steel wire within a sealed argon-field tube (Figure 21.9). The suspension wire and plumb-bob are electrically insulated from the suspension point and energized from a 20 kHz supply. The bob forms the central electrode of a differential capacitance displacement-measuring system formed by two pairs of electrodes mounted orthogonally in the support tube. The electrodes in this particular instance are formed by the single conductors of the MI coupling cables from which the sheath has been removed locally. These cables are locked into grooves cut in the outer surface of a liner tube and the exposed conductors 'see' the plumb-bob through drillings in this tube. The device covers 11 mm of lateral displacement of its lower end, relative to the suspension point of the plumb-bob, representing a small fraction of a degree of arc.

## 21.5. Temperature measurement

The development and production of the MI form of thermocouple cable has been greatly stimulated by the demands of the nuclear industry. This cable is now in widespread use, but some development continues. A high-integrity form has been described by Thomson and Evans (1978).

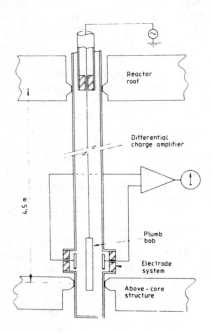

**Figure 21.9.** In pile inclinometer.

The coaxial construction (Figure 21.10(*a*)) was evolved to meet the need for very reliable thermocouples for service in reactor protective systems. The design has been shown to have very high resistance to rapid thermal cycling and to have a fast response with small sample-to-sample variation. Furthermore, because of its double-skinned sheath, not shown in Figure 21.10(*a*), it has a very low probability of failure due to foreign inclusions in the sheath metal. It is also immune to undisclosed faults, i.e. spurious readings must be accompanied by either a conductor open circuit or short-circuit to sheath.

Carrol *et al.* (1982) have described a 'side-welded' form of grounded-junction

thermocouple. A 1.6 mm diameter sheath cable to this construction (Figure 21.10(*b*)) is quoted to have an exponential time constant of 17 ms (nearly an order of magnitude faster than with a conventional grounded-junction sheath closure) in flowing liquid metal.

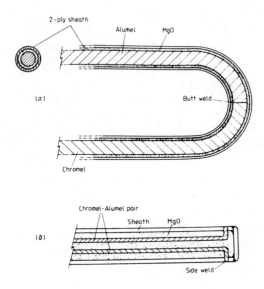

**Figure 21.10.** Coaxial (*a*) and side-welded (*b*) thermocouples.

Tungsten–rhenium thermocouples are used for the arduous duty of reactor fuel temperature measurements in the 1600–2000 °C region. Use has also been made of ultrasonic temperature measurement for this purpose. The basic form of an ultrasonic temperature sensor is illustrated in Figure 21.11. An ultrasonic pulse is generated by a magnetostrictive generator/receiver and is propagated along a transmission line to the active sensor length at the remote end. Part of the pulse energy is reflected from the discontinuity at the transmission line sensor junction, the remainder from the open end of the sensor length. The time interval between the two pulses arriving at the generator/receiver is proportional to the velocity of sound in the sensor, which is a function of temperature. The transmission line may be several metres long and the magnetostrictive head can be constructed to operate in the 600 °C region. This device has the particular property of indicating mean temperatures over its active length.

Various aspects of high-temperature thermocouples and of ultrasonic thermometry were discussed at the Petten Colloquium (1974). More recent experiences on specific designs are published by Tasman et al. (1977) and Hun and Weber (1979). Temperature measurement up to 2500 °C with an uncertainty of 30 degrees by ultrasonic thermometry has been reported by the latter.

Temperature changes in a metal usually give related changes in electrical conductivity. This conductivity change can be monitored electromagnetically, using an inductive sensor (Hughes 1972). McCann (1975) used this principle to monitor fast thermal fluctuations in a flowing stream of sodium–potassium alloy at 850 °C. A smaller version of this type of device was used by Dean (1977) to measure temperature fluctuations of the order of 10 ms rise time in flowing sodium at 550 °C, through the wall of a 5 mm bore stainless-steel containing pocket, the sensor being traversed to observe the values at several vertical positions. The principle is, of course, applicable to metals in solid form.

Another non-contacting device, intended for monitoring for excess temperature, is a form of temperature 'transmitter' shown in Figure 21.12. This device (Duncombe and Winstanley 1982) comprises a 'sender' consisting of a permanent magnet enclosed within a sheath of ferromagnetic material, the Curie temperature of which is chosen to coincide with the required detection level. Below the Curie temperature, the external magnetic field is reduced to a low value by the shunt provided by the magnetic sheath. Above its Curie temperature the sheath loses its ferromagnetic properties, so that the unperturbed external field of the permanent magnet appears. This field can be detected at a point some distance removed from the sender, thereby providing a remote indication of excess temperature. A detector in the form of a fluxgate magnetometer, wound with MI cable on an iron core, has been shown capable of stable operation at up to 600 °C and to readily resolve flux changes of $5 \times 10^{-5}$ T, so that reliable operation typically at 0.4 m distance can be achieved with the detector stationed within the high-temperature environment. One application of this device is to monitor for excess outlet temperatures in fuel subassemblies in a reactor without the need for thermocouples mounted over the outlets.

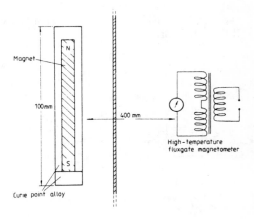

**Figure 21.12.** Temperature threshold transmitter.

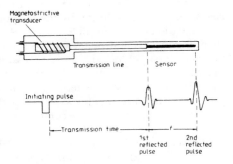

**Figure 21.11.** Ultrasonic thermometer.

The principle has been proposed for temperature monitoring in centrifuges for radioactive liquids.

## 21.6. Pressure measurement

Fluid pressure is a variable usually easily measured outside the process containment. If required within a high-temperature environment, pressure measurement presents particulary difficult problems in that very few materials have good elastic properties in the 500 °C and upward region of temperature. For steady-state measurements, one simple solution is to balance the pressure existing across a flexible diaphragm with a known adjustable external gas pressure, using an electrical contact to determine the balance point. For other than occasional measurements, a more satisfactory method is to use the arrangement outlined in Figure 21.13. A flexible diaphragm isolates the measured fluid from a filling liquid which transmits the applied pressure to a conventional pressure-measuring instrument stationed at a point removed from the high-temperature environment. The filling liquid can conveniently be the sodium-potassium eutectic which is fluid at room temperature, has a small vapour pressure at the 600 °C region and is compatible with most high-nickel alloys. This type of measuring system is available in differential form.

For the best transient response, the pressure sensor must be stationed within the environment. Sensors capable of operating at temperatures in the 600 °C region are available using capacitance or electromagnetic means of sensing the deflection of the measuring diaphragm. The latter form can be of the gauge, absolute or differential pressure type. The inherent response speeds are better than $10^{-3}$ s.

Billeter (1972) has described an ingenious combined pressure and temperature sensor, based on a cylindrical microwave resonant cavity with a flexible end-window. Dilation of the diaphragm by external pressure, or of the cavity dimensions by temperature gives changes in the resonant modes, excited and detected via waveguides. A major drawback to this type of sensor is the need for waveguides penetrating the process containment and so forming a safety hazard.

The pressure being developed within a hermetically sealed vessel has been monitored (Humphreys and Caldwell-Nichols 1981) by measuring hoop strain of the vessel wall using strain gauges of CERL/Planar capacitance type, a method capable of extension to high-temperature ducts and vessels for *ad hoc* measurements.

## 21.7. Flow measurements

Flow-measurement problems with hot aggressive fluids usually lend themselves reasonably well to solution by the adaption of conventional pressure-difference or eddy-shedding types of flowmeter, i.e. by solving the steady-state or dynamic pressure-measuring sensor problem. Where this is not possible (e.g. where significant pressure drops are unallowable), then the choice of flowmeter becomes more restricted.

Electromagnetic flowmeters of the Faraday type are useful for conductive fluids, provided electrodes and insulating flowtube liners can be obtained which will withstand chemical attack and thermal

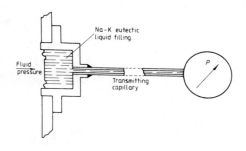

**Figure 21.13.** Liquid-filled pressure gauge.

cycling, and that the flows are not too small. For liquid metals, with their very high electrical conductivities, and for the alkali metals in particular, simple constructions energized from permanent magnets may be made using metal ducts with welded-on potential taps. Such simple flowmeters are extensively used in the sodium-coolant circuits of fast breeder reactors. The head loss with liquid-metal flowmeters, however, is not negligible. Liquid-metal flowmeters on large ducts (i.e. greater than 200 mm bore) become prone to non-linearities due to distortion of the applied fields (Thatcher et al. 1970. Thatcher 1974).

The various forms of ultrasonic flow-measuring methods lend themselves to low-head-loss applications and can be used on very large ducts. The main problem is that of designing suitable high-temperature acoustic transducers or of providing cooled mounting stubs or acoustic waveguides.

Flow measurement based on the transit time of thermal perturbations between thermocouples spaced along the direction of flow (Bentley and Dawson 1966) ideally should be applicable to virtually any fluid. In practice, accurate measurement is difficult and the terminal cross-correlation equipment elaborate although no longer unduly expensive. Bentley (1980) applied this technique to the measurement of 400 °C sodium flows in inaccessible ducts in the Dounreay Fast Reactor. The presence of bends between measuring stations was shown to introduce serious uncertainties. Using corrections derived from experimental measurements in replica ducts, individual measurements to 6% of flow were obtained. The Bentley (1980) paper is one amongst many presented at a meeting devoted to discussing electromagnetic, ultrasonic and transit-time analysis flowmeter techniques applied to liquid-metal flows in large ducts. Further developments of cross-correlation methods and the use of other noise phenomena and sensors have been discussed by Beck (1981).

The flow-measuring principles mentioned above are well known, especially those for pipeline measurement (Brain and Scott 1982). A less well-known form is the flux distortion flowmeter (sometimes called the eddy-current flowmeter). This electromagnetic flowmeter depends on the distortion of the magnetic field distribution in an electrical conductor when perturbed by the relative motion of this conductor with the applied field. In effect, the field distribution appears to be displaced in the direction of the motion of the conductor, relative to the exciting field, and to a degree directly related to velocity. This field distortion is due to the components of magnetic field produced by the eddy currents generated in the moving conductor by the exciting field.

One arrangement of the flux distortion flowmeter, together with an indication of the sensitivity obtainable with liquid sodium is illustrated in Figure 21.14. A central primary winding on a metal flow duct is excited from a constant-current AC supply. This winding is flanked by two similar, symmetrically disposed windings connected in series opposition. Displacement of the magnetic field of the primary winding due to motion of the sodium modifies the effective mutual inductances of the secondaries with the primary, resulting in a difference EMF appearing at the secondary terminals, very nearly linearly related to sodium velocity within the electrical 'skin-depth' of the sodium.

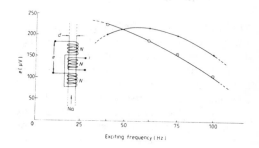

**Figure 21.14.** Flux distortion flowmeter, $d = 100$ mm (bore); $i = 400$ mA; $N = 18$ turns; sodium flow rate = 2.85 m s$^{-1}$. Sodium temperature: ⊙. 200 °C; +. 325 °C.

With refractory MI cable windings, cooled to below 1000 °C, the main temperature limitation is that of the material of the flow tube.

An alternative construction is to have the flow outside the windings. A construction of this type for insertion into a pocket projecting into the measured stream (Davidson and Duncombe 1966) has been further developed into a form wound in MI cable, suitable for direct immersion in flowing sodium (Dean et al. 1970). Three examples of the Dean design have operated in 450 °C sodium within the core of the Dounreay Fast Reactor for 10 years, demonstrating the reliability of the construction. Unhappily, although optimum design parameters can be predicted (Thatcher et al. 1976), an actual calibration is required with the working fluid. Another form of this flowmeter has been described by Wiegand (1972). In this design, the distortion of the field from a permanent magnet is observed, using a high-temperature fluxgate magnetometer. Although this basic form of flowmeter (Lehde and Lang 1948) was used to measure the velocity of seawater, a high-temperature construction would be unlikely to give satisfactory performance with fluids having resistivities of the order of those of the liquid alkali metals, i.e. less than about $2 \times 10^{-8}$ $\Omega$ m.

## 21.8. Acoustic measurement techniques

Although acoustic techniques are not always easy to apply, they not only enlarge the range of physical measurement methods that can be exploited, but can also provide suitable alternatives to the use of specific physical sensors (Crecraft 1983). The main obstacle to the use of acoustic methods at high temperatures is the difficulty of designing sensors and transducers to operate reliably and well under the desired conditions. Magnetostrictive transmitters and receivers wound with MI cable have been developed by

Hans and Podgorski (1978). The usable bandwidth of about 100 kHz is a substantial limitation with magnetostrictive designs but is adequate for monitoring for ultrasound generated by fluid flow, boiling and by structural vibration. These devices can operate at up to 600 °C, as can the acoustic thermometers mentioned earlier.

The inherent bandwidth of piezoelectric ceramics extends above the 5 MHz region and enables wider applications. The upper temperature limit of the most commonly obtainable material, namely lead zirconate (PZT) ceramic with a Curie temperature of 360 °C, is at the lower end of the temperature range in mind. Lithium niobate (Curie temperature 1200 °C) has a practicable upper temperature limit exceeding 700 °C and hence is the subject of device development in several countries. Papers describing the design and application of high-temperature acoustic devices in liquid-metal systems have been published by Argous et al. (1980), Hoitink et al. (1980) and McKnight et al. (1980). An interesting aspect covered in these papers is that of under-sodium viewing by acoustic signals. More recently McKnight (1983) has operated an under-sodium viewer in the Dounreay Fast Reactor using lead-bonded PZT transducers (Figure 21.15).

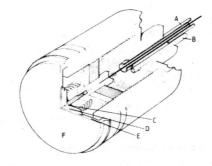

**Figure 21.15.** PZT ultrasonic transducer; $A$, MI cable; $B$, gas filling tube; $C$, backing plate; $D$, crystal; $E$, lead bonding; $F$, active face.

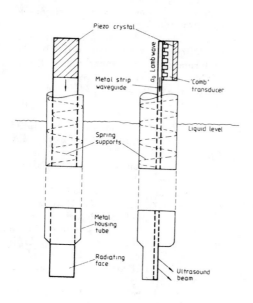

**Figure 21.16.** Ultrasonic waveguide.

background noise. Longer, solid metal waveguides present difficulties arising from energy reflections and refractions from inclusions and inhomogeneities in the metal. For this reason, waveguides filled with alkali metals are being investigated.

Watkins et al. (1982) have described a substantially differing form of waveguide to the simple rod or liquid column. This design (Figure 21.16) is based on the propagation of Lamb waves along along a thin metal strip encapsulated in a tube, using a comb type of transducer having equal half-wavelength mark/space dimensions. Satisfactory operation with water at 2 MHz over a 10 m length has been demonstrated and a similar demonstration in a liquid-sodium facility is being prepared. If this type of waveguide can, indeed, be operated at elevated temperatures then it opens the way for the economic application of a range of ultrasonic fluid level, displacement and inspection techniques within large-scale plant.

Three-dimensional images of the above-core region of the reactor have been constructed from ultrasound signals, with a resolution of 0.5 mm. Although limited to sodium temperatures of 280 °C, there is no reason why this equipment, and other much less sophisticated measurement systems, cannot be operated at much higher temperatures if suitable sensors can be engineered. Examples of lithium niobate transducers for experimental purposes are already available.

The need for high-temperature acoustic sensors and transducers can be avoided by the use of waveguides. Short waveguides have been used in ultrasonic pipe flow measurements, e.g. by Uno et al. (1980). McKnight (1979) has also used short angled waveguides to detect and measure gas bubbles down to 10 $\mu$m in diameter in flowing sodium at 600 °C, the principle being to excite the bubbles into resonance and to use the Doppler frequency shift to distinguish the resonance from the

## 21.9. Conclusions

Although invariably more difficult than in passive moderate-temperature environments, acceptable standards of performance can be obtained with sensors exposed to hot corrosive conditions. Inductive and capacitance devices for operation up to 600 °C are commercially available, or can be constructed without undue difficulty, to monitor process variables and dimensional changes. Experience in the nuclear industry has shown that acceptable long-term reliability can be achieved, particularly with devices based on MI cable windings, although such constructions are necessarily somewhat clumsy, always expensive and electrically inefficient. It is likely that current developments in sensors and in waveguides will eventually allow ultrasonic methods to be used with confidence in remote measurement stations within aggressive environments to give a substantial expansion in the range of possible measurements under these circumstances.

# References

Argous J. P., Brunet M., Baron J., Lhuillier C. and Segut J. L. (1980) Immersed acoustical transducers and their potential use in LMFBR. *2nd Int. Conf. on Liquid Metal Technology in Energy Production* (Conf-800401) (Richland, USA: Am. Nuc. Soc.) pp. 423–9.

Beck M. S. (1981) Recent developments and the future of cross-correlation flowmeters. *Proc. BHRA Conf. Adv. in Flow Measurement Techniques. September 1981, University of Warwick, UK.* Paper K.1, pp. 242–52.

Bentley P. G. (1980) Primary circuit flow measurement in the Dounreay PFR measurement in large LMFBR pipes. *Proc. IAEA Specialists Meeting on Sodium Flow Measurements in Large LMFBR Pipes, Berglisch Gladbach, Germany* (Vienna: IAEA) pp. 73–7.

Bentley P. G. and Dawson D. G. (1966) Fluid flow measurement by transit time analysis of temperature fluctuation. *Trans. Soc. Inst. Tech.* **18**, 183–93.

Billeter T. R. (1972) Composite temperature–pressure measurement instrument for fast reactors. *IEEE Trans. Nucl. Sci* **NS-19**, 814–9.

Brain T. J. S. and Scott R. W. W. (1982) Survey of pipeline flowmeters. *J. Phys. E: Sci. Instrum.* **15**, 961–72.

Carroll R. M., Carr K. R. and Shepard R. L. (1982) Studies of sheathed thermocouples, construction and installation in thermowells to obtain faster response. *Proc. Symp. Temperature: Its Measurements and Control in Science and Industry, Washington, USA, April 1982* (to be published).

Catling E., Wallace F. G. and Ingersol A. (1966) In-pile creep measuring techniques. *Proc. Int. Symp. In-pile Irradiation Equipment and Techniques, AERE, Harwell.* Paper 0.3.

Crecraft D. I. (1983) Ultrasonic instrumentation: principles, methods and applications. *J. Phys. E.: Sci. Instrum.* **16**, 181–9.

Dacey R., Davidson D. F., Leece J. and Harrison E. (1971) Sodium pump instrumentation. *Nucl. Engng Int.* **16** (177), 208–10.

Davidson D. F. and Duncombe E. (1966) Some transducer techniques for use at elevated temperatures. *Trans. Soc. Inst. Tech.* **18**, 51–9.

Dean S. A. (1977) Unpublished work. Nuclear Power Development Laboratories, UKAEA Risley, Warrington.

Dean S. A., Harrison E. and Stead A. (1970) Sodium flow monitoring. *Nucl. Engng Int.* **15** (174), 1003–7.

Downe B., Fidler R., Noltingk B. E., Procter E., Williams J. A. and Phillips L. S. (1981) Performance and application of the CERL-Planer strain transducer. *Proc. BSSM–SESA Int. Conf. Measurements in Hostile Environments. August 1981,* pp. 15–20.

Duncombe E. and Winstanley J. P. (1981) Temperature threshold detectors. *UK Patent Application* 8123184.

Fossheim K. and Holt R. M. (1978) Broad band tuning of helical resonant cavities. *J. Phys. E: Sci. Instrum.* **11**, 891–3.

Hans R. and Podgorski J. (1978) Development of magnetostrictive and piezoelectric high-temperature resistant sensors, a condition for reliable reactor diagnosis. *Proc. IAEA Symp. Nuclear Power Plant Control and Instrumentation* IAEA-SM-226/1 (Vienna: IAEA) pp. 421–37.

Hoitink M. C., Horn J. E., Michaels T. E., Sheen E. M. and Yatable J. M. (1980) Under-sodium viewing development for FFTF. *2nd Int. Conf. on Liquid Metal Technology in Energy Production* (Conf-800401-P1) (Richland, USA: Am. Nuc. Soc.) pp. 437–43.

Hughes G. (1972) Detection of rapid electrical fluctuations in high temperature liquid metals. *J. Phys. E: Sci. Instrum.* **5**, 349–53.

Hugill A. L. (1982) Displacement transducers based on reactive sensors in transformer ratio bridge circuits. *J. Phys. E: Sci. Instrum.* **15**, 597–606.

Humphreys P. and Caldwell-Nichols C. J. (1981) Unpublished work. Nuclear Power Development Laboratories, UKAEA Risley, Warrington.

Hun A. L. and Weber T. (1979) *Proc. Review Group Conference on Advanced Instrumentation for Reactor Safety.* NUREG/CP-0007 (Washington DC: US Nuc. Reg. Soc.).

Irvine W. H. (1962) Extensiometer. *British Patent* 970170.

Lehde H. and Lang W. T. (1948) *US Patent* 2435043.

McCann J. D. (1975) Fast response temperature sensor for use in liquid sodium. *AERE Report* AERE R 7972.

McKnight J. A. (1979) Improvement in or relating to the detection of bubbles in a liquid. *British Patent* 1556461.

McKnight J. A. (1983) The use of ultrasonics for visualising components of the Prototype

Fast Reactor whilst immersed in liquid sodium. *Paper to be presented at the Ultrasonics Int. Conf. Halifax, Canada, July 1983.*

McKnight J., Bishop J., Cartwright D. K. and Diggle W. R. (1980) The applications of ultrasonic technology under sodium. *2nd Int. Conf. on Liquid Metal Technology in Energy Production* (Conf-800401) (Richland USA: Am. Nuc. Soc.) pp. 116–22.

Massey L. M. (1977) High temperature capacitive transducer techniques. *Proc. IEE Colloquium 'Transducers for Use at High Temperature'* Digest No. 1977/58 (London: IEE).

Muller S. and Thun G. (1980) Performances of permanent magnet flowmeter probes for instrumentation of LMFBRS. *Proc. 2nd Int. Conf. on Liquid Metal Technology in Energy Production* (Conf-800401-P1) (Richland, USA: Am. Nuc. Soc.) pp. 444–451.

Petten J. C. R. (1975) *Int. Colloquium on High Temperature In-pile Thermometry Petten, Netherlands* (Eur. 5395).

Pettinger D. S., Duncombe E. and Harrison E. (1968) Position indicating devices. *British Patent* 1101058.

Pugh H. (1980) Capacitance strain gauge. *Patent Application* 12849 DR.

Roach P. F. (1981) Sliding plate transducer. *British Patent Application* 8130428.

Roach P. F. and Dacey R. (1981) Liquid level measuring instruments. *British Patent Application* GB 2061517A.

Sharp W. N. Jr (1975) Strain gauges for long-term high temperature strain measurements. *Proc. Soc. Exptl Stress Anal. (USA)* **32**, 482–8.

Strong J. T. and Procter E. (1981) Strain measuring systems and protections for adverse environments. *Proc. BSSM–SESA Inst. Conf. Measurements in Hostile Environments, Edinburgh, August 1981* (to be published).

Sydenham P. H. (1980) *Transducers in Measurement and Control* (Bristol: Hilger).

Tasman H. A., Schmidt H. E., Richter J., Campana M. and Fayl G. (1977) The Treson Experiments: Measurement of temperature profiles in nuclear fuels by means of ultrasonic thermometers. *High Temp. –High Pressures* **9**, 387–406.

Thatcher G. (1974) *Electromagnetic Flowmeters for Liquid Metals, Modern Methods of Flow Measurement* (Stevenage: Peter Peregrinus) pp. 359–90.

Thatcher G., Bentley P. G., McGonigal G. (1970) Sodium flow measurement in PFR. *Nucl. Engng Int.* **15** (173), 822–5.

Thatcher G., Dean S. A. and Roach P. F. (1976) Flux distortion flowmeter development. *Proc. IAEA Specialists Meeting on the In-core and Primary Circuit Instrumentation of LMFB Reactors, Warrington, England* (Vienna: IAEA) pp. 307–32.

Thomson A., Cladwell-Nichols C. J. and Roach P. F. (1982) Inclinometer for fast reactors. *European Patent Application* 92303010.1.

Thomson A. and Evans R. A. (1978) High integrity thermocouples for in-pile applications. *Proc. IAEA Symp. Nuclear Power Plant Control and Instrumentation.* IAEA-SM-226/109 (Vienna: IAEA) pp. 401–19.

Uno O., Araki H., Horikoshi S., Ozaki Y. and Oda M. (1980) Research and development of the ultrasonic flowmeter for LMFBR. *Proc. IAEA Specialists Meeting on Sodium Flow Measurements in Large LMFBR Pipes Berglisch Gladbach, Germany* (Vienna: IAEA) pp. 95–104.

Walton H. (1977a) Transducers for nuclear reactor measurements. *Proc. IEE Colloquium 'Transducers for Use at High Temperatures'* Digest No. 1977/58 (London: IEE).

Walton H. (1977b) Capacitance transducers in severe environments. *Proc. 'Transducer 1977' Conf.*

Watkins R. D., Deighton M. O., Gillespie A. B. and Pike R. B. (1982) A proposed method for generating and receiving narrow beams of ultrasound in the fast reactor liquid sodium environment. *AERE Report* AERE-R-9965.

Wiegand D. E. (1972) The magnetometer flow sensor. *Argonne Nuclear Laboratory Report* ANL-7874.

Yancey M. E. and Kelsey P. V. (1980) Irradiation effects upon selected ceramic cement and ceramic insulated wires for radiation resistant transducers. NUREG/CR/1335 (Washington DC: US Nuc. Reg. Soc.).

Paper 22

# The impact of the accident at Three Mile Island on plant control and instrumentation philosophy

*F. Catlow*

Instrumentation systems as defined in this book do not provide direct inputs to control systems but provide data for recording or display. The display, or so-called man–machine interface, is a key part of instrumentation systems which is easily neglected to the detriment of overall performance. Consideration of the accident at the Three Mile Island nuclear power plant returns us to an issue raised in the very first paper, the credibility of instrumentation displays. The operator must be in a position to construct an accurate mental picture of conditions and events within the plant, simply on the basis of instrument readings. When these readings conflict, the operator is in the unenviable position of having to decide which are correct. (Eds.)

Independent commissions which were appointed to evaluate the causes of the accident at the Three Mile Island nuclear power plant in the USA exposed major weaknesses in the man–machine interface which they felt might be common to other similar plants. Strengthening this link is regarded as twofold: educating the man to enhance his understanding of plant processes, and improving the machine interface.

The paper reviews suggested improvements in instrumentation which would aid the control of a nuclear plant. These comprise mainly: the application of human factors engineering principles to control design in order to make the 'machine' more manageable, and improved data feedback so that the operator can make an accurate assessment of plant status at any instant.

The author considers that there is a likelihood that the general philosophy of the man–machine interface being applied to the nuclear industry could be applied to some extent to conventional power plants and even other industries.

Originally published in *Trans. S.A. Inst. Elect. Engrs* (August 1983).

## 22.1. Introduction

Fear of nuclear power is one of the major 'hang-ups' of our time. Assuring the public that the energy contained within a nuclear reactor can be controlled with adequate safety to inspire confident acceptance of nuclear power is an extremely sensitive issue. It is for that reason that the accident which occurred at the Three Mile Island (TMI-2) nuclear power plant provoked such widespread reaction. Paradoxically, the harmful effect to the public was trivial although the reactor itself was severely damaged.

Partial melting of the core took place due to high temperatures attained in the absence of adequate cooling. The fact that the radioactive material released by an accident of such magnitude was contained without causing any detectable harm to a single person is regarded by the proponents of nuclear power as vindication of existing reactor design, protection and safeguards. The opponents of nuclear power, on the other hand, regard the accident as a warning that the outcome of future accidents may not be so fortunate.

The various commissions (Kemeny et al. 1979, Rogovin et al. 1980) that were established to investigate the accident all arrived independently at the same conclusion: that the major cause of the accident was due to the operators themselves who due to an inadequate understanding of the plant process not only failed to take corrective action but actually nullified the automatic protection. In this way the operators contributed greatly to the accident which otherwise would have been a minor incident.

To some individuals the problem is principally one of lack of competence on the part of the operators and can be remedied by improved standards of training. Others argue that the operators were understandably confused and that it was impossible for them to evaluate the true state of the plant due to the misleading information that was presented in the control room.

The major recommendations to come out of the TMI investigations may be summarized as follows:

> Improved training, procedures and technical counselling should be provided for reactor operators in order to increase their competence to control the plant during normal and abnormal conditions.
> Control and instrumentation requires to be improved in order to facilitate the operator's task of controlling the plant.

The intention of this paper is to review some of the proposed improvements in control and instrumentation for nuclear power plants. It is my belief that at least some of the recommendations will impact on conventional power plants and on industries requiring a high degree of safety such as the chemical industry.

As the operator may be regarded as being at the heart of the TMI problem most of the proposals concern improvements to the interface between man and machine.

It is important to bear in mind that the purpose of the TMI investigations was to establish the causes of the accident at TMI and to recommend what action should be taken to avoid a repetition of a similar accident taking place.

The challenge may be regarded as the need to prevent, mitigate and control abnormal operating situations. Early detection of potential problems may be regarded as the first line of defence whilst maintaining maximum plant availability. Preventing minor incidents from becoming major problems may be regarded as the second line of defence. Controlling a major problem is the third line with its overall aim being that of public safety.

## 22.2. Man/machine interface

The operator may be regarded as closing the control loop (Figure 22.1).

The information from the plant requires to be presented in such a way that he can readily assimilate it and make prompt decisions based on his knowledge of plant processes.

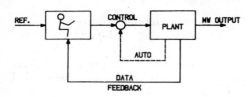

**Figure 22.1.** The operator forms part of the control loop.

The operator is expected to regulate the plant output to the desired level and to compensate for any uncommanded changes of state. Some of these will be in the form of large rapid step functions caused by external disturbances such as reduction or total loss of generator load due to tripping of a feeder or generator circuit breaker. Others will be internal plant disturbances such as tripping of a pump or closure of a valve, failure of a component or circuit.

In each case the operator requires to know:

what changes have taken place and how important they are
what effect those changes will have on other items of plant
what remedial action he should take and how soon
what will be the consequences of his actions

In many instances, the disturbance will take place too quickly for the operator to make an adequate assessment. There is therefore a strong case for more automation. This is generally favoured by certain European countries. In those countries the operator is not permitted to intervene (contrary to TMI) for a certain minimum period of time (10–30 min) after initiation of automatic protection and safeguards.

Automatic systems can be developed to a very high degree of reliability using current fault-tolerant techniques. The logic process however must be pre-established and design of such systems requires analysis of a wide spectrum of possible faults.

Apart from automatic action, which effectively removes the operator, at least for a period of time, from the decision-making process all the proposed improvements enhance the operator's ability to make a decision.

There are two sides to harmonizing the man–machine interface. One is to increase the 'man's' understanding of the machine and the other is to make the 'machine' more readily understandable to the man at all times.

The former can be achieved by providing the operations and maintenance staff with increased competence to control the plant under all conditions, particularly abnormal situations and emergencies through:

improved training to deal with all modes of operations: theoretical in the classroom and practical on a plant simulator. The operators should be encouraged to adopt an 'open-minded' approach to problem solving so that they can analyse each situation, diagnose the problem and respond accordingly

better procedures: particularly to deal with emergencies

back-up support and technical counselling from outside the control room from supervisory staff and specialists. A technical support centre is proposed which will be supplied with plant information and which will provide the operators with outside expertise whilst avoiding large numbers of people congregating in the control room.

The latter can be achieved by improving the control and instrumentation so that the operator has clear and unambiguous information regarding the plant status at all times for all modes of operation. This information on the plant status should be presented in such a way that it can be quickly assimilated and evaluated. Methods which assist the operator to interpret data or make it more intelligible, such as better displays or diagnostic techniques, constitute some of the improvements. A prerequisite to any such methods is that the raw data fed back from the plant is true. There is a psychological limit to the performance of any operator. His ability to cope with large amounts of information, his response, concentration and ability to make decisions under stress must be taken into account. Human factors are therefore an additional consideration in the presentation of information and controls in the control room.

A review of these subjects is dealt with in the following sections. This includes noise diagnostic techniques which can provide a valuable aid in the early detection of potential problems.

## 22.3. Presentation of reliable information

The operator cannot be expected to make sound decisions unless he is supplied with dependable information from the plant.

Unfortunately, false signals occur in most plants at some stage and it is normal for the operator to exercise a certain amount of judgement in distinguishing between true and false information. If the frequency of false alarms or other faulty signals becomes high, however, there is a grave danger that the operator may lose confidence in his instrumentation. Should this happen there is a possibility that he will disregard abnormal signals even when they are true. Apparently, this was one of the problems at TMI; the operators developed a 'mindset' and did not give credence to some of the signals that they were receiving in the control room.

Poor maintenance can be a major cause of false indications. Another important factor is the reliability of the instrumentation. Certain measures have, however, been proposed to increase the credibility of the instrumentation. These are discussed below.

### 22.3.1. Quality
Dependable instrumentation is the product of sound engineering practices such as diligent design with adequate safety

margins, stringent control of manufacturing techniques, extensive testing to cover varied working conditions and careful installation and commissioning. Modern competitive business methods often result in strong pressure being applied to 'cut corners'. To protect against this, a thorough quality-assurance programme is essential.

A high-quality product is more expensive, however, and for computers the cost ratio of equipment built to military specification to normal is about 5 to 1. This means that there may be a 'cut-off' where the same reliability can be achieved through other means (such as providing redundant equipment) at lower cost.

### 22.3.2. Direct measurement
It may not always be possible to make direct measurement of a particular quantity. For instance, the usual method of measuring flow is to measure the differential pressure developed across a constriction to the flow such as an orifice plate. Wherever possible, however, direct measurement is always preferable. In future references TMI may well be quoted as a classic example. The offending pressurizer relief valve which stuck open was indicated in the control room as being closed. This is because the valve indication showed only whether the valve-operating solenoid was energized or not and did not truly monitor the position of the valve. Other values which were not available on pressurized water reactors before TMI and which are included in recommendations are direct methods to measure reactor water level and an indication of the difference between the operating temperatures of the reactor coolant and the saturation temperature at the working pressure. In addition to providing an indication of the coolant water inventory, reactor vessel level instrumentation can also give an indication of the presence of voids within the systems.

### 22.3.3. Validation techniques
Continuous monitoring of measured quantities for their validity is fundamental to

good instrumentation. Various techniques can be used to check instrumentation failures. These are generally based on providing redundant instruments whose output can be compared for discrepancies. In order to identify and isolate a faulty device threefold redundancy is required. Where redundant sensors are not provided, validation is dependent on checking the range of the signal. Signals which transgress upper or lower setpoints can be used to trigger an alarm to warn the operator of instrument failure. Gross faults such as open or short circuits or earth faults can be checked in a similar manner to limit settings through 'live' zeros and 'ceilings'.

Another method is to check instrument loop integrity by checking loop resistance. Noise diagnostic techniques can also be used to check the loop by measuring the loop transfer function through fast Fourier analysis. Such methods will provide an indication of the response and transient performance of the instrumentation.

Advanced methods are currently being researched which are based on analytical techniques (Deyst et al. 1981). Sophisticated methods utilize computer processing of signals and are based on the relationships between different state variables. Calculated values can be obtained which are then checked against measured values to highlight differences. As first-order differential equations are developed to express the signal relationships the method is dynamic and is used to express transient behaviour.

A more crude but nevertheless effective method of checking gross failures, which would have identified the false valve position indication at TMI, would be through a logical process. For example if a pump is running and the line valve open one would logically expect an indication of flow in a pipeline. If flow is indicated and the valve is shown to be closed an inconsistency exists which can be alarmed. In the case of TMI flow of hot coolant through the valve (which was indicated as closed) indicated by resistance temperature detectors downstream of the valve could have been alarmed.

### 22.3.4. Maintenance

Continued reliability of functioning equipment can only be assured through a strictly applied maintenance programme requiring periodic calibration tests in order to maintain accuracy and general performance. In this regard the noise diagnostic techniques mentioned above can be useful.

## 22.4. Human factors

In an emergency such as happened at TMI the operators were apparently overwhelmed and confused by the large number of occurrences, such as alarms, within a short period of time. In order to act effectively, the operator needs to adopt a logical step-by-step approach to restore the plant to a controllable state. As a first step he must be able to correctly diagnose the problem so that he can concentrate on key functions and ignore irrelevant information.

To assist the operator in this decision-making process it has been recommended that the science known as human factors engineering should be applied to control-room design in order to:

reduce operator fatigue and thereby encourage his alertness
make the plant operation more manageable.

### 22.4.1. Control-room layout and design

According to human factors specialists the plant control room should be humanized, that is, the designers should dimension the panels to suit the operator. Such questions as the length, height and shape of the control panels should be carefully considered as well as whether it is better for the operator to be standing or seated (Figures 22.2 and 22.3).

A strategy is necessary to determine which controls and instruments are required on the main control desk and which should be mounted on the back panels or even outside the control room altogether.

Instruments and controls must be arranged in such a way that the operator can rapidly assess the status of the plant and can respond easily whilst maintaining overall plant surveillance. The layout of devices in functional groupings, use of colour codes and attention to labels, scales and other inscriptions must be considered in order to minimize the possibility of errors and to present information clearly and unambiguously.

Sufficient space is required to allow for the logical positioning of additional devices or future modifications.

**Figure 22.2.**   The back panel of a 900 MW PWR control room.

**Figure 22.3.**   The main operating desk of a 900 MW PWR control room.

Without a coordinated design each contractor will invariably apply his own standards to the detriment of design uniformity.

Many pre-TMI nuclear plant control rooms tended to be large and unwieldy with little miniaturization or use of visual displays (CRT screens). I suspect this conservatism is due firstly to a reluctance on the part of the utilities and industry to deviate from tried and proven devices and secondly because of the difficulties of licensing prototypes which require extensive and expensive test programmes.

It is interesting that it has taken the TMI investigators, many of whom were not engineers, to force the nuclear industry to adopt newer methods of displaying plant information.

A first attempt to apply human factors engineering to nuclear control and instrumentation has been to examine the precedent set by the aircraft industry which has much to offer in the way of cockpit design studies.

### 22.4.2. Alarm systems

Alarm systems are worthy of special consideration since nothing is more likely to strike terror into the operator than hundreds of lights flashing simultaneously accompanied by the wail of sirens and klaxons (Figure 22.4).

As plants have increased in size so have the alarm systems. Modern plants may have from 500 to 2 000 alarms. There has been a tendency to provide the operator with all the information regarding the plant status and expect him to be able to sort it out.

One of the major recommendations of the TMI investigators concerned the prioritization of alarms and the suppression of those of a lower order. This will undoubtedly lead to greater use of CRT alarm displays and computers to perform alarm analysis and automatic diagnosis (Figure 22.5).

Diagnostic guidance can be made available in the form of alarm fiches which can be called up manually by the operator or displayed automatically when triggered by their particular alarm.

One interesting proposal is to use different audible alarms for alarms of different priorities. The Japanese have

**Figure 22.4.** A large number of alarms occurring simultaneously can confuse the operator.

developed a system of synthesized voice alarms which are apparently successful.

### 22.4.3. Colour cathode ray tube (CRT) displays (Figures 22.6 and 22.7)

A good CRT display can be used to supplement conventional instrumentation. The

**Figure 22.5.** More recent plant designs utilize CRT alarm displays to supplement conventional annunciators.

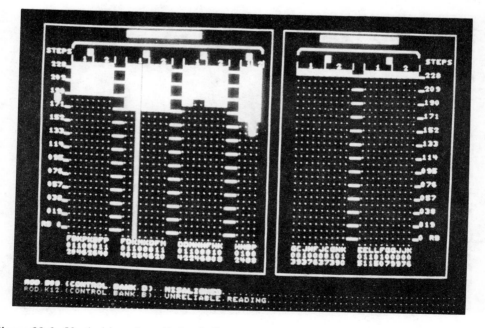

**Figure 22.6.** Vertical bar chart display indicates the position of the control rods of a PWR.

CRT screen is extremely versatile and can display information in the same format as almost any conventional instrument i.e. digital, analogue, instantaneous, historical, etc. In addition it can display mimics with live data which can either be an overview of a particular system or can be 'zoomed' in to display detail such as a single pump with suction and discharge flows, bearing temperatures, lubrication etc.

Discrepancies between actual and demanded state of devices e.g. pumps, valves, etc, as well as other alarm conditions can easily be shown on such mimics.

Pictorial representation of information can be more quickly assimilated by the operator and understood than discrete analogue or digital information.

In addition, the plant computer can be used to calculate the expected values of certain key parameters based on plant operating conditions and these can be displayed alongside actual values to provide an instantaneous comparison. With such representation it is easy to see the dynamic value of the 'margins of safety' between the measured quantity and its

thresholds and hence the operator can take anticipatory action before an alarm condition is reached.

Although the most popular representation of measured quantities is on a bar graph, they can also be displayed against time similar to the trace obtained from a strip chart recorder. The operator can also obtain a hard-copy should he desire to retain any information for reference.

Colour provides an added dimension to such displays and can be used for coding and to highlight or to suppress information.

### 22.4.4. Fixed mimics

In my opinion there is still a place for conventional mimics in a modern control room although diligent layout of components could lessen the need to some extent. Nevertheless, a fairly large mimic which is easily read from anywhere in the control room can give the operator an overview of the plant at a glance and is an advantage (Figure 22.8). The mimic should be kept as simple as possible and should cover only the primary systems. A common failing is

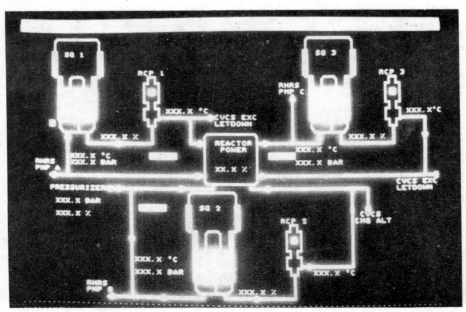

**Figure 22.7.** Animated graphic display of the reactor coolant system of a PWR.

to provide too much detail on mimics so that the operator's reaction time is reduced due to the time taken to search for information. Grouped alarms can be indicated by light-emitting diodes (LEDs) which should not be more than 50 say, in number and major parameters can also be indicated on the mimic. These could either be hard-wired or fed by digital to analogue converters from the computer.

All sub-system mimics can be displayed on the CRT screens and the operator can relate the information on these to the overall system mimic if he needs to do so.

**Figure 22.8.** A 900 MW PWR control room. The fixed mimic can be seen above the rear panel.

*22.4.5. Hierarchy of control*
Automation of certain controls such as sub-groups does not reduce the operator's overall control of the plant but can remove some of the lesser important chores leaving him free to concentrate on fewer major items.

One such chore, although hardly a sub-group, is manual control — on earlier designs of pressurized water reactors — of steam generator water level over the range of 0 to 15% power. On some plants it requires three operators (on a three-loop plant with three steam generators) to regulate the feed flow in order to maintain the correct water level during power raising. Due to the time lags in the system this is a sensitive process and frequently results in reactor trip. Automatic control has eliminated these problems without lessening the operator's responsibility.

## 22.5. Early warning systems — noise diagnostic techniques

These systems are used to provide an early detection of failures and therefore provide the operator with an advanced warning of potential problems. They can serve three functions:

assess the structural integrity of major mechanical and hydraulic circuits by detecting abnormal vibration, the presence of loose parts or leaks;
assess the functioning of measurement channels by checking the transfer function of instrumentation paths;
detect anomalies in the thermo/hydraulic process e.g. blocked channels, excessive voids, etc.

These systems produce stochastic information which requires analysis by statistical methods. They work on the principle that actual measurements are compared with reference patterns or signatures and abnormalities can be highlighted.

To date, the systems in operation on existing plants are checked off-line on a periodic basis. Future systems are likely to operate on-line. Pattern recognition will be performed on-line by computer and the operator will be warned immediately of any potential failures.

Most failures are the result of gradual deterioration (assuming adequate performance at the time of commissioning) and by employing such techniques the operator will be able to take corrective action before a problem reaches crisis proportions.

## 22.6. Accident monitoring

The proposals discussed thus far have been intended to assist the operator to maintain the plant within the correct operating range during normal operation, to warn him of developing situations and to identify problem areas. Once an accident does occur much of the normal instrumentation may no longer be valid since the reactor, turbine and other devices will have tripped.

At TMI the operators were confronted with a situation where:

some of the instruments were not giving meaningful data because they were out of range (e.g. in-core thermocouples — since these had only been designed for normal operating conditions);
some instruments had ceased to function because they had not been designed to operate at the higher temperatures and other environmental conditions to which they were being subjected;
certain additional instruments were needed to measure safety parameters which are unique to accident conditions.

When an accident occurs the normal objectives related to maintaining a specific electrical output are no longer relevant. The operator's objectives then become related to shutting down the plant safely, that is suppressing the fission process and dissipating the residual heat of the core in such a way as to:

firstly, ensure the safety of the public and other personnel by preventing the release of radioactive materials to the environment;
secondly, minimize damage to the plant.

This change in purpose shifts the emphasis in controlling the plant and the instrumentation to be monitored.

### 22.6.1. Instrumentation
During an accident, temperatures, pressures, radioactivity levels etc may exceed their normal design values and it is necessary to provide certain extended-range instruments. This could mean either extending the range of the normal instruments or providing instruments which are solely used for accident monitoring.

If instruments are required to operate under accident conditions they have to be capable of withstanding much harsher environmental conditions, higher than normal design temperature, pressure, humidity, radioactivity and so on. This requires extensive testing for the sensor and its associated connectors and cables under simulated accident conditions. Ageing tests are also required in order to prove the instruments will be reliable even should an accident occur at the end of life of the plant.

The specification for each of these instruments including the degree of redundancy required must be carried out on a case-by-case basis. For instance, certain sensors which are required to imitate a tripping function may only be required to operate until the start of the accident whereas others are required to operate for a considerable period of time under extreme conditions. This period of time may be several minutes or several months depending on the function of the sensor.

### 22.6.2. Symptoms approach
The traditional method of guarding against accidents has been to prepare for certain postulated accidents of a catastrophic nature. This was presumably on the basis that if the worst accidents can be protected against this will provide an 'umbrella' which will cover all lesser accidents.

The accident at TMI did not fit into any of these scenarios and the operators were bewildered. The actions they took in the first few hours did not produce the expected results.

Post-TMI analysis has shown that the event-orientated approach of the past lacks completeness since it is not feasible to provide the operator with specific procedures to cover all possible plant situations which could occur. A new symptoms approach has therefore been proposed which is based on the analysis of physical parameters and their trends. This is a more basic approach since it monitors fundamental concepts such as mass and energy balance.

One system that can assist the operator in this regard is an on-line plant-wide disturbance analysis and surveillance system (DASS).

### 22.6.3. Safety parameters display system

A first step towards a disturbance analysis and surveillance system may be regarded as the safety parameters display. This was one of the recommendations to come from the TMI investigations. Basically the purpose is to monitor the major plant parameters which are fundamental to safe operation and to display them on a screen which is always visible to the operator. The magnitude of each parameter is compared to its expected value for a particular mode of operation and also its relative value compared to other related parameters. Any change from expected conditions will be identified by the operator.

Several levels of display are available to the operator. The top display is an overview of the most important parameters such as coolant loop temperature, pressure, flow, steam pressure, feed water flow etc., which may be related to mass and energy balance. The second level is synoptics of major plant systems which control the basic parameters. Deeper levels give greater details of plant systems and subsystems so that by descending from the top display the operator can 'zoom' in on the cause of the problem.

### 22.6.4. Disturbance analysis and surveillance system (DASS)

The disturbance analysis and surveillance system (DASS) receives inputs such as plant variables, status and alarm information and provides visual displays in the control room as for the safety parameters display system. Whereas the latter is solely a display system providing information in a more convenient format, the DASS system is intended to be a diagnostic tool which operators should find useful to identify potential problems and to cope with unexplained transients. The plant data is used in computer models of the plant vital systems and is intended to alert the operator to any potential problems. Its main purpose is to assist the operator in decision-making when confronted with an abnormal situation.

More specifically it is intended to aid the operator in preventing, terminating and mitigating undesired events. It is intended to achieve this by:

surveillance of the plant status and parameters and alert the operator to potential problems;
detection and analysis of plant disturbances.

Rumancik et al. (1981) describe various levels of functions for DASS as follows:

*Overhead functions:*

identification of the mode of operation;
proper interpretation and validation of process variables;
integrated display of all DASS output.

These functions are required to support the other DASS functions and are essential regardless of how many of the remaining functions are implemented.

*Surveillance functions:*

status of critical safety requirements;
surveillance of subsystem configuration;
verification of automatic control and protective actions;
surveillance of margin to technical specifications.

These functions will provide the operator with an awareness of plant state at a relatively high level. A comparison can be made of plant actual and expected performance and any potentially dangerous trends or reduction of safety margins can be brought to the operator's attention.

In addition a passive check of plant configuration can be made to check mode of operation and any anomalies alarmed — e.g. valve closed which should be open, etc.

*Diagnosis functions:*

disturbance detection by parameter analysis;
determine the cause of the disturbance.

These functions can assist the operator in the correct diagnosis of a problem.

*Corrective action functions:*

determine best corrective action;
assist in monitoring normal and off-normal operating procedures.

These functions require complex system models.

*Predictive functions:*

predict future propagation of disturbances;
evaluation of possible control actions prior to initiation.

These functions would require extensive simulation capability.

The Electric Power Research Institute (EPRI) in the United States originally initiated the DASS project to improve nuclear power plant availability but subsequent to the TMI accident this system is currently being developed to assist the operator to deal with actual or potential plant disturbances.

A similar system is being developed in Germany which has already been successfully tested on the Halden project in Norway.

## 22.7. Conclusion

In summarizing it may be said that none of the proposed changes is particularly new since all were known before the accident at Three Mile Island. What is new is the overall philosophy and the emphasis that is currently being stressed to improve the man-machine interface. Control rooms are unlikely to be automated to such a degree in the foreseeable future (if ever) as to make the operator redundant. In the meantime the possibility of undesired events due to operator action or equipment malfunction can be reduced by assisting the operator to cope with unexpected circumstances. The control room can be made more manageable and checks and diagnostic aids can be provided to support him in his task and to supplement the conventional instrumentation.

It is interesting that probably the greatest impact that the accident at Three Mile Island has had on nuclear power-plant control and instrumentation is to emphasize the need for greater use of computer technology in order to provide an assurance of enhanced safety. This is in an industry which has traditionally been ultraconservative largely because of the prime importance of safety.

Many of the improvements suggested for nuclear power plants are also being considered for conventional power plants where the need for safety may not be so strong. However, for a normally operating plant, safety, availability and optimum performance are synonymous since a plant with frequent malfunctions and therefore a poor safety record cannot have a high availability. Therefore if any of the above proposals can be shown to improve availability while increasing safety there is a strong possibility of their adoption not only for conventional power plants but also for other industries as well.

Finally, although the TMI accident may be considered history, the changes and improvements which were suggested in order to obviate a repetition are still very much in a state of flux three and a half years later. Improvements which were proposed only 12 months ago are already out of date. The immediate post-TMI reaction was to produce short-term solutions in the way of additions and backfits. Then as similarities of purpose were realized there was a tendency to integrate the various solutions. Today there is a search for something more basic than hardware additions and I believe that a more fundamental and philosophical approach exists.

## References

Kemeny, J. G. et al. (1979) Report of the President's commission on the Accident at Three Mile Island October.

Rogovis, M. et al. (1980) Three Mile Island. A Report the Commissioners and the Public Vol 1. Jan.

Current Nuclear Power Plant Safety Issues. (1980) *IAEA Conference Proceedings Stockholm.* Oct. Vols. 1 and 3.

Need for Nuclear Education (1982) *Financial Mail* 7 May.

Post TMI Plant Design (1980) *Power Eng.*

Livingston, W. L. (1980) The legacy of TMI-2 for the nuclear power plant operations system *Combustion* Nov.

Deyst, J. J. Kanazawa, R. M. and Pasquenza, J. P. (1981) Sensor validation: A method to enhance the quality of the man/machine interface in nuclear power stations. *IEEE Trans on Nuclear Science.* Vol NS-28. No. 1 Feb.

Clark, R. N. and Campbell, B. (1982) Instrument fault detection in a pressurized water reactor pressurizer. *Nuclear Technology.* Vol. 56 January.

Wensley, J. H. (1982) Fault tolerant techniques for power plant computers. *IEEE Trans on power apparatus and systems.* Vol. PAS-101. No. 1 January.

Human factors evaluation of control room design and operator performance at Three Mile Island-2. (1980) *Nureg/CR-1270* Jan.

Bagchi, C. N. and Gotilla, S. C. (1981) Application of human engineering criteria to annunciator display systems in a large fossil power station. *IEEE Trans on Power Apparatus and Systems* Vol. 1 Pas-100. No. 6 June.

Seminara, J. L. and Parsons, S. O. (1980) Survey of control room design practices with respect to human factors engineering. *Nuclear Safety.* Vol. 21. No. 5 Sept.–Oct.

The human side: Training and tools to help the operator (1981) *IEEE, Spectrum* April.

Don't forget the operator: he's only human (1981) *Electrical Review* Vol. 208. No. 11 **20** March.

Meijer, C. H. Pucar, J. L. and O'Connell, T. J. (1981) Applied human engineering to improve the man precess interaction in a nuclear power plant. *IEEE Trans on Nuclear Science.* Vol. NS-28. No. 1 Feb.

Gandrille, J. L. Kerboul, M. and Oliot, A: Symptoms characterization as a basis for post-accident procedures. *Framatome. Paris.*

A computerized disturbance analysis and surveillance system (1981) *Nucleonis Week* **17** Sept.

Human factors acceptance criteria for the safety parameter display system. (1981) *Nureg 0835* Oct.

Rumancik, J. A. Easter, J. R. and Campbell, L. A. (1981) Establishing goals and functions for a plant wide disturbance analysis and surveillance system (DASS). *IEEE Trans on nuclear science.* Vol. NS-28. No. 1. February.

Disturbance analysis and surveillance system scoping and feasibility study. (1982) *EPRI NP-2240 Project 891-3 Final Report* July.

# Index

Accelerometers, miniature, 68–74, 111–2, figs. 8.1–8.5, 12.3–12.4
design, 73–4
Accident monitoring, 250–2; DASS, 251–2; instrumentation, 250; safety parameters display system, 251; symptoms approach, 250–1
Accident, nuclear, at Three Mile Island, 240–53, figs 22.1–22.8; early warning systems—noise diagnostic techniques, 249; human factors, 244–9; major recommendations, 241; man/ machine interface, 241–3; presentation of reliable information, 243–4
Acoustic sensors; fibre-optic, 127–8, figs 14.2–12.5; nuclear reactors, 236–7, figs 21.15–21.16
Adjustability, smart pressure transmitters, 42–3
Air humidity measurement, 63–7, figs 7.1–7.5; definitions, 63–4; methods, 65–7
Alarm systems, 246–7, figs 22.4–22.5
Amplifier configurations, 203–4, figs 19.4–19.5
Amplitude modulation, 162–3, fig 16.5
Analytical instruments, 6–8, 91–101, figs 11.1–11.4; automated chemical analysers, 99; electrochemical methods, 97–8; gas chromatography, 92–3, fig 11.1; infrared absorption, 93–5, fig 11.2; mass spectrometry, 93; neutron-based techniques, 97; physical parameter measurements, 98–9, fig 11.4; visible and ultraviolet absorption, 95–9; X-ray fluorescence and absorption, 96, fig 11.3; other techniques, unmet needs, 100
Anatomy of sensors, 13–14
Angular digital encoders, 25–8, figs 3.1–3.3
Atomic absorption, 100

Bell provers, 193–4, fig 18.15
Bimorphs, 148, fig 15.14

Cables, 155–6; fibre-optic, 164–5; for hostile environments, 224
Calibration; density-measuring instruments, 54; flowmeters, 184–99, figs 18.1–18.19; points to watch, 192–3, 196–7; with gases, 193–7; with liquids, 185–93. See also Errors and inaccuracies
CAMAC, 157–8
Capacitance sensors, fig 2.8
CRT displays, 246–8, figs 22.5–22.7
Colour measurement, 219, fig 20.15
Communications and transmission systems, 8–9, 77, 151–70, figs 16.1–16.7; density measurement, 52–4, figs 5.15–5.18; digital transmission, 156–63; electrical analogue, 153–6; fibre optic, 164–5; future, 169–70; networks, 105–9; pneumatic transmission, 152–3
Computer systems; buses, 157–8; interfaces, 160–2, 165
Control room layout and design, 244–6, figs 22.2–22.3
Control systems, 148, fig 15.14. See also Communications and transmission systems; Displays
Converters, 30–4, figs 3.8–3.13
Coriolis forces, in mass flow measurement, 55–62, figs. 6.1–6.10; density measurement, 62; general operation, 58–60; gyroscopic precession, 57–8; mass flow meter, 56–7; mechanical configuration, 60–1; timing diagram, 61–2
Current and flux balance, 141–2, fig 15.6

Data links. See Communications and transmission systems
Density and level measurement, using non-contact nuclear gauges, 82–90, figs 10.1–10.10
Density meters, 45–52, figs 5.1–5.15; flowing liquids, 46–9, 62; food industry, 219; gravity, 46; inertia, 49–50; inferential-type, 50–2

Density of liquids, measurement, 45–54, figs 5.1–5.18; calibrating the instruments, 54; measuring instruments, 45–52; transmitting the measurements, 52–4
Depth transducers, 36
Dielectric constant density meters, 51, fig 5.11
Digital feedback instruments, 147–8
Digital transducers, 24–38, figs 3.1–3.14; angular digital encoders, 25–8; depth transducers, 36; laser systems, 37–8; linear displacement transducers, 28–30; magnetic systems, 36–7; radiation transducers, 37; synchro-resolver conversion, 30–4; variable frequency devices, 34–6; vortex transducers, 37
Digital transmission, 156–63, figs 16.3–16.5
Dimension gauging, 220–1, figs 20.16–20.17
Displacement and strain measurement in nuclear reactors, 226–31, figs 21.1–21.9
Displacement balance instruments, 146, fig 15.11
Displays, 4, 240, 243–4, 246–9, 251–2, figs 22.2–22.8
Disturbance analysis and surveillance system (DASS), 251–2
Doppler shift; phase-balance instruments, 146–7, fig 15.12; ultrasound, 79

Electrical analogue transmission, 153–6, fig 16.2
Electrical current fibre-optic sensors, 130, figs 14.8–14.9
Electrical noise suppression and prevention, 200–7; amplifier configurations, 203–4; grounding configurations, 204–6; noise generation, 200; noise reduction, 201–3; quantitative evaluation of noise suppression, 206
Electrochemical methods, 97–8
Ergonomics, 4, 240–53, figs 22.1–22.8; accident

monitoring, 250–2; early
warning systems—noise
diagnostic techniques, 249;
human factors, 244–9; man/
machine interface, 241–3;
presentation of reliable
information, 243–4
Errors and inaccuracies, 171–83,
figs 17.1–17.6; measurement
errors, 173–8; the statement of
errors, 178–9; specification of
an instrument's errors, 179–81;
examples of instrument
inaccuracy, 181–2. See also
Calibration

Feedback, 134–50, figs
15.1–15.14; current and flux
balance, 141–2; displacement
balance, 146; force and torque
balance, 139–41; general
properties of feedback systems,
135–8; heat-flow balance,
142–3; instrument control
systems, 148; instruments with
digital outputs, 147–8; phase
balance, 146–7; radiation
balance, 147; temperature
balance, 145; types of feedback
measuring system, 138–9;
voltage balance, 143–5
Fibre-optics communications,
8–9, 164–5, fig 16.6; density
gauges and transmission
systems, 53–4, figs 5.16–5.17;
flowmeters, 126, fig 14.1;
gyroscopic sensors, 130–2, figs
14.8–14.9; hydrophones, 127–8,
figs 14.2–14.5; liquid-level
sensors, 126; magnetic field
sensors, 129, fig 14.7; sensors,
16, 23, 123–33, figs 2.12, 13.14,
14.1–14.9; spectrophones,
128–9, fig 14.6; temperature
sensors, 126–7
Flowmeters, 40; calibration,
184–99, figs 18.1–18.19;
calibration using liquids,
185–93; calibration using gases,
193–7; doppler, fig 9.11; fibre
optic, 126, fig 14.1; food
industry, 213–4; nuclear
industry, 234–6, fig 21.14;
using Coriolis principle, 55–62,
figs 6.1–6.10; using pressure
balance, 145, fig 15.9
Food process plant
instrumentation, 208–22,
figs 20.1–20.17; level (or vessel
contents), 215–7; liquid flow,
213–4; moisture, 210–3;
temperature, 209–10; weighing
of dry materials, 214–5; other

standard physical
measurements, 217–9; special
transducers, 219–21; the future,
221–2
Force balance instruments,
139–41, figs 15.4–15.5
Force transducers, food industry,
218
Foreign-body detection, food
industry, 219–20
Frequency division multiplexing,
163
Frequency modulation, 163,
fig 16.5
Function-generator converters,
32–3, figs 3.10–3.11
Future instrumentation. See
Instrumentation industry

Gas chromatography, 92–3,
106–7, figs 11.1, 12.1
Gases, use in flowmeter
calibration, 193–7, figs 18.14–
18.18
Glass-to-silicon bonding, 104–6
Gravimetric tanks, in calibration,
185–9, figs 18.2–18.7
Gravimetric gas-meter calibrator,
194–5, fig 18.17
Gravity, density meters, 46–9,
figs 5.1–5.7
Gyroscopes, fibre-optic, 130–2,
figs 14.8–14.9
Gyroscopic mass flow meter,
55–62, figs 6.1–6.10; density
measurement, 62; general
operation, 58–60; gyroscopic
precession, 57–8; mass flow
meter, 56–7; mechanical
configuration, 60–1; timing
diagram, 61–2

Half-duplex, 160
Harmonic oscillator converters,
34, fig 3.13
Heat-flow balance instruments,
142–3, fig 15.7
Heatscan, 220–1, figs 20.16–20.17
Hermetic seals, silicone sensors,
104–6
Hierarchical networks, 166
High-temperature environments,
instrumental techniques. See
Nuclear reactors, transducers
Highway networks, 166–7
Histograms, 175, fig 17.5
Hostile environments,
instrumental techniques. See
Nuclear reactors, transducers
Humidity measurement, 63–7,
figs 7.1–7.5; definitions, 63–4;
methods, 65–7; in food
industry, 218–9, fig 20.14

Hyde prover, 194
Hydrometer, 47, fig 5.4
Hydrophones. See Acoustic
sensors
Hygrometry. See Humidity
measurement

Inaccuracies. See Errors and
inaccuracies
Incremental shaft encoders, 27–8
Inductance change sensors, fig 2.6
Inertial-mass density sensors,
49–50, figs 5.8–5.10
Inferential-type density meters,
50–2, figs 5.11–5.14
Infrared; absorption, 93–5,
fig 11.2; analysis, 7
Instrument inaccuracy:
specifying, 179–81; some
examples, 181–2, fig 17.6
Instrumentation industry, 1–11,
figs 1.1–1.4; analysers 6–8;
developments, 1928–1978,
10–11; instrument hardware,
9–10; profile, 2–3; sensors, 3–6;
signal transmission, 8–9
Integrated sensors, 15–16,
102–13, figs 12.1–12.4;
accelerometer formed from a
cantilever, 111–2; and gas
chromatography, 106–7;
micromachining, 103–4;
pressure sensors, 107–11;
sealing, 104–6; thermal
conductivity, 107; unifying the
sensor field, 113
Interfaces, computer-based
systems, 160–2, 165
Interferometry, 119–21, 130,
148–9, figs 13.7–13.10
Interpolation systems, 30,
figs 3.6–3.7
Invasiveness, 4
Ion implant sensor, 43–4, fig 4.4

Laser techniques, 7, 37–8, 100,
114–9, figs 13.1–13.6;
laser-based position sensors
and alignment devices, 116–9;
summary of laser
characteristics, 116
Level (vessel contents)
measurement, food industry,
215–7, figs 20.12–20.13
Level and density measurement
using non-contact nuclear
gaufes, 82–90, figs 10.1–10.10
Light-sensitive detectors, 117,
table 13.2
Limiting mean, 175
Linear displacement transducers,
28–30, figs 3.4–3.7
Liquids; density measurement,

45–52, figs 5.1–5.14; fibre-optic sensors, 126; nuclear gauges for density and level measurement, 82–90, figs 10.1–10.10; ultrasonic measurement, 75–81, figs 9.1–9.12. *See also* Flowmeters
Local area networks (LAN), 8

Magnetic digital systems, 36–7
Magnetic fields, fibre-optic sensors, 129, fig 14.7
Man/machine interface. *See* Ergonomics
Mass flow measurement using Coriolis, 55–62, figs 6.1–6.10; density measurement, 62; general operation, 58–60; gyroscopic precession, 57–8; mass flow meter, 56–7; mechanical configuration, 60–1; timing diagram, 61–2
Mass spectrometry, 93
Master meters, 190–1, 196, figs 18.10–18.11; combination of master meter and calibrator, 191–2, fig 18.12
Measurement, 171–3, figs 17.1–17.2. *See also* Errors and inaccuracies
Mechanical input and output sensors, figs 2.1–2.2
Mechanical properties, sensors for, 12–23, figs 2.1–2.11; anatomy, 13–14; applications, 13 classification, 14; mechanical measurands, 13; output signals, 14; technological trends, 15–16, 23; transduction techniques, 14–15; types, figs 2.1–2.11
Microprocessors, 2–3, 9; and analysis, 7–8; smart pressure transmitters, 39–44, figs 4.1–4.4
Micromachining, 103–4
Microwave absorption in gases, 100
Mimics, 248–9, fig 22.8
Modems, 162–3
Moiré fringe techniques, 30, fig 3.5
Moisture measurements, food industry, 210–13, figs 20.6–20.8
Multiplexing, 158–9, 163, 165, fig 16.3

Networks, 8, 165–9, fig 16.7
Neutron-based techniques, analytical instruments, 97
Noise; diagnostic techniques, accident prevention, 249; suppression and prevention, electrical, 200–7, figs 19.1–19.9;

amplifier configurations, 203–4; grounding configurations, 204–6; noise generation, 200; noise reduction, 201–3; quantitative evaluation of noise suppression, 206. *See also* Acoustic sensors
Normal distribution, 176, fig 17.5
Nuclear gauges, 82–90, figs 10.1–10.10; density detection and transmission systems, 54, fig 5.18
NMR, 100
Nuclear reactors, transducers, 223–39, figs 21.1–21.16; principles of measurement at high temperatures, 224; acoustic measurement techniques, 236–7; flow measurement, 234–6; pressure measurement, 234; sensor construction techniques, 224–5; strain and displacement measurement, 226–31; temperature measurement, 231–4. *See also* Accident, nuclear, at Three Mile Island
Nucleonic sensors, fig 2.11

Open systems interconnection (OSI), 8
Optical fibres. *See* Fibre-optics
Optical gratings, 29–30, fig 3.4
Optical measurement methods, 114–24, figs 13.1–13.15; classical interferometry, 119–20; diffraction, 121–3, figs 13.11–13.13; hographic and speckle interferometry, 120–1; mechanical measurements, 16, 23. *See also under* Fibre-optics
Optical shaft encoders, 26–7, figs 3.2–3.3
Optoelectronic sensors. *See* Fibre-optics, sensors
Output signals, 14

Parallel data communication, 157–9, fig 16.3
Phase; balance instruments, 146–7, fig 15.12; modulation, 163, fig 16.5; shift converters, 32, figs 3.8–3.9
Piezoelectric sensors, 16, fig 2.6; accelerometer, 68–74, figs 8.1–8.5; noise suppression and prevention, 200–7, figs 19.1–19.9; pressure, 107–11, fig 12.2
Pipe provers, 189–90, figs 18.8–18.9

Piping, for pneumatic transmission, 152–3
Pitot; traversing, 196, fig 18.19; tube calibration, 197, fig 18.20
Pneumatic; sensors, fig 2.13; transmission, 152–3, fig 16.1
Population mean, 176–8
Potentiometers, 148, fig 15.14
Presentation of reliable information, 243–4
Pressure measurement; flow meter using pressure balance, 145, fig 15.9; food industry, 217–8; nuclear industry, 234, fig 21.13; silicon sensors, 107–11, fig 12.12; smart transmitters, 39–44, figs 4.1–4.4
Proway, 168–9
PVT system for calibrating gas-meters, 195–6, fig 18.18
Pyrometry. *See* Temperature measurement

Quality control, 7–8

Radiation; absorption density meters, 51, fig 5.13; balance instruments, 147, fig 15.13; basics, 82–5, figs 10.1–10.2; detectors, 85–6, figs 10.3–10.5; gauges, level and density measurement, 82–90, figs 10.1–10.10; transducers, 37
Raman spectroscopy, 7; laser, 100
Random errors, 174–5, fig 17.3; randomising systematic uncertainties, 178
Readings. *See* Displays
Redundancy, 10
Refractive index; density meters, 51, fig 5.12; process control, 98–9
Reliability, 4; maintenance, 244; presentation of reliable information, 243–4; smart pressure transmitters, 43–4, fig 4.3
Remote adjustability, 42–3
Resistance change sensors, figs 2.4–2.5
Resolvers; synchro-resolver conversion, 30–4, figs 3.8–3.13
Ring-type networks, 167

Safety parameters display system, 251
Sagnac interferometer, 130, figs 14.8–14.9
Sampling, 7
Sealing silicon sensors, 104–6
Serial data communication, 159–63, figs 16.4–16.5
Shaft encoders, 25–8, figs 3.1–3.3

Signal transmission. *See*
Communications and
transmission systems
Silicon technology and integrated
sensors. See Integrated sensors
Simplex, 160
Smart pressure transmitters,
39–44, figs 4.1–4.4; performance,
40–1; rangeability, 41–2;
reliability, 43–4; remote
adjustability, 42–3
Soap film burettes, 193, fig 18.14
Sonar, 78, figs 9.8, 9.9
Sound-speed density meters,
51–2, fig 5.14
Specification of instrument
inaccuracy, 179–82, fig 17.6
Spectrophone, 128–9, fig 14.6
Standard deviation, 175–6
Standards; digital networks, 8,
157–8, 160–2, 168; electrical
analogue transmission, 154,
155. *See also* Calibration
Star networks, 166
Strain and displacement
measurement, in nuclear
reactors, 226–31, figs 21.1–21.9
Successive approximation
converters, 34
Surface waves, 80, fig 9.13
Synchro-resolver conversion,
30–4, figs 3.8–3.13
Systematic errors, 173–4;
randomising systematic

uncertainties, 178
Tachometers, figs 2.5, 2.7, 15.11;
digital, 28
Temperature measurement; fibre-
optic sensors, 126–7; food
industry, 209–10, figs
20.1–20.5; nuclear reactors,
231–4, figs 21.10–21.12;
temperature-balance
instruments, 145, fig 15.10
Thermal conductivity detection,
98–9, 107
Thermocouples, 231–3, fig 21.10
Three Mile Island. *See* Accident,
nuclear, at Three Mile Island
Time division multiplexing, 163
Timing diagrams, 61–2, fig 6.10
Torque-balance instruments,
139–41, fig 15.5
Tracer methods for calibrating
gas meters, 196
Tracking converters, 33–4, fig
3.12
Transmission systems. *See*
Communications and
transmission systems
Tree networks, 166
Trends, 1–11, 15–16, 23

Ultrasonic sensors, 75–81, figs
2.10, 9.1–9.12; generation and
detection, 76–7; non-invasive
techniques, 75–6;
thermometers, 233, fig 21.11;

using, 77
Ultraviolet absorption, 95–6
Unity–gain feedback, 137–8

Validation techniques, 243–4
Variable frequency devices, 34–6,
fig 3.14
Velocity sensors, figs 2.5, 2.7;
calibration, 196–9, fig 18.20;
ultrasound, 77–8
Venturi-nozzles, sonic, 194, fig
18.16
Vibrating, string, beam, cylinder
transducers, 35–6, fig 3.14
Vibration-type sensors. *See*
Acoustic sensors; Piezoelectric
sensors
Viscometers, 221
Visible and ultraviolet
adsorption, 95–6
Voltage-balance instruments,
143–5, fig 15.8
Volumetric tanks, in calibration,
185, fig 18.1
Vortex transducers, 37

Wave analysers, 148, fig 15.14
Weigh feeders, 214–5, figs
20.9–20.11
Windings, high-temperature
inductive, 225
X-rays; diffraction, 100;
fluorescence and absorption,
96, fig 11.3